Landscapes of material are also landscapes of meaning: praxis is itself symbolic, and all landscapes are symbolic in practice. *Ideology and landscape in historical perspective* draws together fifteen historical geographers to examine landscapes as messages to be decoded, as signs to be deciphered. The range of examples is wide in terms of period, from the medieval to the modern, and of place, embracing the USA, Canada, Palestine, Israel, South Africa, India, Singapore, France and Germany. Each essay addresses a specific problem, but collectively they are principally concerned with the ideologies of religion and of politics, of church and state, and their historical impress upon landscapes. The book is introduced by an essay which explores the dialectical understanding of landscapes, and landscapes as expressions of the connection of an ideology to a quest for order, to an assertion of authority and to a project of totalization. The issues raised by landscapes and their meanings – issues of individual and collective action, of objective and subjective knowing, of materialist and idealist explanation – are fundamental not only to historical geography but to any humanistic study, and render the geographical study of landscapes of interest to scholars in many disciplines.

Cambridge Studies in Historical Geography 18

IDEOLOGY AND LANDSCAPE
IN HISTORICAL PERSPECTIVE

Cambridge Studies in Historical Geography

Series editors:
ALAN R. H. BAKER   J. B. HARLEY   DERYCK HOLDSWORTH

Cambridge Studies in Historical Geography encourages exploration of the philosophies, methodologies and techniques of historical geography and publishes the results of new research within all branches of the subject. It endeavours to secure the marriage of traditional scholarship with innovative approaches to problems and to sources, aiming in this way to provide a focus for the discipline and to contribute towards its development. The series is an international forum for publication in historical geography which also promotes contact with workers in cognate disciplines. For a list of titles in the series please see end of book.

IDEOLOGY AND LANDSCAPE
IN HISTORICAL PERSPECTIVE

Essays on the meanings of some places in the past

Edited by

ALAN R. H. BAKER
University Lecturer in Geography
and Fellow of Emmanuel College, Cambridge

and

GIDEON BIGER
Lecturer in Geography at Tel Aviv University

CAMBRIDGE
UNIVERSITY PRESS

CAMBRIDGE UNIVERSITY PRESS
Cambridge, New York, Melbourne, Madrid, Cape Town, Singapore, São Paulo

Cambridge University Press
The Edinburgh Building, Cambridge CB2 2RU, UK

Published in the United States of America by Cambridge University Press, New York

www.cambridge.org
Information on this title: www.cambridge.org/9780521410328

First published 1992
This digitally printed first paperback version 2006

A catalogue record for this publication is available from the British Library

Library of Congress Cataloguing in Publication data

Ideology and landscape in historical perspective: essays on the
meanings of some places in the past / edited by Alan R. H. Baker and
Gideon Biger.
     p.   cm. – (Cambridge studies in historical geography: 18)
Includes index.
ISBN 0 521 41032 0
1. Landscape assessment. 2. Historical geography. I. Baker,
Alan R. H. II. Biger, Gideon. III. Series.
GF90.I34 1992   91–40968
304.2–dc20   CIP

ISBN-13 978-0-521-41032-8 hardback
ISBN-10 0-521-41032-0 hardback

ISBN-13 978-0-521-02470-9 paperback
ISBN-10 0-521-02470-6 paperback

Contents

Figures

The paintings in chapter 11 are reproduced by courtesy of the McMichael Canadian Collection, Kleinburg, Ontario.

Contributors

ALAN R. H. BAKER is Lecturer in Geography and Fellow of Emmanuel College at the University of Cambridge

YOSSI BEN-ARTZI is Lecturer in Geography at Haifa University

GIDEON BIGER is Lecturer in Geography at Tel Aviv University

ROBIN A. BUTLIN is Professor of Geography at the University of Technology, Loughborough

DIETRICH DENECKE is Lecturer in Geography at Göttingen University

RUSSEL L. GERLACH is Professor of Geography at Southwest Missouri State University

PETER G. GOHEEN is Professor of Geography at Queen's University, Kingston, Canada

LEONARD GUELKE is Professor of Geography at the University of Waterloo

RUTH KARK is Professor of Geography at the Hebrew University of Jerusalem

D. W. MEINIG is Professor of Geography at Syracuse University

HANS-JÜRGEN NITZ is Professor of Geography at Göttingen University

BRIAN S. OSBORNE is Professor of Geography at Queen's University, Kingston

REHAV RUBIN is Lecturer in Geography at the Hebrew University of Jerusalem

GRAEME WYNN is Professor of Geography at the University of British Columbia

BRENDA S. A. YEOH is Lecturer in Geography at the National University of Singapore

Preface

These essays were originally among those presented as papers to a conference, Ideology and Landscape in Historical Perspective, being the Seventh International Conference of Historical Geographers, held at the Hebrew University of Jerusalem in July 1989. The meeting was organized by Professor Y. Ben-Arieh, Professor Ruth Kark and Dr R. Aaronsohn (the Hebrew University of Jerusalem), Dr Y. Ben-Artzi (Haifa University), Dr G. Biger (Tel Aviv University) and Dr Y. Katz (Bar-Ilan University) on a theme originally suggested by Dr Alan R. H. Baker (University of Cambridge). The editors thank Mrs P. Goodman for assistance with the preparation of the chapters for publication and Dr A. J. Crowhurst for preparing the index.

Introduction: on ideology and landscape

ALAN R. H. BAKER

People, period and place

Napoleon stood five feet six and a half inches in his stockinged feet: about average height for a Frenchman for his day ... The most distinctive feature of his body was the broad chest, covering lungs of exceptional capacity ... Napoleon's shoulders were broad, and his limbs well-made. But they were not particularly muscular. His thighs, for instance, lacked grip. He sat his horse like a sack of potatoes, had to lean well forward to keep his balance and in the hunting field was often thrown. His was an energetic but not a powerful physique ... Napoleon had a head of average size; he took, in today's measurements, a number 7 hat. The head seemed large because his neck was short. His feet were small: twenty-six centimetres long, that is size 6. His hands were small and beautifully made, with tapering fingers and well-formed nails. Small too were the penis and testicles ... Napoleon used to say that his heart-beat was less audible and less pronounced than most men's, but his doctors could find no evidence of this. His pulse rate ranged between fifty-four and sixty beats a minute. So the rhythm of metabolism appears to have been average. No physical peculiarity can explain the speed at which his mind worked.[1]

In Vincent Cronin's biography of Napoleon, which runs to some 470 pages, only these sentences and a few others describe Napoleon's physique: Cronin's concern is much more with Napoleon's attitudes and actions than with his appearance, even though it contributed significantly to public perception of his personality and even though the image of Napoleon became familiar through drawings and prints to many ordinary people not only in France but throughout Europe. A biographer seeks not so much to describe the appearance of a person as to understand a whole personality and to situate that individual within a society. A graphic pen-and-ink sketch of Paul Vidal de la Blache's countenance drawn five years before he died undoubtedly enlivens T. W. Freeman's biographical essay[2] and enables us to connect vicariously with the founding-father of the French school of regional geography, but it is Vidal de la Blache's ideas – their origins, evolution and impact – and not his physiognomy which retain our

1

attention. Those ideas cannot be taken at face value but have to be evaluated critically within the context of social practice and social thought in France during the nineteenth century, for Vidal de la Blache was working in a period of political and intellectual turmoil which both influenced his own thinking and was itself influenced by it.[3]

People may, in this regard, be taken as a metaphor for place: a sketch of a landscape may capture or caricature its essence but it cannot be expected to reveal very much about its whole, complex, character.[4] Even the appearance of a place is not easily to be described[5] but a landscape also needs to be situated within its own natural and cultural history – its own ancestry and upbringing – if it is to be properly understood. Places have been perceived as people: Vidal de la Blache recognized that a region has a 'geographical personality'; Fernand Braudel that a country has an 'identity'; and Marvyn Samuels that a landscape has a 'biography'.[6] A place (like a person) needs to be recognized in terms not only of individuality but also of contextuality, as a product both of nature and of nurture. A landscape is a social construction, both intentionally and unintentionally. Landscape is not only, as Dennis Cosgrove has described it, 'a way of seeing';[7] it is also 'a way of thinking and a way of doing' which involves what David Harvey terms a process of 'creative destruction'.[8] The explicit 'idea of landscape' might only date from the Renaissance but both what have come subsequently to be termed 'landscapes' and also ideas about landscapes existed centuries earlier. The 'idea of landscape' has a more restricted historical and geographical currency than does 'the landscape of ideas'.

Mainly on ideology

A landscape is a resultant of attitudes and actions; but to the extent that actions are themselves outcomes of attitudes the latter deserve – but have by no means always been granted – a privileged status over the former in historico-geographical studies.[9] The process of landscape creation and reformation was captured tangentially in Marx's view of history as being specific to particular places: at every period of history there is 'a material outcome ... a historically created relationship to nature and of individuals towards each other', a sum total of production forces 'that is transmitted to each generation by its predecessor' and 'on the one hand is modified by the new generation but on the other itself prescribes its own living conditions and imposes upon it a definite development, a special character of its own – so that, in other words, circumstances make men just as men make circumstances'.[10] Historical specificity can thus be envisaged as underpinning the individuality of a landscape. Such an approach to landscape change allows for, indeed requires, consideration both of intended and unintended consequences of actions, and of material and non-material motivations. It recognizes the possibility of a false consciousness on the part of people who have enacted historical roles assigned to them by forces which they did not understand and which can only be comprehended by an observer, while it also allocates a role to ideology as a system of ideas that aspires both to explain the world and to change it. Marx insisted that:

a distinction should always be made between the material transformation of the economic conditions of production which can be determined with the precision of the natural science, and the legal, political, religious, aesthetic or philosophical – in short ideological – forms in which men become conscious of this conflict and fight it out. Just as our opinion of an individual is not based on what he thinks of himself, so we cannot judge such a period of transformation by its own consciousness; on the contrary, this consciousness must rather be explained from the contradictions of material life, from the existing conflict between the social forces of production and the relations of production.[11]

Historical studies of landscapes must be grounded in an analysis of material structures: they are properly concerned with tangible, visible expressions of different modes of production, with hedgerows and field systems, with canals and factory systems. But such material structures are created and creatively destroyed within an ideological context: such studies must therefore also acknowledge that landscapes are shaped by mental attitudes and that a proper understanding of landscapes must rest upon the historical recovery of ideologies.

'Ideology' is an imprecise term, a concept which has acquired a number of different meanings and, as Georges Duby has pointed out, the negative, pejorative connotations which it has developed in a political context have obscured its utility for the purposes of historical analysis which will be best served by the acceptance of the widest of definitions.[12] The two conventional uses of 'ideology', as distinguished by J. B. Thompson, are as 'the lattice of ideas which permeate the social order, constituting the collective consciousness of each epoch' and as 'a consciousness which is in some way "false" and which fails to grasp the real conditions of human existence': Derek Gregory conveniently refers to these respectively as a 'generalized' and as a 'distorted' system of ideas.[13] But Thompson, dissatisfied with the first definition because it is too wide and with the second because it is too narrow, argues for ideology as 'a system of signification which facilitates the pursuit of particular interests' and sustains specific 'relations of domination'.[14] As Gregory has emphasized, this formulation draws critically upon both A. Giddens' structuration theory and J. Habernas' version of critical theory.[15] Giddens has suggested, for example, that 'to analyse the ideological aspects of symbolic orders' is 'to examine how structures of signification are mobilised to legitimate the sectional interests of hegemonic groups'.[16] Duby, as a social historian, adopted Althusser's definition of ideology as 'a system (possessing its own logic and structure) of representations (images, myths, ideas or concepts) existing and playing a historical role within a given society'.[17] Ideology, then, involves systems and structures of signification and domination: any landscape is likely to contain all manner of ideological representations so that a description of its appearance must also logically be 'thickened'[18] into an interpretation of its meaning.

Of the many characteristics of ideology, just three will be addressed here because of their especial relevance to landscape creation and so to landscape interpretation: the connections of an ideology to a quest for *order*, to an assertion of *authority*, and to a project of *totalisation*.[19] An ideology is seen by Edward Shils as 'the product of man's need for imposing intellectual order on the world. The need for an ideology is an intensification of the need for a cognitive and moral map of the universe.'[20] Ideologies offer ordered, simplified visions of the world; they substitute a single certainty for a multiplicity of ambiguities; they tender to individuals both an ordered view of the world and of their own place within its natural and social systems. The function of an ideology in this regard is to furnish assurance; it does so, paradoxically, either by highlighting perfect patterns in the present or by promising utopian forms in the future. In both cases, the concept of order comes to be represented in landscapes, both unintentionally and intentionally. In the case of cities, for example, Yi-Fu Tuan has emphasized that:

The city is a place, a centre of meaning, par excellence. It has many visible symbols. More important, the city itself is a symbol. The traditional city symbolized, first, transcendental and man-made order as against the chaotic forces of terrestrial and infernal nature. Second, it stood for an ideal human community ... It was as a transcendental order that ancient cities acquired their monumental aspect. Massive walls and portals demarcated sacred space. Fortifications defended a people against not only human enemies but also demons and the souls of the dead. In medieval Europe priests consecrated city walls so that they could ward off the devil, sickness, and death – in other words, the threats of chaos.[21]

More particularly, Paul Wheatley has shown how the ordered spatial structure of the ancient Chinese city 'afforded a ritual paradigm of the ordering of social interaction at the same time as it disseminated the values and inculcated the attitudes necessary to sustain it'[22] and Dennis Cosgrove has demonstrated how the Italian Renaissance ideal city, with its geometrically ordered proportions and its monumental buildings, expressed and communicated not only the symbolic order of humanism but also the symbolic authority of the Italian nobility.[23]

Ideologies involve the assertion of authority, transcendental or earthly. Consequently, they are concerned with struggles for power between conflicting interest groups. Ideologies compete with each other, so a given society and landscape may have several different systems of symbolic representations existing within it simultaneously and antagonistically. Ideologies are inevitably political (but not necessarily, of course, party political) and some of the most intense power struggles have been between proponents of allegedly transcendental ideologies. Conversely, even an earthly ideology concerns itself with the sacred because it reaches out to sanctify an entire life-world by bringing every part within its compass: hence the emphasis on the sacred and the profane, light and dark, inside and outside, 'us' and 'them' in any ideology, whether nominally religious or anti-religious, and the avowed importance of converting the latter into the former in each case. Whether authority (as identified by Max Weber) has its locus in tradition and custom, in reason and knowledge, or in the claims of a charismatic leader,[24] the exercise of power is expressed forcefully in language and in a wide range of written texts whose function it is to defend and to propagate the particular system of ideas and values upon which a given ideology is based.

Reconstructing ideologies thus involves identifying, linking and interpreting a mass of archival, documentary sources, ranging from legal codes to personal correspondence. But ideologies exert their authority and find expression not only in language but also in landscape. Non-verbal 'documents' in the landscape can be powerful visual signs, conveying messages forcefully, and their importance is enhanced when we are dealing with historical but (often) predominantly illiterate societies. An historical landscape was, for example, rich in signs of identity, linking an individual with a social group: coats of arms, flags, totems, uniforms, decorations, tattooings, hairstyles, trademarks. It was also replete with expressions of social codes – protocols, rituals, fashions and games – which were elements of communication: gestures and dances, customs and ceremonies, parades and processions, eating and drinking, all provided opportunity for a social message to be communicated, for authority to be asserted, for prestige and power to be proclaimed. This argument is without closure because 'everything is a sign: presents, our houses, our furniture, our domestic animals'.[25] Ideologies create, unintentionally as well as deliberately, a landscape as a system of signification, expressive of authority.

An ideology employs individual, sacred, symbols to signify its own holistic character. Ideologies are, as Duby puts it, 'complete systems, fulfilling, by nature, a *globalising* function; they also claim to offer an overall representation of a society, its past, present and future, integrated into a complete *Weltanschauung*'.[26] Shils argues that an 'ideology contends for the realization of a state of affairs which, its proponents allege, either never existed previously or existed in the past but exists no longer' and that 'ideologies impel their proponents to insist on the realization of the ideal, which is contained in the sacred, through a "total transformation" of society'.[27] Within that 'globalising function' and 'total transformation' landscapes were both products and process, structured and structuring. One way in which an ideology seeks to achieve its hegemony is through total conquest, be it of the physical and/or the human environment; another way is through total withdrawal, so that the ideology's essence can be isolated from the polluting influences of its existing environment. Either way, the reorganization of a landscape signifies a realignment of authority.

Ideological impulses are constantly modifying landscapes. Ideologies, moreover, contain within themselves a contradiction: they seek permanence and continuity but argue for change, for the realisation of fundamental principles through praxis. Ideologies perform a stabilizing role while perfecting transformation; they are simultaneously loaded with inertia and pregnant with change. Some ideologies use systems of representation to protect the privileges of an established authority; others use them to promote new power relations. But 'the ideal social organisation to which the most revolutionary ideologies aspire is still perceived, beyond the battles fought and won in the cause, as the establishment of something permanent; no utopia preaches permanent revolution'.[28] Both the conflicts engaged in explicitly by ideologies and their own internal contradictions are enmeshed in the constant, but not continuous, process of landscape change.

Mainly on landscape

The description and interpretation of landscapes has been a long and honourable tradition within geography: along with the locational school and the ecological school, it stands as one of the main 'deviations' from geography's central concern with regions, places and areas. Like the term 'ideology', that of 'landscape' has been given a number of different – and debated – definitions: it has been most usually employed to refer to the impression conveyed by an area, to the objects producing that impression, and to the area itself.[29] But the character of the landscape school of geography has itself been redefined – and debated – in recent years as part of the humanistic renaissance of modern geography.[30]

The practice (but not the contention) of geography as the study of the transformation of 'natural landscapes' into 'cultural landscapes', of 'changing landscapes', owes much to the research and writings of Carl Sauer and the Berkeley school of cultural geography, to Clifford Darby and the British school of historical geography, and to the writings of Roger Dion in France.[31] Each of those influences was, of course, constructed upon foundations laid by earlier scholars and each derived some of its strengths from liaisons with cognate disciplines, especially archaeology, anthropology and history (particularly economic history). Moreover, their impact has been magnified in association with a concern for landscape history by non-geographers like J. B. Jackson in North America, W. G. Hoskins in Britain and Fernand Braudel in France.[32] Furthermore, the lineage of such studies descends through to the present day in works such as Michael Conzen's *The making of the American landscape*, R. A. Dodgshon and R. A. Butlin's *An historical geography of England and Wales* and Jean-Robert Pitte's *Histoire du paysage français*.[33] These three works are unambiguously 'children' of the founding-fathers of landscape geography, but they have certainly matured and acquired a sturdy independence from them, exhibiting a remarkable diversity of approaches to the study of landscape.

That there are at least 'ten versions of the same scene' has been elucidated by Donald Meinig who argues that different observers of the 'same' prospect might see the landscape before them, depending upon their 'perspectives', as representing *nature* (emphasizing the insignificance of people), *habitat* (as people's adjustment to nature), *artefact* (reflecting people's impact on nature), *system* (a scientific view of interacting processes), *problem* (for solving through social action), *wealth* (in terms of property), *ideology* (revealing cultural values and social philosophy), *history* (as a record of the concrete and the chronological), *place* (through the identity that locations have), and *aesthetic* (according to some artistic quality possessed).[34] More succinctly, Conzen has codified landscape studies into four principal approaches, reflecting respectively concerns with *environmental awareness*, with *symbolic representation*, with *landscape design*, and with *landscape history*.[35] In practice, as Conzen admits, these four approaches collapse into one because:

to view the landscape historically is to acknowledge its cumulative character; to acknowledge that nature, symbolism, and design are not static elements of the human record but change with historical experience; and to acknowledge too that the geographically distinct quality of places is a product of the selective addition and survival over time of each new set of forms peculiar to that region or locality.[36]

Studies of landscape, like those of ideologies, necessitate an historical perspective. Because landscapes have histories they possess a compelling human significance.

'Actual' landscapes are constructions, 'ideal' landscapes are conceptualizations. At the same time, 'actual' landscapes are moulded by ideologies and ideologies are themselves fashioned by 'actual' landscapes: the relationship is reciprocal, the product is a dialectical landscape which is a resolution of nature and culture, of practice and philosophy, of reason and imagination, of 'real' and 'symbolic'. This 'way of seeing' landscapes was not the one adopted by most traditional cultural and historical geographers who focussed their attentions upon landscape expressions of material culture and tended to ignore, or at least to neglect, the *mentalité* of the people who created them. Cultural landscape forms were seen in materialist terms, interpreted straightforwardly as products of work, as people laboured to meet their basic needs for food, water, clothing and shelter and transformed landscapes in so doing – clearing woodland, draining marshland, reclaiming heathland, building farmsteads, erecting barns, constructing railways and so on. This approach to landscape, fundamentally utilitarian and economistic, was grounded in a weak conceptualization of culture which separated off economy from consciousness, action from ideology. Alternatively, we can agree with Sahlins that economy should be regarded as 'a category of culture rather than behaviour, in a class with politics or religion rather than rationality or prudence; not the need-serving activities of individuals, but the material process of society'.[37] Landscapes are indeed products of, *inter alia*, the struggle for survival and the need to 'make a living' but that activity is engaged in by conscious human beings. As Cosgrove – drawing in part upon Vidalian reserves – has aptly put it: 'Any mode of production is a "mode of life", a *genre de vie*, constituted by men [*sic*] and symbolic *ab initio*.'[38] Landscapes of material are also landscapes of meaning: praxis is itself symbolic, all landscapes are symbolic in practice.

A shift away from narrowly materialistic interpretations of landscapes to a broader concern with the meaning of landscapes has not, however, necessarily meant a drift towards idealist interpretations,[39] because material needs and constraints remain as integral components of culture and of landscape. Such a move has already enriched studies of even 'ordinary' landscapes[40] and is increasingly infusing broader syntheses. For example, the essays edited by Conzen on the making of the American landscape are explicitly 'rooted' in the fertile traditions of landscape study and some deal with familiar themes like 'the clearing of the forests' and 'the settlement of the American grassland', but most of the essays bring forth newer fruits like 'French landscapes in North America', 'democratic utopian landscapes', 'landscapes of private power and wealth' and the construction of a 'national landscape'.[41]

The discovery of meaning in landscapes is a challenging but difficult task, both in theory and in practice.[42] Perhaps for that reason it has been described using a great variety of metaphors most of which serve to emphasize the duality of the 'apparent' and the 'hidden' in landscapes: 'surface/depth' figures-of-speech abound. Pitte's likening of a landscape to the *visible tip of an iceberg*[43] was echoed in the use of Lawren Stewart Harris' oil painting *Icebergs, Davis Strait* by Cosgrove and Daniels to illustrate the dust-jacket of their edited collection of essays on the iconography of land[*sic*]scape.[44] Mixing his metaphor, Pitte also likened landscape to a *snapshot* (*pellicule*) of geographical reality[45] (which in turn recalls to mind Marc Bloch's description of the retrogressive method in history as like unwinding a spool of film to reveal earlier and earlier frames).[46] For Pitte, a landscape is also 'an act of liberty: it is a *poem* written on a blank sheet of climax vegetation'.[47] For Audrey Kobayashi, landscape is like a *dance* because it, too, is

a process totalized, the extension and collection of human activity in a material setting. It is organized, choreographed, although seldom so expressively as in the dance; like the dance, it can produce profound emotional effects on the beholder whose senses are awakened to its presentation. Both landscape and the dancer require a practised eye to appreciate their finer qualities. Both are irreducible to their constituent elements, both transcend their moments of expression, yet neither exceeds its physical features: buildings or props, movements, expressions in flux.[48]

Others view landscape as *theatre*, or as *carnival* or as *spectacle*.[49] Alternatively for Kobayashi landscape is 'a form of *language*, and it is necessary to know its history, its structures and its syntax if it is to be understood'.[50] Many other interpreters of landscapes have employed *geological* metaphors, referring to the 'sedimentation' of meaning in landscapes, to 'layers' and 'strata', and to 'underlying structures' of meaning. More frequently, however, landscape is being likened to a *written document* to be read critically as a 'deeply-layered text'.[51] Like texts, landscapes are being opened to Derrida's strategy of deconstruction and, to borrow J. B. Harley's phrase, 'searched for alternative meanings'.[52] To deconstruct (a text or a landscape) is, in Terry Eagleton's alternative metaphor, 'to reverse the imposing tapestry in order to expose it in all its unglamorously dishevelled tangle the threads constituting the well-heeled image it presents to the world'.[53] Easily stated but difficult to accomplish because the landscape, as Roland Barthes has emphasized, is the richest of our system of signs.[54] A common metaphor currently in use in landscape studies is that of a landscape as a *message which has to be decoded*. But while a landscape is composed of a multiplicity of signs it may also contain a plurality of meanings: it is not a simple matter of 'one landscape/one message'. So, to the important tasks of reconstructing 'actual' landscapes of the past and tracking their changes through time has to be added that of deciphering their signs, of discovering their meanings.

Considerable movement towards a redefinition of landscape and towards a theory of cultural landscape change has been made by Cosgrove, in a series of studies initiated some years ago in his paper on 'place, landscape and the dialectics of cultural geography'[55] and by Kobayashi in her recent 'critique of dialectical landscape' which – as its title signals – draws heavily but critically upon some of the ideas of Jean-Paul Sartre.[56] There has been a palpable shift in landscape studies in recent years. Nonetheless, continuities can also be detected running from Vidal de la Blache's concern with the concept of *pays* through the ideas of structuralism to those of deconstruction.[57] For example, Kobayashi's argument that 'the geographical theory of landscape can provide the third component in a triad of action, discourse and object, for a comprehensive, dialectical understanding of social history'[58] provides clearly audible echoes of concepts enunciated deep within the salons of the French geographic tradition.[59] This is hardly surprising because, as Cosgrove has remarked 'the issues raised by landscape and its meanings point to the heart of social and historical theory: issues of individual and collective action, of objective and subjective knowing, of idealist and materialist explanation'.[60] Such issues are fundamental to a humanistic geography, indeed to any humanistic study, and together with the holistic character of landscape, explain why the geographical study of landscape must involve many disciplines in addition to geography.

Place, period and people

Such excursions into the territories of 'theory' have, of course, their own intrinsic interest and value but they are also and ultimately intended to enable us to comprehend more meaningfully landscapes of 'reality'. Better equipped conceptually, we can set to work more effectively in the field, in the record office and in the library. If only it were that simple. In practice, the task of understanding places in their appropriate historical and cultural contexts remains a daunting one because landscape as ideology can take as many forms as people are able to imagine and to create ways of establishing order and authority. The variability of landscape is understandable; landscapes themselves are less easily understood. An intellectual 'high priest' can impose his authority and distinguish between landscapes of dominant cultures and alternative landscapes of residual (relict) features, of emergent cultures and of excluded cultures but such a classification cannot – indeed, does not in this (Cosgrove's) case – claim to be inclusive or of objective validity.[61]

No such order is imposed upon the chapters in this volume. Each author addresses a specific problem but collectively the chapters can be seen to be principally concerned with ideologies of religion and of politics, with ideologies of church and of state. Major ideological changes during the last two thousand years – like the transition from feudalism to capitalism, the secularisation of society, and the emergence of nation states (and dependent colonies) – had profound effects upon landscapes and some of them are touched upon in these chapters. Other important ideological concerns – those which allocate central roles to gender, or to class or to ethnicity and those which value highly a reverence for nature or for the past – are not addressed here. There was never an intention – either at the conference which initially attracted these contributors or in the minds of the editors when coralling some of the conference papers into a book – to be comprehensive. The more limited purpose was that of putting ideology and landscape higher on the agenda of historical geography and perhaps of signalling to cognate disciplines the nature of geographical interest in these trans-disciplinary themes.

NOTES

1 V. Cronin, *Napoleon* (Newton Abbot 1972) 179–80
2 T. W. Freeman, *The geographer's craft* (Manchester 1967) 44–71
3 G. Nicolas-O and C. Guanzini, *Paul Vidal de la Blache: géographie et politique* (Lausanne 1987); H. F. Andrew, The early life of Paul Vidal de la Blache and the makings of modern geography *Transactions, Institute of British Geographers* (1986) 174–82
4 For some field sketches of Alpine landscapes by Emmanuel de Martone (son-in-law to Paul Vidal de la Blache), see J. R. Pitte, *Histoire du paysage français* 2 vols. (Paris 1983) vol. I between 15 and 16. As Pitte points out, these sketches exaggerate geomorphological features at the expense of the biogeographical and cultural aspects of the landscapes
5 H. C. Darby, The problem of geographical description *Transactions, Institute of British Georgaphers* 30 (1962) 1–14; D. W. Meinig, Geography as an art *Transactions, Institute of British Geographers* 8 (1983) 314–28
6 P. Vidal de la Blache, *Tableau de la géographie de la France* (Paris 1903): the first four chapters for part I of this work, with the title 'Personalité géographique de la France'; F. Braudel, *L'Identité de la France: espace et histoire* (Paris 1986); M. S. Samuels, The biography of landscape, in D. W. Meinig (ed), *The interpretation of ordinary landscapes* (Oxford 1979) 51–88
7 D. Cosgrove, *Social formation and symbolic landscape* (London, 1984) esp. 1–38
8 D. Harvey, *The condition of postmodernity* (Oxford 1989) 258
9 A. R. H. Baker, On ideology and historical geography, in A. R. H. Baker and M. D. Billinge (eds.) *Period and place: research methods in historical geography* (Cambridge 1982), 233–43
10 Quoted in H. Fleischer, *Marxism and history* (New York 1973) 21–2
11 Quoted in R. Aron, *Main currents of sociological thought* vol. II (Harmondsworth 1968), 120
12 G. Duby, Ideologies in history, in J. Le Goff and P. Nova, *Constructing the past: essays in historical methodology* (Cambridge 1985) 151–65
13 J. B. Thompson, *Critical hermeneutics* (Cambridge 1981); D. Gregory, Ideology, in R. J. Johnston, D. Gregory and D. M. Smith (eds.), *The dictionary of human geography* 2nd edn (Oxford 1986) 214–15
14 J. B. Thompson, *Studies in the theory of ideology* (Cambridge 1984)
15 Gregory, Ideology, 214
16 A. Giddens, *Central problems in social theory: action, structure and contradiction in social analysis* (London 1979)

17 Duby, Ideologies, 152

18 The metaphor is derived from C. Geertz, Thick description: toward an interpretative theory of culture, in *The interpretation of cultures: selected essays* (New York 1973), 3–30

19 This section draws upon E. Shils, The concept and function of ideology, in D. L. Sills (ed.), *International encyclopedia of the social sciences* vol. VII (New York 1968) 66–76

20 Shils, Concept and function, 69

21 Yi-Fu Tuan, *Space and place: the perspective of experience* (London 1977) 173

22 P. Wheatley, *The pivot of the four quarters: a preliminary enquiry into the origins and character of the ancient Chinese city* (Chicago 1971) 478

23 D. E. Cosgrove, Problems of interpreting the symbolism of past landscapes, in A. R. H. Baker and M. D. Billinge (eds.) *Period and place: research methods in historical geography* (Cambridge 1982) 220–30

24 P. Claval, *Espace et pouvoir* (Paris 1978) 28–30, drawing upon M. Weber, *Economie et société* (Paris 1971) 219–307

25 This section draws upon P. Guiraud, *Semiology* (London 1975). The quotation is from p. 90. For specific examples, see W. Zelinsky, O say can you see? Nationalistic emblems in the landscape *Winterhur Portfolio: A Journal of American Material Culture* 19 (1984) 277–86; D. Upton, White and black landscapes in eighteenth-century Virginia *Places* 2 (1985) 59–72; V. Konrad, (ed.) Focus: nationalism in the landscape of Canada and the United States *Canadian Geographer* 30 (1986) 167–80

26 Duby, *Ideologies*, 153

27 Shils, *Concept and Function*, 67

28 Duby *Ideologies*, 153

29 M. W. Mikesell, Landscape, in D. L. Sills, (ed.), *International encyclopedia of the social science* vol. VIII (New York 1968) 575–80

30 For recent 'progress reports' on the place of landscape studies in geography, see D. Ley, Cultural/humanistic geography *Progress in Human Geography* 5 (1981) 249–57, 7 (1983) 267–5, 9 (1985) 415–23; L. Rownstree, Cultural/humanistic geography *Progress in Human Geography* 10 (1986) 580–6; Cultural/humanistic geography *Progress in Human Geography* 11 (1987) 558–64; D. Cosgrove, A terrain of metaphore: cultural geography 1988–89 *Progress in Human Geography* 13 (1989) 566–75; D. Cosgrove, ... Then we take Berlin: cultural geography 1989–90 *Progress in Human Geography* 14 (1990) 560–8

31 C. O. Sauer, The morphology of landscape *University of California Publications in Geography* 2 (1925) 19–54; H. C. Darby, The changing English landscape *Geographical Journal* 117 (1951) 377–94; R. Dion, *Essai sur la formation du paysage rural français* (Tours 1934)

32 See, for example; J. B. Jackson, *Landscapes: selected writings of J. B. Jackson* (Amherst, Mass. 1970); W. G. Hoskins, *The making of the English landscape* (London 1955); F. Braudel, *L'Identité de la France: espace et histoire* (Paris 1986)

33 M. Conzen (ed.), *The making of the American landscape* (Boston 1999); R. A. Dodgshon and R. A. Butlin (eds.), *An historical geography of England and Wales* (London 1990); J. R. Pitte, *Histoire*

34 D. Meinig, The beholding eye: ten versions of the same scene *Landscape Architecture* 66 (1976) 47–54. See also P. Lewis, Learning from looking: geographic and other writing about the American cultural landscape *American Quarterly* 35 (1983) 242–61

35 Conzen, *American landscape*, 3–4

36 *Ibid.*, 4

37 M. Sahlins, *Stone Age economics* (London 1974) xii. See also his *Culture and practical reason* (Chicago 1976)

38 For an extended account of the idealist alternative, see L. Guelke, *Historical understanding in geography* (Cambridge 1982)

39 Cosgrove, Symbolism of past landscapes, 221

40 Meinig, *Interpretation of ordinary landscapes*

41 Conzen, *American landscape*

42 For some discussion of the problems see, in addition to literature already cited here, C. Rose, Human geography as text interpretation, in A. Buttimer and D. Seamon (eds.), *The human experience of space and place* (London 1980) 123–34; D. Cosgrove, Geography is everywhere: culture and symbolism in human landscapes, in D. Gregory and R. Walford (eds.), *Horizons in human geography* (Basingstoke 1989) 118–35; I. Hodder, Converging traditions: the search for symbolic meanings in archaeology and geography, in J. M. Wagstaff (ed.) *Landscape and culture: geographical and archaeological perspectives* (Oxford 1987) 134–45

43 Pitte, *Histoire*, 23

44 D. Cosgrove and S. Daniels (eds.) *The iconography of landscape* (Cambridge 1988)

45 Pitte, *Histoire*, 23

46 M. Bloch, *The historian's craft* (Manchester 1954) 46

47 Pitte, *Histoire*, 24

48 A. Kobayashi, A critique of dialectical landscape, in A. Kobayashi and S. Mackenzie (eds.), *Rethinking human geography* (Boston 1989) 164–83. The quotation is from p. 164.

49 D. Cosgrove and S. Daniels, Fieldwork as theatre: a week's performance in the Venice region *Journal of Geography in Higher Education* (1989); P. Jackson, 'Street life': the politics of carnival *Environment and Planning D: Society and Space* 6 (1988) 213–28; D. Ley and K. Olds, Landscape as spectacle: world's fairs and a culture of heroic consumption *Environment and Planning D: Society and Space* 6 (1988) 191–212; D. Cosgrove, Spectacle and society: landscape as theatre in pre- and post-modern cities, in P. Groth (ed.), *Vision: culture and landscape* (Berkeley 1990) 221–39

50 Kobayashi, Dialectical landscape, 164–5

51 J. Duncan and N. Duncan, (Re)reading the landscape *Environment and Planning D: Society and Space* 6 (1988) 117–26

52 J. B. Harley, Deconstructing the map *Cartographica* 26 (1989) 1–20

53 T. Eagleton, *Against the grain* (London 1986) 80

54 R. Barthes, *L'Empire des signes* (Geneva 1970). Cited in Pitte, *Histoire*, 20

55 D. Cosgrove, Place, landscape and the dialectics of cultural geography *Canadian Geographer* 22 (1978) 66–72

14   Alan R. H. Baker

Kobayashi, Dialectical landscape
57 P. Claval, De Vidal de la Blache au structuralisme, in P. Claval and J. P. Narady, *Pour le cinquantenaire de la mort de Paul Vidal de la Blache, Annales Littéraires de l'Université de Besançon* 93 (1968) 115–25
58 Kobayashi, Dialectical landscape, 182
59 For a survey, see A. Buttimer, *Society and milieu in the French geographic tradition* (Chicago 1971)
60 Cosgrove, *Symbolic landscape*, 38
61 Cosgrove, Geography is everywhere, 127–33

1

Ideology and landscape in early printed maps of Jerusalem

REHAV RUBIN

Cartography, the art and science of map-making, is thought by many to be a scientific and objective method of portraying the morphology of lands and seas throughout the world. Believed to be so, maps have been judged often as being 'true or false', 'accurate or inaccurate'. Actually, this concept of cartography and maps is misleading. In most cases maps should be treated as media of communication,[1] or a kind of language.[2] Maps should be critically studied and judged in the same way as literary verbal sources and documents are by historians and other scholars. While literary documents use words and phrases in order to promote ideas, or to influence beliefs, maps are designed to do just the same, but using graphic figures instead of words. Maps, especially before the modern, measurement-based, topographical ones, were often value-laden images and were used in many cases to create political and social images.[3]

Moreover, maps have unique characteristics which give them an advantage over literary documents. Using graphic figures and creating visual images they have a strong influence, sometimes even stronger than verbal or documentary sources in presenting values and transferring messages. By supplying a visual image the map gives another dimension to the image it represents, and restricts, or even overtakes, the freedom of its reader to create an image of his own. Thus, we believe that as a medium of communication maps have a stronger influence than the literary text.

Jerusalem has been conceived in the Christian world as the Holy City and the centre of the world. Therefore, there was an immense output of Christian thought, art and literature dealing with the Holy City, and cartography was part of this.

The number of maps of Jerusalem, created in Europe between the invention of the printing press in the middle of the fifteenth century and the early measured maps from the nineteenth century,[4] is estimated at about 250–300 items.[5]

This large volume of maps has produced two consequences. On the one hand, it brought about a large body of material for research – many maps made by different people of various countries, cultures and religious ideas. On the other hand, almost none of the maps and none of their makers were objective or indifferent towards Jerusalem as an object for mapping. For all, the city was an object for worship viewed through ideology and historical sentiments. Therefore these maps provide us with a special opportunity to study the relation between ideology and maps.

The maps of Jerusalem – the Holy City – were iconographic by nature. The map-makers, while creating these maps, could not avoid expressing their own thoughts, feelings and associations referring to the concept of the Holy City. Indeed, many of these maps are reflections of ideology, concepts and propaganda, rather than realistic portraits of the city. They are much more depictions of the Jerusalem which existed in the hearts and minds of Christian European map-makers and map-readers, than portraits of Jerusalem of earth, stone and dust.

The aim of this chapter is to use these early printed maps of Jerusalem as a series of case-studies, showing how Christian–European values, ideas and propaganda were transferred through maps. We will show that it was not only political ideology that used maps, in the way many European maps of countries and regions did,[6] but religious ideology – a Christian one – which was so important in the case of Jerusalem.

The early printed maps of Jerusalem are perhaps an appropriate case-study because at that time, the sixteenth to eighteenth centuries, Jerusalem was a remote and unimportant town under Ottoman Turkish rule, a city that most Europeans could think of, dream of, and pray for, but did not visit and see with their own eyes. Therefore map-makers had the freedom to depict it almost as they wished, and as a result they played an important role as image makers.

Reviewing the numerous maps of Jerusalem brings about a distinction, suggested by Ben-Arieh and Elhasid, between two main categories of maps.[7] The maps in the first category described an imaginary historical-Biblical Jerusalem, and were based mainly on the Holy Scriptures and the works of Josephus Flavius. Most of these maps were drawn by artists and scholars who had never visited the city. These imaginary maps are conceptual – ideological documents which reflect the Christian images of Jerusalem in the time of Christ rather than the realistic landscape of the city. This can easily been shown by comparing the topography of the city with the historical imaginary maps like the ones drawn by Adrichomius (Fig. 1.1), Villalpandus or the Dutch Visscher family, which were popular and were copied often during the period under discussion.[8]

The study of the ideological element in these maps is an almost endless task. They are extremely detailed and every item needs to be checked to assess its historical and ideological value. These maps are actually repre-

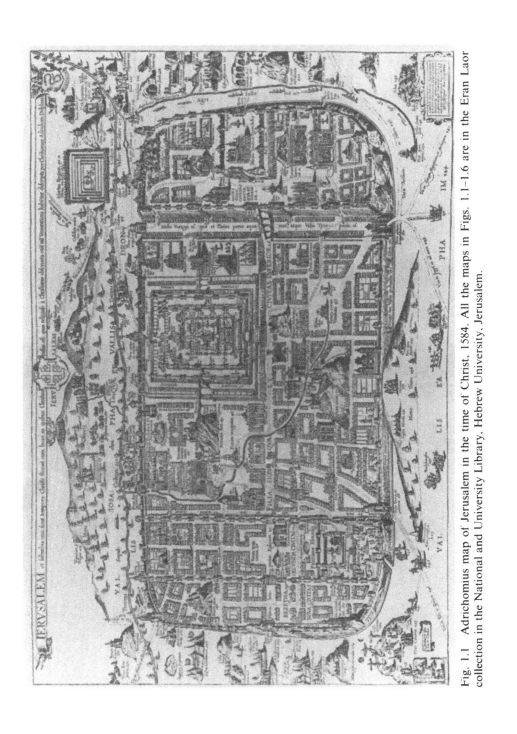

Fig. 1.1 Adrichomius map of Jerusalem in the time of Christ, 1584. All the maps in Figs. 1.1–1.6 are in the Eran Laor collection in the National and University Library, Hebrew University, Jerusalem.

Fig. 1.2 Quaresmius map of Jerusalem, 1630.

sentations of what their makers thought of Biblical Jerusalem, its buildings, people, monuments and the events described in the Holy Scriptures.

The second group of maps, defined as the 'realistic' ones, claimed to portray contemporary Jerusalem. Some were drawn by pilgrims and travellers, based on their own impressions, while others are copies and imitations drawn in Europe by people who had never seen the city but who used eye-witnesses' maps as a basis for their work.[9]

In this discussion we prefer not to delve into the study of the imaginary maps of the first group, where the conceptual and ideological aspects are obvious. We shall rather look at those maps which were drawn by people who had actually visited Jerusalem, and whose maps were based upon direct eye-witness impressions, and which can justly be defined as 'realistic', for example, the Quaresmius map of Jerusalem (Fig. 1.2). The study of these maps shows that even the most reliable ones, drawn by people who stayed in Jerusalem for several months, and even for years, contain important ideological elements. The following case-studies identify examples of these ideological elements in maps drawn between the fifteenth and the eighteenth centuries.

The orientation

Many of the early maps of the Holy Land were drawn facing east. It was not a rule and there are examples of maps with different orientations, but the eastern orientation was the most common one.[10] But in the case of Jerusalem, most of the maps were drawn from the top of the Mount of Olives, thus facing west. This was for practical and ideological reasons: the Mount of Olives is the best vantage point to look at the city, and at the same time it is the traditional site where Jesus Christ stood and prophesied upon the city (Luke 19: 41–4), and the site of his Ascension. But beside these reasons for the western orientation, map-makers used orientation as one of their tools for emphasising sacred sites, the best example perhaps being Breydenbach's map (Fig. 1.3).

In 1483 Bernard von Breydenbach, a noble clergyman from Mainz, Germany, went on a pilgrimage to the Holy Land.[11] He took with him a Flemish artist, Erhard Reuwich, who created many illustrations and a large detailed map of the whole country, with Jerusalem in its centre. This map was drawn facing east even showing the pilgrims' ship anchoring near Jaffa. But Jerusalem itself was drawn from the top of the Mount of Olives, looking from the east westward, and thus rotated 180° in relation to the rest of the map.

In the centre of Jerusalem, the Church of the Holy Sepulchre is depicted with its facade facing the viewer. In reality this facade faces south and can not be seen from the east. Therefore, the church was rotated in the map 90°

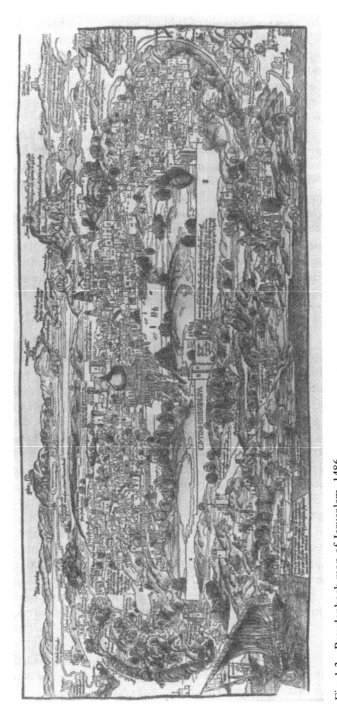

Fig. 1.3   Breydenbach map of Jerusalem, 1486.

counter-clockwise, in such a manner that instead of facing south, as it does in reality, it was drawn as if it were facing east, so that it could be clearly viewed by the reader of the map. Thus, we can see three different orientations in the same map: that of the whole country, that of Jerusalem and that of the Church of the Holy Sepulchre. All this was done in order to emphasise the Church of the Holy Sepulchre within Jerusalem, and the Holy City within the entire country, because of their holiness in the map-maker's conception.

Religious symbols

In Jerusalem religious symbols are an integral part of the landscape. Mosques' minarets and churches' belfrys and spires with their crescents and crosses decorate the city's skyline. These symbols were used throughout the city's history as elements in the competition between Christianity and Islam. A famous example of this competition is the huge golden cross erected by the Crusaders on the top of the Dome of the Rock, which was destroyed by Saladin, who replaced the cross by a crescent immediately after he reconquered the city.[12]

This competition between crescents and crosses was depicted on the maps. In some maps of Jerusalem, crescents on top of mosques and minarets were indeed drawn[13] (Fig. 1.4), while others ignored them. But none of the artists and map-makers could ignore the minarets which were too prominent, and these were used by the artists to symbolise the oriental character of the city's landscape and skyline.

At the same time, among the reliable maps there are none with crosses depicted above all the churches of Jerusalem. Some maps put crosses above the Church of Holy Sepulchre, while some did not draw crosses at all. Only one map – the De Angelis map of 1578 – which is very important and reliable, put an impossible cross on top of the al-Aksa mosque defined as the site of the presentation of the Blessed Virgin Mary.[14]

It seems that under Muslim rule there were crescents on top of every minaret but no crosses on top of churches. Yet many of the map-makers purposely ignored that prominent symbol of Islam in Jerusalem. This attitude towards the crescents as Islamic symbols is a clear case of Christian conceptual preference and perhaps we should use here the term propaganda.

Nomenclature and anachronism

The early printed maps of Jerusalem contain many terms and names written in the map body or in the legend. These terms identify sites and connect them with historical traditions. Therefore these terms were a suitable means to emphasise ideas and ideologies.

Fig. 1.4   Munster map of Jerusalem, 1550.

An interesting case is the names of the mosques on the Temple Mount in maps of the sixteenth to eighteenth centuries. In that period Jerusalem was under Ottoman Muslim rule. In the eyes of Islam, the Temple Mount was the most sacred area in Jerusalem, and the third in its holiness in the world. Non-Muslims were not even permitted to enter it.

Nevertheless, many of the maps include anachronistic terminology, calling the Dome of the Rock and the al-Aksa mosque by their Crusader names – Templum Domini and Templum Salomonis – respectively.[15] Other maps identified the al-Aksa Mosque as Templum Simeon[16] or as the site of the 'presentation of the Blessed Mary'.[17] The first terms are anachronistic, using in the sixteenth to eighteenth centuries a Crusaders' (twelfth-century) nomenclature. The other terms represent Christian traditions which relate to the Temple Mount, referring to it as part of the Christian holdings in the city and ignoring the Muslim rule of these holy sites.

Use of these anachronistic terms, and especially the terms Templum Domini and Templum Salomonis, is not a 'mistake'. These terms were used through a long series of maps from the early printed ones in the late fifteenth century to the nineteenth century. Here, like in the case of the crescents, it was part of an ideological tendency to ignore the Islamic rule of Jerusalem and its holy places and particularly of the Temple Mount. Using these Crusader terms is a way of commemorating the glory of this past period when the city was ruled by Christians.

Multi-period time concept

Modern maps are usually updated once in a while, and the date of their last revision is printed in the legend. This is part of the general concept that

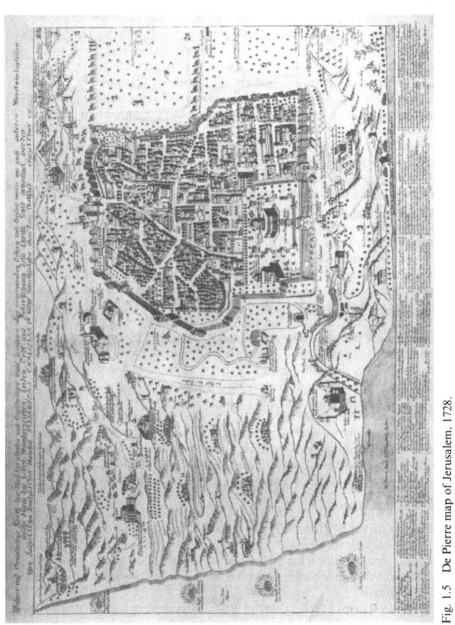

Fig. 1.5 De Pierre map of Jerusalem, 1728.

maps should be, for practical reasons, time-bound documents. The application of this concept of time to maps was uncommon in the early maps of Jerusalem. We shall use, as an example of the dual or multi-period concept of time in these maps, a map of Jerusalem printed in 1728 by De Pierre, a person whose origin and life are almost unknown (Fig. 1.5).[18]

In this map, the city itself was depicted fairly accurately, but the area around the city was drawn rather freely without applying the rules of perspective. Large parts of the Judean desert and Judean hills are portrayed in this free manner, and many sites in these areas are shown and identified by captions.

Among the sites, we can find on the one hand Biblical traditions from the Old Testament, such as Adam and Eve, Cain and Abel, Abraham and the three angels, David and Goliath, together with traditions from the New Testament, like the well of the three Magi, the suicide of Judas Iscariot. On the other hand, side by side with these Biblical traditions, we find daily scenes from the author's own time, such as the Arab horsemen, the camel caravans entering Jerusalem or the Greek Orthodox monk from St Sabas monastery in the Judean desert.

The concept of the Holy City and the Holy Land was so dominant in the map-maker's eyes and mind that it overtook the concept of time in such a way that all these items could be shown side by side on the same map. This concept of time, actually the multi-period concept of time, made the map-readers feel that they were actually studying the real Bible Land, where the historical Biblical events are really part of the landscape and can be seen, side by side with elements of the contemporary landscape and daily life seen by the eighteenth-century map-maker.

This attitude towards time and traditions is quite common in other maps, such as the maps of the Holy Land by van Doetecum, copied later by Fairborn and by Sayer.[19] In these maps the whole country was portrayed with many small figures presenting scenes from the Holy Scriptures, in a beautiful, somewhat naive picturesque way (Fig. 1.6).

Reality and images

Another aspect of dualism in portraying Biblical traditions is the appearance of realistic elements of the landscape side by side with figures and scenes representing religious traditions. Having described above the dualism of the time concept, we will refer now to the dualism in portraying the landscape.

In several maps of Jerusalem the Biblical scenes were portrayed as if they were a real part of the landscape and could really be seen. In the Jerusalem map of Hermanus van Borculus, printed in Utrecht in 1538,[20] two religious scenes are depicted in the foreground. The one on the left shows Jesus

Fig. 1.6   Details from the maps of Allard, c. 1700, and Sayer 1786.

Christ's entrance into the city on Palm Sunday, and the one on the right portrays the stoning of St Stephen next to the city gate. Both scenes are drawn in the foreground of a map which is reliable, realistic and full of details, and are copied in the later maps of van Doetecum and Sayer.[21]

The same kind of mixture of reality and traditional images exists in other maps. In the Quaresmius map of Jerusalem (1639),[22] Judas Iscariot hanging from a tree, and the stoning of St Stephen are portrayed. In the same map there are two bridges across the Kidron Brook. The northern one is next to Gethsemane and the southern is next to Absalom's Tomb. Both bridges are indeed realistic portrayals and did exist.[23] In fact, the main road to Jericho still passes over the northern bridge, and remains of the southern one can still be seen. But next to the southern bridge, into the realistic context, the place where Jesus Christ had crossed the Kidron Brook is portrayed, and his footprints are depicted on the map as if they were still there! This depiction of the bridge and the site of Christ's crossing were also copied in many other maps. Indeed these scenes and many other similar ones were drawn as if they could really be seen there at the time of the map-maker, because this was the image of the city in the minds of the map-maker and his readers, and because it represented their conceptual world, cultural and religious.

Statement and silence

In most of the realistic maps, sites are identified by captions within the map or by legends in their margins. Some legends are fairly short, containing about twenty items or less, and naturally most of these items are Christian holy sites.

In some maps there are extremely detailed legends, like those in the maps by De Angelis (90 items) (Fig. 1.7)[24], Amico (66 items),[25] Quaresmius (115 sites)[26] and De Pierre (about 200 items in the legend and in captions within the map). We will use these maps of Jerusalem as rich examples for content analysis.

As Harley claimed, the cartographical contents of a map should not be studied only positively, through what is mentioned, but also through the silence, the blank areas in the map, the things that are missing, that were ignored or overlooked by the map-maker.[27] Analysis of the most detailed legends from this viewpoint demonstrates the attitudes and values of the map-makers of the most reliable maps of Jerusalem.

Antonio De Angelis, a Franciscan monk who spent eight years in Jerusalem, published a map of the city in 1578.[28] In his map there is a legend containing ninety items. Two of these ninety items relate to the Jews in Jerusalem, then a minority in the city; seven refer to the Muslims, then the majority and the rulers of the city; eight items refer to the city's walls

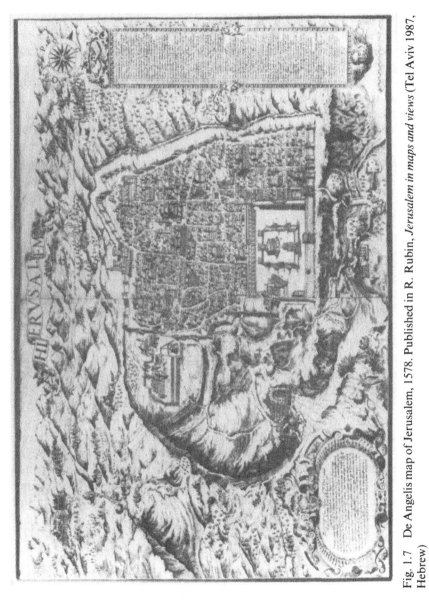

Fig. 1.7   De Angelis map of Jerusalem, 1578. Published in R. Rubin, *Jerusalem in maps and views* (Tel Aviv 1987, Hebrew)

and gates and its daily life; and all the remaining *seventy-three* items deal with Christian holy sites and traditions. Moreover, even the few sites mentioned in regard to the Jews and the Muslims are often in a negative connotation, like in De Angelis map item, no. 72: 'the place where the Jews live, but they are a few'. On the other hand, even geographical objects are described in their religious context and not as topographical features, as for example, 'the Siloam pool where the Lord enlightened the eyes of the blind man' (no. 34), or 'the spring where the Blessed Virgin Mary washed the clothes' (no. 86, as a description of the Gihon spring).

The same attitude and proportion will be found in other maps with such detailed legends like the above-mentioned maps by Quaresmius (1639), Amico (1620) and others. It seems, therefore, that the map-makers emphasised the Christian sites and subjects and kept silent almost entirely in regard to the Muslims, Jews and even aspects of contemporary daily life in Jerusalem.

Conclusion

In this chapter we demonstrate that there was a strong ideological element in the early printed maps of Jerusalem. As Jerusalem was in this period a small, remote unimportant town in the Ottoman Islamic empire, there is a contradiction between the large number of printed maps and the immense attention that the city attracted, given its political and economic importance at the time. The maps with realistic contents were at the same time conceptual portraits of the city. So these maps are an interesting series of cases of the preference of ideology over reality. The use of several case-studies emphasizes that even those maps which were drawn by reliable people, who had visited Jerusalem or stayed in the city for a long time, who knew the city and its reality, were full of devotion and emotion, and religious motives.

These maps served as medium of communication which transferred mainly ideology, concepts and symbols of the Holy City as it was in the eyes of the European Christian map-makers and their readers. On one hand, these 'realistic' maps reflect the topographical reality of the city's landscape, but at the same time, and perhaps much more than that, they reflect the ideological concept of the Holy City.

NOTES

1 N. Thrower, *Maps and man* (New Jersey 1972) 1

2 J. B. Harley, Maps, knowledge, and power, in D. Cosgrove and S. Daniels (eds.), *The iconography of landscape* (Cambridge 1988) 277–312 and esp. 278

3 J. B. Harley, Silence and secrecy: the hidden agenda of cartography in early modern Europe *Imago Mundi* 40 (1988) 57–76

4 The first map of Jerusalem based on actual measurements was printed in 1818. It is known as Sieber, 1818. See Y. Ben-Arieh, The first surveyed maps of Jerusalem *Eretz-Israel* vol. XI (M. Dunayevsky Memorial Volume, Jerusalem 1973, Hebrew) 64–74

5 E. Laor, *Maps of the Holy Land, a cartobibliography of printed maps 1475–1800* (New York and Amsterdam 1986)

6 J. B. Harley, Silence and secrecy: the hidden agenda of cartography in early modern Europe *Imago Mundi* 40 (1988) 57–76

7 J. Ben-Arieh and N. Elhassid, Some notes on the maps of Jerusalem 1470–1600, in A. Cohen (ed.), *Jerusalem in the early Ottoman period* (Jerusalem 1979, Hebrew) 121–51

8 C. Adrichomius, *Jerusalem et suburbia eius, sicut tempore Christy floruit* ... Coloniae Agrippinae (Cologne 1584), J. B. Villalpando, Vera Hierosolymae veteris imago a Ioanne Baptista Villalpando ... (Rome 1604); see also R. Rubin, *Jerusalem through maps and views* (Tel Aviv 1987, Hebrew)

9 For a general discussion on original and copied maps of Jerusalem see R. Rubin, Original maps and their copies: cartogenealogy of the early maps of Jerusalem, from the 15th to the 18th century *Eretz-Israel* vol. XXI (Amiran Volume, Jerusalem 1991, Hebrew)

10 N. Kadmon, The orientation of maps of Eretz-Israel in general and old Arabic maps in particular, in N. Kadmon (ed.), *Cartography in Israel 1985* (Tel Aviv 1985, Hebrew)

11 Bernard von Breydenbach, *Peregrinatio in terram sanctam* (Mainz 1486)

12 See S. Shein, *The history of Eretz Israel under Moslem and Crusader rule (634–1291)* (Jerusalem 1981, Hebrew) 196, 328

13 For a map with crescents, see for example S. Munster's map of Jerusalem. The Munster map is entitled 'Jerusalem civitas sancta', and appears in Munster's book *Cosmographia universalis* (Basle 1550) 1016–17. For a map which ignored the crescents, see for example R. Rubin, The map of Jerusalem by Hermanus Borculus (1538) and its copies *The Cartographic Journal* (1992). Many other maps are photographed in Rubin, *Jerusalem*

14 R. Rubin and M. Levy, The legend of the De Angelis map *Cathedra* 52 (1989) 112–19 (Hebrew)

15 For examples of the terms Templum Salomonis and Templum Domini on maps see those of Braun and Hogenberg (1575); P. Ligorio (1559); L. Des Hayes (1629) De Pierre (1728) and even as late as the map of N. Whittock (1839). For all these maps see Rubin, *Jerusalem*

16 Breydenbach, *Peregrinatio*

17 A. Moldovan, The lost De Angelis map of Jerusalem, 1578 *The Map Collector* 24 (September 1983) 17–24; R. Rubin and M. Levy, The legend of the De Angelis map *Cathedra* 52 (1989) 112–19 and note 2 (Hebrew)

18 De Pierre, Map of Jerusalem 1728. See Laor, *Maps of the Holy Land* 145–6; R. Rubin, A chronogram dated map of Jerusalem *The Map Collector* (1992)

19 For the Van Doetecum map see K. Nebenzhal, *Maps of the Holy Land* (New York 1986) 120–1. The Sayer map is part of the Laor collection. See Laor, *Maps*

of the Holy Land 162, 185. The Fairburn map is kept in the British Library. See *The British Catalogue of printed maps, charts and plans* (London 1967) vol. VIII, 131, no. 705 (26)

20 R. Rubin, Map of Hermanus Borculus

21 See notes 14 and 19

22 F. Quaresmius, *Historica et moralis Terra Sanctae* (Antwerp 1639) For a photo-graph of the map see Rubin, *Jerusalem* 125

23 R. Rubin, Old maps of Jerusalem as historical–geographic sources *Studies in the geography of Eretz Israel* 12 (1986) 52–64

24 See note 17 above

25 Bernardino Amico, *Trattato dell piante et immagini de sacri edifizi di Terra Santa* (Florence 1620); B. Amico, *Plans of the sacred edifices of the Holy Land* trans-lated by T. Belloni and E. Hoade, preface and notes by B. Bagatti (Jerusalem 1953)

26 Quaresmius, *Terra Sanctae*

27 Harley, Silence, esp. 57–8

28 See R. Rubin, The De Angelis map of Jerusalem (1578) and its copies *Cathedra* 52 (1989) 100–11 (Hebrew) and also note 17 above

2

Ideological contexts and the reconstruction of Biblical landscapes in the seventeenth and early eighteenth centuries: Dr Edward Wells and the historical geography of the Holy Land

ROBIN A. BUTLIN

In the early years of the eighteenth century Dr Edward Wells, a clergyman of the Church of England and rector of a church in south Leicestershire, published four volumes on the historical and geographical background to the places, landscapes and events of the Old and New Testaments, using the term 'historical geography' in their titles.[1] In these works he directly and indirectly reflected a range of past and then contemporary theological and geographical ideologies and perspectives, and through these portrayed the geography behind the events and tenets of belief in the Bible. Wells had graduated in arts and divinity and had taught geography, rhetoric, and mathematics at Christ Church, in the University of Oxford, from 1697 to 1701. He had not visited the Holy Land himself, and his historical geographies of the Old and New Testaments were based on his own analysis of the contents of the Bible and use of a wide range of secondary sources, ranging from the accounts of the classical Greek and Roman geographers to seventeenth- and early-eighteenth-century maps and accounts of travel, together with contemporary Biblical and theological treatises.

Although there were many European accounts of the 'sacred geography' of the Holy Land published in the seventeenth and eighteenth centuries, the best, including Wells' works, are of considerable interest, reflecting as they do a fascinating amalgam of complex issues and ideologies pertaining to the changes in Western Christian understanding of the interrelationships between people, environment and divine purpose. They are located at the juncture, in the late seventeenth century in particular, of theologically and scripturally informed knowledge, and show the tension between revealed knowledge (based on sacred texts) and experimental knowledge (based on mathematics, navigation and exploration) that was increasing at this time and was central to the development of geographical thought. The best of these works cannot simply be dismissed as 'armchair geographies', and should be seen as important and revealing accounts and symbols of a series of cultural and intellectual trends which were to continue to influence geographical and theological thought well into the nineteenth century. Wells' historical geographies provide interesting possibilities for the analysis of the relationship between ideologies of the late seventeenth and early eighteenth centuries and the perception of the places and landscapes of the Holy Land.

This chapter will examine, therefore, the contexts of Wells' descriptions of Biblical landscapes in the late seventeenth and early eighteenth centuries, and pay particular attention, through analysis of his texts, to the nature and content of contemporary geography (especially historical geography) which they portray; the very broad issue of 'sacred geography' as a major topic of European theological interest, both Catholic and Protestant; and the types of evidence, including maps and travellers' accounts, used by him and other geographers of the period.

Edward Wells: a brief biography

Edward Wells (1667–1727), the son of Edward Wells, vicar of Corsham in Wiltshire, was admitted as a scholar at Westminster School in 1680, and was elected as a student to Christ Church, Oxford, in 1686, taking his BA in 1690 and his MA in June 1693, and his Bachelor and Doctorate degrees in Divinity in April 1704. He was tutor and lecturer in rhetoric at Christ Church from 1697 to 1701. In this capacity, Wells was bound up in an ancient system of education mainly based on classical studies (Greek and Roman literature being seen as a model for the training of the mind and as a basis for sound moral and social principles) and was required to lecture on the work of Aristotle. The system of education at the college was based on oral debate and disputation, lectures and related set exercises. As a tutor he had broader pastoral responsibilities. Logic was an important part of undergraduate studies, and included mathematics, which also functioned as an introduction to such natural sciences as astronomy and optics. Geography was an heir to both traditions, logical and mathematical and classical, and Wells' work indicates this quite clearly, especially his *Treatise of antient and present geography* (1701).[2]

One of the common features of the careers of tutors of the college was that livings for its students were often obtained through the connections and families of the tutor's pupils in the college. Hence 'in 1702 Edward Wells was appointed to the living of Cotesbach in Leicestershire by the patron St John Bennett, to whose son Thomas Bennett had been tutor from 1693 to 1696, and in 1716 he was appointed to the rectory of Bletchley in Buckinghamshire by his old pupil Browne Willis, who held the patronage'.[3]

It appears that he held both livings at the same time, but preferred to live in Cotesbach, where he died and is buried. Cotesbach is a small parish in south Leicestershire, one mile south of Lutterworth, described by Nichols in his *History and antiquities of the county of Leicester* (1807) as an area of rich land, with good pasture for sheep and cattle, but depopulated as a result of enclosure in about 1626. Nichols records that 'Dr Wells, a noted mathematician, was rector here; and during his incumbency built the neat parsonage here opposite the church; to which his successor John Fanshawe, DD, student of the same college, made great improvements, by enlarging the house, new laying out and walling the gardens.'[4] Wells' pastoral duties at Cotesbach cannot have been very extensive, and it was during his period here that the *Historical geographies* were written.

Little is known of his life in Cotesbach and Bletchley, except for a rather scurrilous account by Thomas Hearne, the Oxford antiquary, bibliographer and diarist, who was not without bias in his thoughts about Wells. Thus, in his diary for Monday 24 July 1727, Hearne wrote:

Above a week since died Dr Edward Wells, Rector of Cottesbach in Leicestershire and of Blechley in Buckinghamshire ... When he was of Christ Church he was a noted Tutor, and was looked upon as a sincere religious man, but after he came to live in the country, he shewed himself an hypocrite, & grew into contempt among those that before respected him, particularly after he became (upon my declining it, upon account of the oaths, when 'twas offered me [Hearne was an eminent Jacobite scholar, who had refused in 1716 to take the oath of allegiance to George I]) Rector of the rich living of Blechley, which he shamefully neglected & seldom went to, but lived at Cottesbach, where he taught school also, & writ & scribbled many books, on purpose to scrape up money, tho' he was a single man, & was otherwise very rich. The truth is, he was a man of great industry, but the books he hath writ & published, as they are very many, so they are inaccurate, & contain very little that is curious.[5]

Wells had undoubtedly got into difficulties at Bletchley: T. Cooper's entry in the *Dictionary of national biography* states that he had attacked his patron and former pupil, Browne Willis, from the pulpit, and that Willis had responded with a tractate, *Reflecting sermons considered; occasioned by several discourses delivered in the parish church of Bletchley*. Cooper confirms that he had both livings when he died on 11 July 1727, and adds 'He was esteemed one of the most accurate geographers of his time.'[6] One of Wells' additional interests was the building and restoration of churches, on which he published a book in 1717 (reprinted with an introduction by Henry Newman in 1840), and on which subject he introduces an interesting polemic *en passant* in Chapter 3 of the first part of *An historical geography of the New Testament*.

Texts: the historical geographies of the Old and New Testaments

An historical geography of the New Testament was initially published in 1708, there being two parts in one volume. This was followed by *An historical geography of the Old Testament* in three volumes, published in 1710, 1711 and 1712 respectively. Both parts of the *New Testament* book and the three volumes of the *Old Testament* were dedicated to royal and aristocratic ladies, starting with Queen Anne, and followed by Lady Crew, Lady Essex Mostyn, Lady Dowager Wray and Lady Mansel of Margam. These were not the first books in or on geography that he had published, nor the first that contained reference to what was termed Bible or sacred geography. In 1701 he had published a *Treatise of antient and present geography*, for the use of students, for whom he had also published a revised edition of the *Orbis descriptio* of Dionysius. In 1701 he also published *A new sett of maps, both of antient and present geography*.[7] In addition, he published a range of pamphlets and books on mathematics, various theological topics, editions of Xenophon and had provided maps for editions of Josephus' history of the Jews.

The historical geographies of the Old and New Testaments were written for a particular purpose, clearly enunciated in their respective title-pages and prefaces: they provide geographical and historical backgrounds to and explanation of the significant places and events recorded in the Bible. The work on the New Testament is divided into two parts, the first part being concerned with the 'journeyings' of Christ, the second with the travels of St Paul, the whole work

Being a Geographical and Historical Account of all the places mention'd, or referr'd to, in the Books of the New Testament; Very useful for understanding the History of the said Books, and several particular texts. To which end there is also added a chronological table. Throughout is inserted The Present State of such Places, as have been lately Visited by Persons of our own Nation, and of unquestionable Fidelity; whereby the work is rendred very Useful and Entertaining. Illustrated and Adorned with MAPS, and several copper-plates; wherein is represented the Present State of the Places now most remarkable.

He proceeds to the interesting statement that

And as Geography is acknowledged to be One Eye of History in general, so nothing can more conduce to illustrate the History of Our Lord, than giving as it were a Plan of those places which made up the Scene, whereon the Particulars of his Holy and Unspotted Life were Transacted, especially if the Geographical Description be rang'd after an Historical Method, or according to the Series of Time, wherein the Places were Visited by our Blessed Lord.[8]

He goes on to say that his principal sources of information for the ancient condition of places were the Jewish historian Josephus (c. AD 37–100) and the account of the journey from Aleppo to Jerusalem in 1697 by Henry Maundrell, chaplain to the Levant Company Factory at Aleppo.[9] He enthuses on the importance of maps and illustrations for geographical treatises, and concludes his preface with this statement:

'Tis here to be further remarked, that I have not contented my self with giving a Bare Geographical *Account of Places*; but have also taken notice of such famous Persons or Actions or Other Circumstances, as the Places are Memorable for in History, or at least deserve our Present Observation. And this I have done to the End, that this work might be Useful, but also Pleasant and Entertaining to the Reader. On this Historical Account, as also by reason of the historical Method I have made use of both in this and the other Part, I have given to this Work the Name of An Historical Geography of the New Testament.[10]

Chapter 1 deals with 'the Holy Land in general, and its principal divisions; as also of such other countries and places, as lay without the Holy Land, and are mentioned or referred to in the four gospels'. In some respects this is the most interesting chapter in the first part of the book, the rest primarily being a kind of Biblical historical and geographical gazetteer. He commences with a description of the provinces of the Holy Land, starting with Judea, Samaria and Galilee, paying particular attention to the location and general characteristics of the provinces and their principal cities, rivers and physical features.

Subsequent chapters treat chronologically of the history and geography of Christ's life, the last chapter (7) dealing with 'the places honoured with Our Lord's presence after his Resurrection', and contain detailed descriptions of the major towns and places so involved, including Nazareth, Bethlehem and Jerusalem. The detailed descriptions of such places are a mixture of Wells' reading of the New Testament and of a range of authorities, both ancient and modern. Thus, for example, his accounts of Nazareth, Bethlehem, Jerusalem, Tyre, Sidon and Jericho are acknowledged to derive much from Henry Maundrell's *Journey from Aleppo to Jerusalem at Easter, AD 1697*, and from Josephus. Thus, speaking of the Lake of Galilee:

This lake takes its name from the country that surrounds it, which is fruitful and agreeable to imagination. No plant comes amiss to it; besides that it is improved by the skill and industry of the inhabitants to the highest degree; and by a strange felicity of climate every thing prospers there; as nuts, palms, figs and olive-trees . . . Such a delicious country was the land of Gennesareth in the time of Josephus, who lived in the same age with our Saviour.[11]

It is interesting to note in passing that the same passage from Josephus, from a different edition, is quoted by George Adam Smith in *The historical geography of the Holy Land* (1894) in his description of the Plain of Gennesaret.[12]

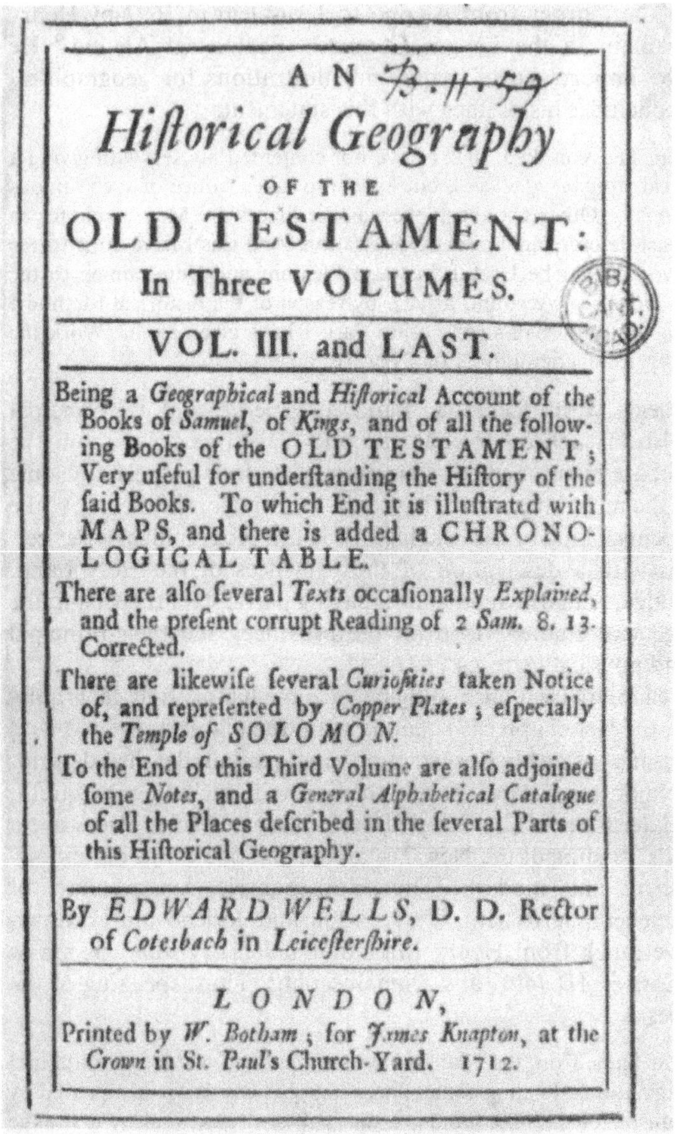

A N *B. H. 59*

Hiſtorical Geography

OF THE

OLD TESTAMENT:

In Three VOLUMES.

VOL. III. and LAST.

Being a *Geographical* and *Hiſtorical* Account of the
Books of *Samuel*, of *Kings*, and of all the follow-
ing Books of the OLD TESTAMENT ;
Very uſeful for underſtanding the Hiſtory of the
ſaid Books.   To which End it is illuſtrated with
MAPS, and there is added a CHRONO-
LOGICAL TABLE.

There are alſo ſeveral *Texts* occaſionally *Explained*,
and the preſent corrupt Reading of 2 *Sam.* 8, 13.
Correƈted.

There are likewiſe ſeveral *Curioſities* taken Notice
of, and repreſented by *Copper Plates* ; eſpecially
the *Temple of SOLOMON.*

To the End of this Third Volume are alſo adjoined
ſome *Notes*, and a *General Alphabetical Catalogue*
of all the Places deſcribed in the ſeveral Parts of
this Hiſtorical Geography.

By *EDWARD WELLS*, D. D. Reƈtor
of *Coteſbach* in *Leiceſterſhire.*

LONDON,

Printed by *W. Botham* ; for *James Knapton*, at the
Crown in St. *Paul's* Church-Yard.   1712.

Fig. 2.1   Title-page of *An historical geography of the Old Testament* vol. III (1712).
By permission of the Syndics of Cambridge University Library.

The second part of *An historical geography of the New Testament* follows similar guidelines to those of the first. He does, however, specify in more detail the sources of information he has used in the second part. These include the works of Sir Paul Rycaut, Henry Maundrell's journey, and illustrations from the work of a Monsieur Bruyn. The second part is, in fact, very heavily derived from the work of Maundrell, especially his detailed description of Damascus and other important towns. Although the source is fully acknowledged, these descriptions seem to lie uncomfortably in the text, especially as Wells cites directly, using the first person plural ('we') in the account. The second part is well illustrated by copperplates and maps.

An historical geography of the Old Testament (Fig. 2.1) was published in three volumes, the first dealing with the geography of the Book of Genesis, the second with the Books of Exodus, Leviticus, Numbers, Deuteronomy, Joshua, Judges and Ruth, and the last with the remainder of the Old Testament. Its general design is similar to that of the *Historical geography of the New Testament*, 'the most observable difference', according to Wells in his preface, 'between one work and the other, is this, that in my Geography of the Old Testament, I have found it requisite, to have frequent recourse to the Hebrew Language'.[13]

The first volume deals with the location of the Garden of Eden, the structure of Noah's Ark and the place where it came to rest, the place of the original 'plantation' or settlement after the Flood, and the travels of the patriarchs Abraham, Isaac and Jacob. The first, rather lengthy, chapter is concerned with the 'Places of the antediluvian world – viz. the Garden of Eden, the land of Nod, and the city of Enoch'.[14] The principal purpose is the identification of the location of these places, with the aid of a wide range of scholarly authorities. Wells concludes, after some interesting reflections on the nature of the tides and coastlands of the northern end of the Persian Gulf (largely courtesy of Strabo), that 'it will not be difficult to assign the very situation of the country of Eden, wherein God planted the garden of Paradise. For it is evident, from the words of Moses, that it lay on the single channel, which is common to all the four rivers' (Pishon, Gihon, Hiddikel and Euphrates).[15] Southern Mesopotamia was Wells' preferred location (Fig. 2.2), but although this type of scholarly search continued well into the nineteenth century, the current view is that although Mesopotamia might have been a possibility, the primitive and mythological geographical ideas incorporated in the Genesis narrative make it impossible to be certain.

Chapter 2 is a similar locational detective puzzle, this time in relation to the mountains of Ararat, and is followed by a long, prolix attempt to reconstruct the size of Noah's Ark and the location of the animals within it. Chapter 3 opens with the interesting statement that 'The Sacred historian,

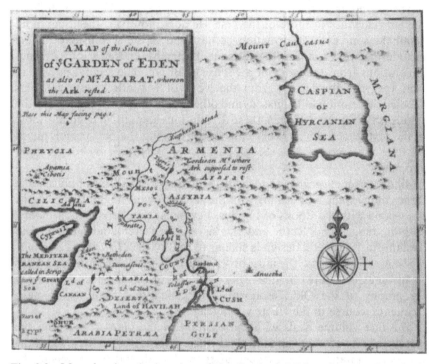

Fig. 2.2 Map showing the location of the Garden of Eden, from *An historical geography of the Old Testament* vol. I (1710), opposite p. 1. By permission of the Syndics of Cambridge University Library.

having inform'd us, how the world was dispeopled by the Flood, proceeds to inform us next, how it was repeopled by the posterity of Noah.'[16] In this and the two ensuing chapters we are firmly exposed to the prevailing religious and geographical beliefs of the seventeenth century, including the idea of an all-exterminating flood,[17] the notion that the Germans, Gauls, Welsh and Anglo-Saxons are descendants of Japheth (one of Noah's three sons), and the strict Biblical interpretation of the demise of the cities of Sodom and Gomorrah. In addition, the chronological table that terminates the first volume indicates Wells' view that the age of the earth at the birth of Christ was 3,950 years, though he does mention another opinion (presumably that introduced by Archbishop Ussher) that the true date of origin was 4004 BC.

The second and third volumes of *An historical geography of the Old Testament* deal chronologically with the remainder of the Old Testament. Of particular interest are the lengthy accounts of the pyramids of Egypt, of Alexandria and the early journeyings of the Israelites.[18] In volume III there is a long account of the ancient state of the city of Jerusalem described in an itemized fashion, starting with the layout of the old city, then moving to descriptions of the gates, the hills and mountains about Jerusalem, and the rivers and valleys. The material is from early and contemporary authorities.[19]

The overall impression, therefore, is of an interesting mixture of Biblical exegesis and the use of secondary authorities to reconstruct the geographical and historical landscapes and environments of the Old and New Testaments. The limitations are obvious, especially when compared for example with the contemporary account of Palestine by Maundrell, and later accounts such as that of George Adam Smith. Wells' works lack the ring of first-hand experience, but they do exemplify the widespread scholarly 'sacred' geography of the seventeenth and eighteenth centuries (examined at greater length below), are also of interest for their early use of the term 'historical geography', and provide interesting insights into a number of aspects of the geographical and theological writings of their period.

These works seem to have had a common and popular currency at least to the early part of the nineteenth century. Thus, R. Greaves, in his account of religion in the university of Oxford in the period 1715–1800, indicates that what he calls Wells' *Geography* was one of a number of 'Anglican classics' which were published in cheap editions, one per year, from about 1790 onwards, the Wells volume printed in 1801, with a re-engraving of the map of Palestine 'according to D'Anville' (J. B. B. d'Anville, an important French geographer of the eighteenth century). Here, in these popular editions, according to Greaves, 'was a body of religious knowledge for the young cleric or theological student that was orthodox, learned and widely acceptable'.[20] This combined work was translated into German, appearing as *Historische Geographie des Alten und Neuen Testaments* (Nuremberg, 1765). Wells' *Historical geographies* also attracted the attention of Karl Ritter, being listed without comment in the bibliographical section of his volume on the geography of Palestine dealing with 'a certain class of works, which, although not referring directly to the results of personal observation, are yet valuable, as digests of what had been observed by others, and as studies preparatory to the prosecution of personal enquiry'.[21]

Contexts

The ideological and cultural contexts reflected in Wells' historical geographies include: the changes in the nature of geographical understanding, including the meaning of historical geography, related to old and new geographical information (including maps), theory and methodology; the tradition, dating from medieval times, of 'sacred geography', and the closely linked changes in theological thought, dating mainly from the

Reformation, which it exhibits; and the related search for the reasons for cultural diversity, including notions of the origins and dispersal of the tribes and races of the earth. It is, of course, somewhat arbitrary to separate these aspects one from the other, for they were essentially bound up together in a cosmographic view. As Glacken has indicated:

In ancient and modern times alike, theology and geography have often been closely related studies because they meet at crucial points of human curiosity ... The conception of a designed world, in both classical and in Christian thought, has transcended personal piety. In Western thought the idea of a deity and the idea of nature often have had a parallel history; in Stoic pantheism they were one, and in Christian theology they have supplemented and reinforced one another.[22]

The nature of geographical thought, including historical geography

Geography in the seventeenth and early eighteenth centuries was in an obvious state of transition, reflecting the clear tension between the new scientific empiricism, epitomized in the work of Francis Bacon, and the more traditional descriptive work to be found in conventional geography. Hence

Astronomy, optics and mechanics, for example, gained a distinct advantage from the new theories, but many traditional studies, geography among them, soon came into conflict with the growing demands for certainty and objectivity in science, while the mechanist world view favoured by the rationalist school provided an added complication for inquiries concerned with living organisms, and particularly with man. In many ways geography was seriously hampered by the new frame-work.[23]

Historical geography in the seventeenth century was very much on the traditional side, paying particular reference to the study of the Greek and Roman classical geographical authors, and still dependent to a significant degree on Ptolemaic interpretations of the geography of the world. The term 'historical geography' was rarely used, and when it was, had either the context of the geography behind important historical events, as in Wells' historical geographies, or the study of the history and geography of changes in the territorial possessions and boundaries of states and empires in the past, related to the chronology of royal houses and administrations, particularly in Europe. An early example is the *Nouvelle métode de géografie historique pour aprendre facilement et retenir ... la géografie moderne et l'anciène, l'histoire moderne et l'anciène, le gouvernement des états, les intérêts des princes, leurs généalogies* ... published in Paris in 1697 and written by the Abbé Louis de Courcillon de Danegeau. The term was also used in collections of historical maps, as in P. Duval's *Cartes pour la géographie historique ancienne*, published in Paris in 1657. In the seventeenth century there were still very strong links between geography and history. Wells, in his introduction to *An historical geography of the New Testament*, thus makes reference to the fact that 'Geography is acknowledged to be One Eye of History in general',[24] and refers to the advantages which will accrue in studies of the Holy Land 'especially if the Geographical Description be rang'd after an Historical Method, or according to the Series of Time, wherein the Places were Visited by our Blessed Lord'.[25] This was a common metaphor in the seventeenth and eighteenth centuries: May cites Hakluyt, who speaks of 'Geographie and Chronologie (which I may call the Sunne and the Moone, the right eye and the left of all history)', and Vico, who refers also to geography as 'the other eye of history', and indicates that 'in the centuries immediately preceding Kant, phrases such as "geography is the other eye of history" or "the right eye of history" or "one of its beacons" had become undiscussed clichés, handed down and repeated almost verbatim from generation to generation'.[26] However familiar, the link is important, both in relation to the emergence of the term historical geography, and also of the emergence of geography as a visual ideology.

The context of geographical teaching at Oxford, where Wells studied and taught, has been outlined by J. N. L. Baker, who in his account of academic geography in the seventeenth and eighteenth centuries[27] and of the teaching of geography in the Oxford colleges in particular, gives, as also does Gilbert,[28] an impression of a continuous and corporate sense of identity in the teaching of geography at Oxford. Baker shows that: 'In the colleges the undergraduate was given a general account of the world in order that he might correct and supplement the knowledge he acquired from classical authors, and in particular from such works as the *Orbis descriptio* of Dionysius; in short, he was taught what would now be called regional geography.'[29] The classical studies aspect of geography is, according to Baker, illustrated in the works of Wells. Wells' approach to geography is illustrated in his *Treatise of antient and present geography* (Fig. 2.3) published in 1701: 'I shall not spend time' he says in the preface,

in setting forth the great Usefulness of Geography, it being what the Learned and more knowing Part of Mankind is already convinced of. And for the same reason there seems likewise to be no occasion of proving, that (at least) the Design of Maps and Treatise now published by Me is Useful and Beneficial; since it is no other, than to help and further Young Students in acquiring a competent knowledge of Geographical Science.

These students at universities were, according to Wells in the same preface, 'obliged to converse as much (if not more) with Old Authors as with New; it becomes hereupon requisite for them, to have a considerable Insight into Antient as well Present Geography'.[30] Downes, in a study of Georgian

Fig. 2.3   Frontispiece of *A treatise of ancient and present geography* (1701). Coleville Library, Leicestershire County Council Libraries and Information Service.

geography, says that 'partly as a result of Hakluyt's labours, a school of geography arose at Oxford whose influences spanned the period (1599–1712) between the publications of George Abbott, Archbishop of Canterbury, and Edward Wells, Rector of Cotesbach in Leicestershire'.[31] Abbott had become master of University College, Oxford, in 1597, and in 1599 had published *A briefe description of the whole worlde*, which Baker indicates had reached a ninth or tenth edition in 1664, being 'doubtless intended to supply the latest geographical information for the benefit of those who were reading the standard works of ancient authors, and its general plan was copied by a number of writers. Such short treatises continued to be popular in Oxford for more than a hundred years.'[32] Why Downes should see Wells as the last of the representative types of this era of the Oxford school of geography is not entirely clear, though he seems to infer that Wells' books mark the end of a very traditional period of geographical writing, which was ended by the adoption in France and Scotland in particular of new geographical ideas strongly influenced by the philosophy of the Enlightenment: thus 'If, as we may suppose, Scottish colleges were more receptive than their Oxford counterparts to the new geographical ideas from Holland and France, we may regard the eighteenth century geographical grammars, published mostly in Edinburgh and London ... as a marked advance on the works of the Oxford geographers of the previous century.'[33] The organization of the *Treatise of antient and present geography* is mainly regional, after some preliminary chapters on geographical location, latitude and longitude. The regional chapters draw heavily on such classical authors as Ptolemy, and generally treat of the classical accounts of individual countries and regions, contrasted with the modern state of those places. Chapter 25 is titled 'Of the Sacred or Biblical Geography': 'Having gone through the description of Asia according to its principal Divisions &c mentioned by Heathen Writers, it seems convenient to proceed immediately to the Sacred Geography thereof ... it hath therefore been judg'd best to dispatch the whole Bible-Geography in this place, that so it may be the better apprehended, lying all together under one view.'[34] The chapter deals separately with the Old Testament and the New Testament, and is essentially an attempt to identify and locate the tribal and administrative divisions of the Holy Land and adjacent areas, followed by descriptions of the chief mountains, rivers and seas. This link with classical authors is very much characteristic of the chorographic/cosmographic tradition in early modern geography. The organization of this book bears close resemblance, for example, to the organization of the *Introductio in universam geographiam* of Philip Cluverius, published in Latin in Holland in 1624, and translated into English as *An introduction into geography both ancient and moderne*, published in 1657 at Oxford, where Cluverius had spent some time in the early seventeenth century.[35] Cluverius' work followed in the cosmographic traditions of Sebastian Münster, and stressed the importance of comparing the accounts of the classical geographers with that of more modern writers ('comparative geography' – a common feature of the seventeenth century), and generally seems to be reflective of traditions of the sixteenth century, including a strong orientation towards Europe and the Asiatic East, with little importance given to the New World.[36] Wells in his preface cites Cluverius' book as one of the two more useful current introductory geographical works for students (the other, as already noted, being Heylyn's *Cosmographie*, published in 1652). Like Cluverius, Wells avoids mention of the Copernican controversy.

The 'ancient and modern' geography theme is followed in his complementary set of maps, the *New sett of maps, both of antient and present geography* (1701), produced as an atlas comprising pairs of maps for regions of the world, one 'antient', one modern. The maps have very little detail, generally capital cities, seats and sees of bishops, archbishops and patriarchs, university towns, chief rivers, and have only poor physiographical detail. The outlines of the continents are reasonably accurate for their time, but it is clear that this is a student reference atlas, and bears out J. N. L. Baker's observation that at Oxford 'college teaching was intended to give undergraduates a knowledge of the ancient and modern world sufficient to enable them to follow intelligently the writings of classical authors and of such modern writers as they cared to study ... The great Oxford atlas of Moses Pitt (1680–82) and the smaller atlas of Edward Wells ... are additional pieces of evidence to support this theory.'[37] Seventeenth-century atlas-makers frequently made use of historical material from the two principal sources most readily to hand in all major libraries and collections – the classical authors and the Bible – and frequently in doing so compared past with present geography, as had Wells in his atlas and had Philippe Briet in his *Parallela geographiae veteris et novae* (1648–9). An advantage for cartographers short of contemporary geographical data was that by using extensive maps of the past they did not have to be up to date, and these maps could also be drawn at fairly small scales.

This type of geography mainly reflects the *special* geography or topical and regional geography which was one of the two main components of geography in the seventeenth century, the other being the *general* geography concerned with mathematical and astronomical aspects of the world or sphere as a whole. This division is in turn a reflection of the views on geography of the classical authors such as Strabo and Ptolemy, particularly the latter's concept of chorography. The intellectual route of the division of geography into general and special geography or chorography is complex, but may roughly speaking be traced via the classical Greek authors, the Ptolemaic debate of the seventeenth century and the specific work in the sixteenth and seventeenth centuries of Keckermann, Varenius, Carpenter and Heylyn.[38] Wells seems to be following in part the chorological tradition of such Greek geographers as Hecataeus, Herodotus and Polybius, including its very strong chronological and historical leanings. He was, of course, familiar with the writings of the classical Greek philosophers and historians, having himself while a tutor at Christ Church edited the works of Xenophon for publication. Wells' own methodological views are made clear in his own writings. In the introduction to the *Treatise*, for example, he states that 'Geography differs from Cosmography (or the description of the universe) as a Part from the whole: from Chorography and Topography (that is, the description of a particular country or place) as the whole from a greater and lesser part'.[39] Wells seems to be effecting, perhaps subconsciously, a marriage of the distinctive concerns of both 'general' and 'special' geography, but with distinctive leanings towards the latter, in a way not dissimilar to that incorporated in the works of Bernhard Varen (or Varenius). Although according to the recent reading by Warntz, Varenius' *Geographia generalis* was symbolic of an 'unequivocal refutation of the Aristotelian–Scholastic conception of Geography and its enthusiastic acceptance and application of Cartesian philosophy of science',[40] Warntz' reading of Varenius is debatable in the sense that both J. N. L. Baker and Bowen make it clear that Varenius was also committed to special geography. Thus, 'his next statements provide fairly strong evidence of his intention to follow the *Geographia generalis* with a work on special geography, to accompany an earlier venture in that field, his previous publication, the *Descriptio regni Japoniae et Siam* of 1649, a compilation of descriptions by authors'.[41]

It is perhaps unfair to seek too closely for deep methodological purpose in a set of works designed for undergraduate use, but it is important to have some insight into Wells' Oxford geographical (and mathematical – he was a believer in mathematical geography for undergraduates, and wrote several mathematical treatises) writings as a partial basis for understanding his later work in the historical geographies.

Sacred geography

The origins of sacred geography date back, as Rubin has indicated,[42] to the Biblical toponymy titled *Onomastikon*, compiled by Eusebius, bishop of Caesarea early in the fourth century AD, and this Biblical or sacred geography, involving the use of historical and geographical skills to locate and characterize the important Biblical places and events, continued through the Middle Ages to the early modern period (indeed continues in some countries to the present day), frequently incorporating the use of both old and new maps of the Holy Land. Added energy was given to this work by the new emphasis of the Reformation on the authority of the Bible, with a consequent intensification of interest in the detailed topographical and geographical environments of scripture: giving rise to attempts to reconcile revealed knowledge and the authority of the sacred text with the empirical knowledge and authority of the new science, attempts which came to crisis point with Darwin in the nineteenth century. Thus, in addition to the centrality of theology to the study of all the sciences, including geography, was added the special focus on the Holy Land itself, through direct and indirect experience. Though much of this work was, inevitably (given the nature of the Reformation) Protestant, it also had a profound effect on Catholic thinking, especially by the Jesuits. Dainville, in his important study of the reflection of changes in the nature of geography in the study curricula of the Jesuits in France in the sixteenth and seventeenth centuries,[43] has shown how the appeal of the Reformation to the exclusive authority of the Bible launched a vast current of interest in its contents, including the many places of which it speaks. Thus, many cartographers, including Mercator and Finé, launched their careers with chorographies and maps of the Holy Land. The Jesuits, as Dainville states, didn't lag behind this new enthusiasm for sacred geography:

Les Jésuites, comme bien l'on pense, n'entendirent pas rester en marge ou en retard. Leurs constitutions firent même de la défense de la Bible approuvée par l'Église, l'une des raisons de l'étude des humanités. Ils créèrent et developpèrent face aux protestants, à côté de la théologie positive catholique ... étaient l'étude approfondie des trois languages classiques: latin, hébreu et la connaissance de la géographie et de l'histoire.[44]

The Jesuit priest A. Possevin, in his review of the diverse disciplines pertinent to the study and practice of theology (*Bibliotheca selecta*, 1593), included in book XV (*De mathematicis, ubi item de architectura ad religiosos praesertim spectante, deq. cosmographie et geographia*) two chapters on geography, in which geography was seen as that part of cosmography which describes the terrestrial globe and places it before our eyes.[45] His advice to teachers of sacred geography (*la géographie sacrée*) was to begin with such general ideas on the inhabitation of the earth as the earthly Paradise after the Flood and the division of the earth between the children of Noah, these to be followed by a more detailed examination of particular countries and places such as Syria, Palestine, Phoenicia and the frontiers of the promised land. It was necessary to know the places where the main events recorded in the Scriptures happened, including the seas, rivers, towns, fortresses, floods and migrations of peoples. The map of Palestine should be very carefully scrutinised in order to ascertain the exact locations of Biblical places, and the city of Jerusalem with its historic territories, the journeys of St Paul and the other apostles should be given particular attention. At the time when Possevin was writing, there were many sacred geography texts available for teaching, including Thevet's *Cosmographie de Levant* (1554), Montanus' *Apparatus ad literas sacras* (1572), and maps of Palestine by Finé, Ortelius (*Theatrum orbis terrarum* 1570), and Adrichomius (*Theatrum Terrae Sanctae ac biblicarum historiarum* 1590).[46]

Wells was clearly writing within this broad tradition of sacred geography, and although he was working within the reformed Protestant tradition, he drew, especially for illustrations and maps, on the work of Jesuits such as Kircher and Villalpandus, as will be shown later. His awareness of this tradition of sacred geography is clear from the chapter 'Of the sacred or bible geography' (Chapter 15) in his *Treatise of antient and present geography* (see Fig. 2.4), and also from his prefatory remarks to, and arrangement of, the two works on the historical geography of the Old and New Testaments. The first book of *An historical geography of the Old Testament*, for example, approximates very closely, in its outline and purpose, to the sequence advised by Possevin: Wells indicates 'The Principal Particulars described in this Treatise, are, 1. The Situation of the Garden of EDEN; 2. The structure or make of NOAH's Ark, together with the Place whereon it rested; 3. The Original Plantations after the Flood; 4. And Lastly, The Travels of the Three Celebrated Patriarchs, Abraham, Isaac and Jacob.'[47]

Another, related, contextual influence on Wells' ideas and writings is that which is bound up in the search in the sixteenth and seventeenth centuries for a more scientific mode of understanding the diversity of the earth's population, in terms of their physical and cultural (including religious and political) characteristics. They also, however, reflect older and deeper traditions of Biblical exegesis and illustration. Very much related to what has been described as the questing spirit of the Renaissance, manifest in part by the establishment of collections of curios and artefacts in the major cities of Europe, this taxonomic tradition, embraced and pursued by the humanists (though not by the natural scientists until a later period), was extended and energized in the sixteenth and seventeenth centuries by the voyages of discovery and travellers' accounts. Hodgen, in her illuminating study of early anthropology in the sixteenth and seventeenth centuries, shows how this quest for scientific collection (in which she includes the details of the geography of the Holy Land required by the theologian) broadened from artefacts to customs, which were in turn classified and described, and which posed in consequence interesting questions of organization, ideology and methodology.[48] This was an interesting period of tension and transition, during which the old established traditions of dependence on the geographical and cultural information of the classical writers of Greece and Rome, and their methods of organization of such knowledge, together with the Eurocentric perspectives of cosmographers and collectors, was gradually eroded by the impact of travel accounts and

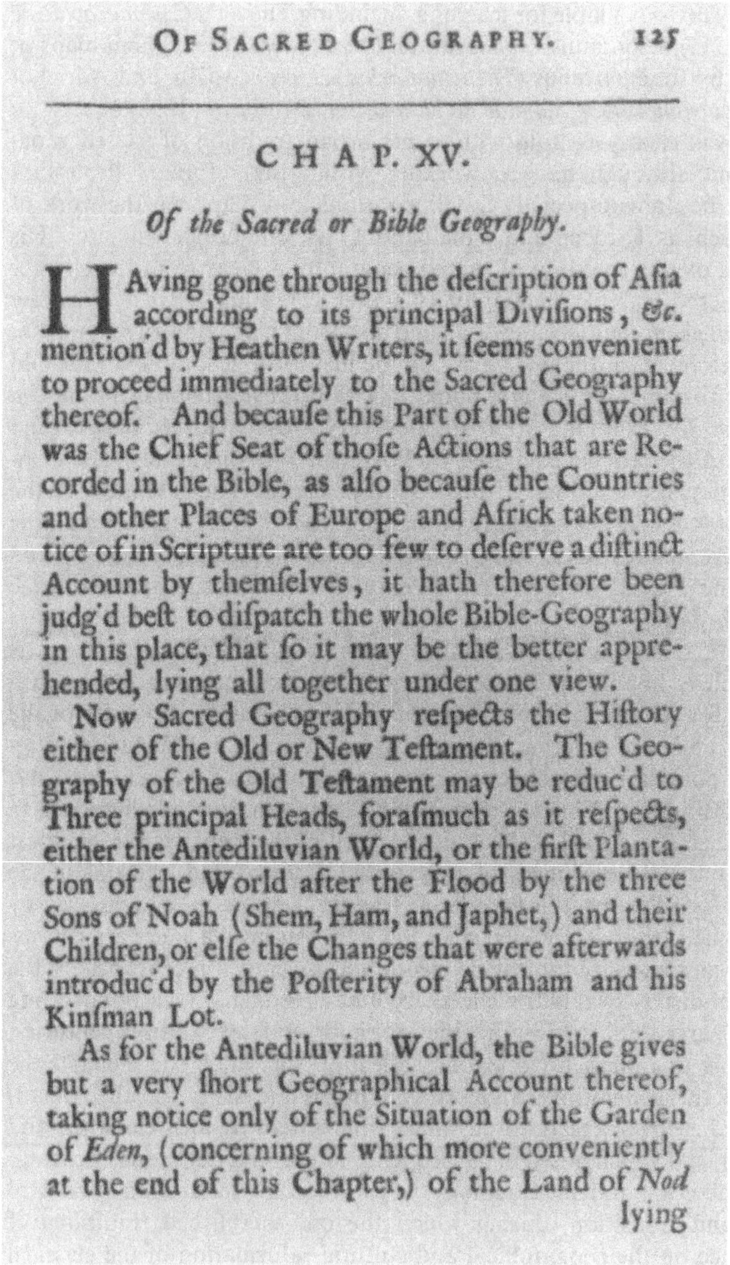

C H A P. XV.

Of the Sacred or Bible Geography.

HAving gone through the defcription of Afia according to its principal Divifions, &c. mention'd by Heathen Writers, it feems convenient to proceed immediately to the Sacred Geography thereof. And becaufe this Part of the Old World was the Chief Seat of thofe Actions that are Recorded in the Bible, as alfo becaufe the Countries and other Places of Europe and Africk taken notice of in Scripture are too few to deferve a diftinct Account by themfelves, it hath therefore been judg'd beft to difpatch the whole Bible-Geography in this place, that fo it may be the better apprehended, lying all together under one view.

Now Sacred Geography refpects the Hiftory either of the Old or New Teftament. The Geography of the Old Teftament may be reduc'd to Three principal Heads, forafmuch as it refpects, either the Antediluvian World, or the firft Plantation of the World after the Flood by the three Sons of Noah (Shem, Ham, and Japhet,) and their Children, or elfe the Changes that were afterwards introduc'd by the Pofterity of Abraham and his Kinfman Lot.

As for the Antediluvian World, the Bible gives but a very fhort Geographical Account thereof, taking notice only of the Situation of the Garden of *Eden*, (concerning of which more conveniently at the end of this Chapter,) of the Land of *Nod*
lying

Fig. 2.4 Beginning of Chapter 15, p. 125, of *A treatise of antient and present geography* (1701). Coleville Library, Leicestershire County Council Libraries and Information Service.

guides and new ways of organizing information. A particular problem, which explains in part the emphasis given in the first book of Wells' *Historical geography of the Old Testament* to the question of Noah's Ark and the repopulation of the earth after the flood, is that which involves the conflict between monogenetic and polygenetic explanations for the cultural diversity of the world's population as it was becoming increasingly evident from the expansion of geographical knowledge. As Hodgen put it:

Though the account in the Book of Genesis was generally accepted among European men of learning in the sixteenth and seventeenth centuries, and with it the monogenetic hypothesis [a single beginning in a single place for the whole of mankind, with subsequent demographic events of repopulation, after the flood, for example], there was some revision of the interpretation of its contemporary ethnological meaning. In the age of faith, or while the medieval population was small, cultural diversification, in so far as it was known to most men, had been restricted geographically to the shores of the Mediterranean, to the vaguely rumoured tribes of Northern Europe, and to a few of the inhabitants of the Near and Far East ... To account for this small human sample, the Biblical story of the Noachian dispersion, with its short muster roll of tribes and empires mentioned as the posterity of the three brothers, seemed circumstantial enough ... But as geographical knowledge, combined with Biblical criticism, began its work upon the intimate fabric of the Book of Genesis, the unrevised sacred history no longer seemed inviolate or universally satisfying.[49]

The responses to such problems varied, and are extremely complex. Wells seems to belong to a category of theologians who sought to use new geographical knowledge and the resources of mathematics to support traditional views, including the monogenetic theory of the origin and spread of the peoples of the world. Thus, in Chapter 3 ('The first plantations after the Flood') of the first book of *An historical geography of the Old Testament*, he says: 'The Sacred Historian, having informed us, how the World was dispeopled by the Flood, proceeds to inform us next, how it was repeopled by the Posterity of Noah',[50] and proceeds to a genealogy of dispersion with the aid of the authority of Josephus and Camden's *Britannia*. Yet later on in the same book he enlists a late-seventeenth-century estimate of the population of London to determine the size of the ancient population of the city of Nineveh: an interesting and contrasting pair of examples of the difficulties which he and many fellow scholars faced in reconciling old and new ideas. The type of conservative world-view exemplified by Wells was also to be seen in other geographical works of the seventeenth century such as Peter Heylyn's *Cosmographie* (1652, with subsequent editions in 1657, 1666, 1670, 1674, 1677, 1682 and 1703), by which Wells was influenced, having cited Heylyn in his historical geographies and his geographical textbook of 1701. Hodgen's view of Heylyn's and similar works in this particular context is that 'They resisted the inclusion of material that might have been considered outright atheism; and they included in their treatment of the problem of cultural differences the usual and repetitious lists of nations descended from the sons of Noah.'[51] Heylyn's work remained in favour in the late seventeenth century, according to Bowen, because the strength of the Church of England over natural science was increased by the reinstatement of an Anglican monarchy at the Restoration of 1660.[52]

An additional problem in this transitional period was that of chronology, which influenced the approach and content of works such as those of Wells. Given Wells' view, already noted above, that the age of the earth was at most 4,004 years before the birth of Christ, from his perspective the period about which he was writing in his historical geographies allowed for a total period of just over 4,000 years for human habitation of the earth from the time of the creation. The implications of this are fairly obvious, and manifest themselves first in the credibility given to the accounts of the cultures and places of the Middle East by Biblical and ancient classical authors as being fit bases for not only reconstructions of their ancient geography but also of their contemporary geography and secondly in the firm support given to the monogenetic theory of human origin and diversity.

Sources of information

The sources which Wells uses directly reflect the major concerns and priorities of the ideologies briefly outlined above. By the late seventeenth and early eighteenth centuries, the volume of material available to students and scholars of sacred or Biblical geography had grown enormously, as had the array of maps to be consulted and used, and Wells, as we shall see, drew on both very early and the latest authorities for his information. His principal sources were: the Bible itself, in the form of the Anglican authorized version of 1622 and presumably, given his linguistic expertise, earlier related Greek, Latin and Hebrew versions and texts; the classical writers of Greece and Rome and the early historians of the Christian Church, including Strabo, Pliny, Josephus, Eusebius, Ptolemy and others; a small number of seventeenth-century travellers' accounts; together with the works of theologians and geographers working on particular aspects of sacred geography in the seventeenth and eighteenth centuries, from which he also drew some of his illustrations and maps.

The two sources which he uses and cites most extensively, apart from the Bible itself, are Flavius Josephus' *Jewish war*, originally written in Aramaic, and published in 77–8 AD, and Henry Maundrell's *Journey from Aleppo to Jerusalem* (1697).

Flavius Josephus was a native of Palestine who fought against the Roman occupation, for part of the time as Governor of Galilee, but who

ultimately became a Roman citizen, receiving a pension which enabled him
to write a series of accounts of the times he lived in. He was clearly a
thoroughly unsavoury and in many respects brutal character, whose
writings, *inter alia*, contain frequent attempts to justify his betrayal of his
fellow Jews, while protesting that he remained a loyal Jew.[53] However, his
Jewish war, written in Aramaic and translated into Greek, comprises a
first-hand account of events in Palestine in the first century, supplemented
by material from other authors, and is an important, if biased, source of
information. The edition which Wells uses is the 1702 translation by Sir
Roger L'Estrange, for which Wells had drawn several maps. Henry Maundrell, the author of Wells' second major source, was chaplain to the Levant
Company factory at Aleppo from 1671 to 1681. Like other chaplains,
particularly his predecessors Charles Robson (Aleppo) and Thomas Smith
(Constantinople), he wrote travel-books about his experiences in the
Levant. Maundrell (1665–1701) was an Oxford graduate, and curate of
Bromley in Kent, who was elected in December 1695 to the chaplaincy of
the Levant Company's factory in Aleppo. His journey to Jerusalem,
together with fourteen other Aleppo residents, began on 26 February 1697.
They stayed in Jerusalem over Easter, and they returned to Aleppo on 20
May. Maundrell died, it is thought of fever, in Aleppo early in 1701. His
book, *A journey from Aleppo to Jerusalem at Easter AD 1697*, was
published in Oxford in 1703; there was a second edition in 1707 and a third
in 1714. There were many subsequent editions and impressions, including
translations into Dutch, French and German.[54]

A journey from Aleppo to Jerusalem at Easter AD 1697 is a lively and
fascinating account of Maundrell's experiences, both of searching for holy
places and of the state of Palestine in the late seventeenth century. It offers
far more than Wells borrows, for Wells for the most part reproduces at
length only those sections dealing with the major cities and important sites,
whereas Maundrell has in addition much to say about the conditions of
travel and the nature of Arab life and agriculture which Wells generally
does not cite. His style of description is graphic and potent. Thus, of
Damascus, he writes:

It is thick set with mosques and steeples, the usual ornaments of the Turkish cities;
and it is encompassed with gardens, extending no less, according to common
estimation, than thirty miles round; which makes it look like a noble City in a vast
Wood, the gardens are thick set with fruit trees of all kinds, kept fresh and verdant
by the waters of Barrady. You discover in them many turrets, and steeples, and
summer-houses frequently peeping out from amongst the green boughs, which may
be conceived to add no small advantage and beauty to the prospect.[55]

He was similarly enamoured of the 'prospect' from the top of Mount
Tabor, overlooking Galilee: 'It is impossible for man's eyes to behold a

higher gratification of this nature. On the North West you discern at a distance the Mediterranean; and all round you have the spacious and beautiful plains of Esdraelon and Galilee, which present you with the view of so many places memorable for the resort and miracles of the son of God.'[56] He is also quite critical of some of the attempts to show visitors to the Holy Land the supposedly exact sites of important Biblical events, and includes the cynical observation 'that almost all passages and histories related in the Gospel are represented by them that undertake to shew where every thing was done, as having been done most of them in Grottos; and that even in such cases, where the condition and the circumstances of the actions themselves seem to require places of another nature'.[57] This work is an interesting piece to read, and worthy, with the companion descriptions of travel by other Aleppo chaplains, of further study in its own right, but it does add life to the otherwise rather dull descriptions of Wells. Maundrell's work also provides some interesting insight into the development of Arabic studies in the seventeenth and early eighteenth centuries.[58]

Wells also made extensive use of a wide range of contemporary authorities in the field of sacred geography, including theologians and geographers, a more extensive study of which could yield a fascinating perspective on the information field of this particular scholar. Given the mode of citation, which generally only includes the name of the author rather than the details of the exact book or work of the author cited, it is quite difficult to be exactly specific in relations to his sources, but bibliographical searching does enable a reasonable estimation to be made. In his preface to the second part of *An historical geography of the New Testament*, he gives an idea of some of these sources:

The antient state of several places, which lay without the Holy Land, is taken from Strabo, who lived in the first century; and the present state is taken chiefly either from Sir Paul Rycaut, or Mr Maundrel. To Mr Maundrel we are beholden for the present state of Damascus and Ptolemais, which he visited in 1697: to Sir Paul Rycaut for the present state of the seven churches in Asia ... which he visited in 1678. As for the draughts [engravings] they are taken from Monsieur Bruyn, and the chronological table from the Reverend and learned Dr Cave.[59]

Sir Paul Rycaut (1628–1700), former secretary to the British ambassador in Constantinople, and subsequently consul to the Levant Company in Smyrna, was an authority on Turkey and the Ottoman empire, had published in 1668 the influential *The present state of the Ottoman empire*, and in 1679 the book which Wells uses as his source for the diffusion of Christianity outside the Holy Land, *The present state of the Greek and Armenian churches*. Rycaut was (and is still used as) an important authority on the Ottoman empire and on the historical doctrines of the Greek and Armenian churches. In a recent study he is described in the following terms:

Fig. 2.5   Nazareth, from *An historical geography of the New Testament* (1708) p. 18. By permission of the Syndics of the Cambridge University Library.

'In many respects he was the archetypical early fellow of the Royal Society, with an informed interest "in all the Observables of Nature and Art", and a diligence in recording evidence that enabled him to make a real contribution in more than one field.'[60]

The 'Monsieur Bruyn' from whom Wells copied the fine engravings of towns such as Nazareth (see Fig. 2.5), Bethlehem and Jerusalem, was most likely Cornelius de Bruyn, a Dutch painter and traveller, who had travelled extensively in the Middle East, starting in 1678, and who in 1698 had published the extensively illustrated *Reizen van Cornelius de Bruyn, Door de vermaardste van Klein Asia ... en Palestina.*[61] The 'Reverend and learned Dr Cave' was William Cave (1637–1713), an English writer on ecclesiastical history, whose major work was *Scriptorum ecclesiasticorum historia literaria* (1698-9).

Wells had also consulted the works of two prominent French theologian–geographers, Bochart and Huet. Samuel Bochart (1599–1667) was a famous Biblical scholar and oriental linguist and sometime pastor at Caen. His major work, undoubtedly the one consulted by Wells, was *Geographia sacra* (1646, 1651), which is mainly concerned with the genealogy of the nations recorded in Genesis 10 and with the Phoenicians, a book extensively illustrated with maps and frequently republished in the late seventeenth and early eighteenth centuries.[62] Dainville records that it was a lecture by Bochart on sacred geography in 1646 that had turned the interest of Pierre Daniel Huet (1630–1721) from law to the study of Hebrew.[63] Huet became bishop of Avranches in 1689, and was also an authority on 'sacred geography'. In his thesis on the location of the Garden of Eden, Wells quotes Huet, presumably *A treatise of the situation of Paradise (Tractatus de situ Paradisi terrestris)*, published in Paris in 1691 and in English in London in 1694.

Another work cited is the study of the world after the Flood – *Geographia conjecturalis de orbis terrestris post diluvium transformatione ex variorum geographorum sententia cui author subscribit* (1675) – and the *Turris Babel* (1679), by the German Jesuit cartographer, philosopher and linguist Athanasius Kircher (1602–80). Wells was clearly familiar with a wide range of English and European authorities on aspects of Biblical history and theology; his information field seems to have been both extensive and up-to-date, within the context in which he had chosen to work.

The books on the historical geography of the Old and New Testaments include a number of maps and illustrations (see Fig. 2.6). Most of the general 'territorial' maps appear to have been drawn especially for these particular books, though it is more than likely that further detailed searching of other contemporary cartographic authorities will reveal that Wells, as in other matters, had extensively culled existing published maps.

The question of the origins and genealogies of maps and illustrations of the Holy Land of this period, including maps from Bibles, is a fascinating subject, on which interesting work has been, and continues to be, undertaken. Delano Smith has shown, for example, that:

the story of maps in bibles is the story of the Reformation in general. In particular, it concerns the religious convictions of leading reformers such as Martin Luther and John Calvin and their followers. Each of the maps involved bear witness to a complex web of contacts which would have been at times covert, dangerous, even fanatical. Most of the books in which these maps first appeared offended the authorities, and many were outlawed and burned, thus accounting for the rarity of some of these early editions.[64]

Although the first printed Bible with illustrations dates from 1455 (illustrated manuscript Bibles are much older, of course), the first map in a printed Bible dates from 1525. Both printed maps and illustrations in Bibles were created to mirror and complement the symbolism and imagery of the written text. The themes of pictures and maps were consistent over long periods of time. The picture illustrations date back in some cases to the fourth century AD, with an acceleration in production in the twelfth and

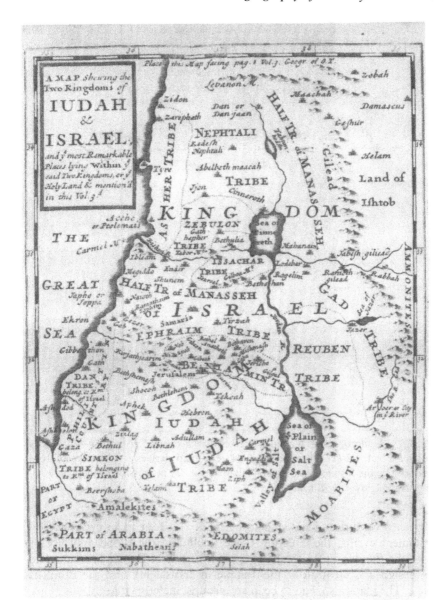

Fig. 2.6 Map showing the kingdoms of Judah and Israel, from *An historical geography of the Old Testament* vol. III (1712), opposite p. 1. By permission of the Syndics of Cambridge University Library.

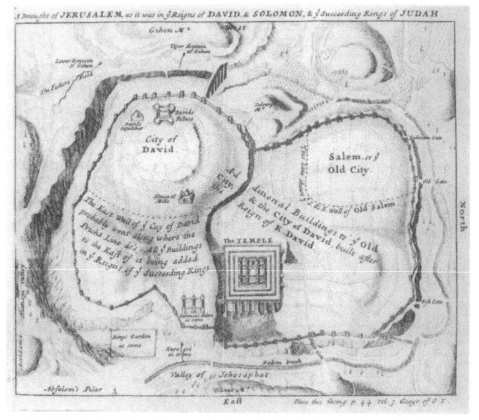

Fig. 2.7 A 'draught of Jerusalem', from *An historical geography of the Old Testament* vol. III (1712), opposite p. 44. The map derives from a map of 1604 produced by the Jesuit Juan Bautista Villalpandus. By permission of the Syndics of Cambridge University Library.

thirteenth centuries. They tend to be related to four main cycles of Old Testament events, the Creation Cycle (including the Flood and Noah's Ark), the Exodus Cycle, the King Cycle (the time of David, Solomon, and the design and creation of the Temple in Jerusalem), and the Prophets Cycle. The maps also tend to be in key groups, including the Exodus maps, later included in a group (of four) containing additionally a map of the division of the Land of Canaan among the Twelve Tribes of Israel, with two New Testament maps, one showing the places mentioned in the gospels, the other the places mentioned in the Acts of the Apostles, including the journeyings of St Paul. There were also maps of the location of the Garden of Eden and of Paradise.[65] All these date from the sixteenth century. It is not surprising to find that maps and illustrations of this period had quite long genealogies of republication[66] making it at times difficult to check their originality.

Wells drew heavily on existing material. The map or 'draught' of Jerusalem 'as it was in ye reigns of David and Solomon, and ye succeeding Kings of Judah' (see Fig. 2.7) which he includes in volume III of *An historical geography of the Old Testament*, has as its source the imaginary plan of ancient Jerusalem seen from the east first published by Juan Bautista Villalpando or Villalpandus (1552–1608), the Spanish Biblical geographer and cartographer who worked in Rome, in volume III (1604) of *Explanationes in Ezechielem et apparatus urbis ac templi Hierosolymitani commentat*.[67] The link evidenced here between Wells and Jesuit sources and thinking is an extremely important one for understanding part of the background of the sacred historical geography in which Wells was engaged.

Villalpandus was one of an important group of Jesuits working within a context of a particular kind of Christian Hermetism, which flowered in the late sixteenth and early seventeenth centuries. According to Taylor:

As it existed in the Renaissance, Hermetism may be compared to an intricate web made up of numerous strands of very different origins and dates. Fundamentally, however, it postulated belief in an astrologically ordered cosmology. According to this view, the universe was geocentric, being divided into three zones – namely the elemental world of man, the celestial world of the planets and fixed stars, and the supercelestial world of God and His angels, known also as the intellectual world.[68]

Villalpandus who was born in Córdoba in 1552, joined the Society of Jesus in 1575, and by 1583 had already indicated a major interest in the theories and nature of construction of Solomon's Temple in Jerusalem. He became a Jesuit architect, and was transferred to Rome in 1592, together with Father del Prado, an older man with whom he collaborated, albeit uneasily, on the production of a commentary on the book of the prophet Ezekiel (whom Taylor describes as 'that most Hermetic of the prophets')[69] which appeared in three volumes, in 1596 and 1604. The second volume, *De postrema Ezechielis prophetae visione* (1604) contains a major and famous reconstruction of the Temple, based on Ezekiel's vision of it, the reconstruction having been encouraged by the architectural interests of Philip II, and stimulated and influenced by the deep-rooted concept, reinforced by the visit of the founder of the Jesuit order to Jerusalem in 1523, of the Temple of Solomon as the symbol and forerunner of the heavenly or celestial Jerusalem to come. The reconstruction was made in classical style. Wells includes copies of these reconstructions in volume III of his *An historical geography of the Old Testament*, and the map produced by Villalpandus for the 1604 volume. Wells also draws strongly, in his descriptions of the Flood and the Tower of Babel, on the work of another Jesuit scholar, Athanasius Kircher, who was the last of the major Jesuit Hermetic scholars, and who attempted to provide greater scientific authority for Hermetic ideas at a time when they were being discredited. His major works in the field of mystic architecture, one of which – on the Tower of Babel – is cited by Wells in volume I of *Historical geography of the Old Testament*, were: *Arca Noë* (1675) and *Turris Babel* (1677). Kircher was a very important figure, who attempted to bridge the gap between the old mysticism and the new science, but never really succeeded. As Taylor puts it: 'The tragedy of Father Kircher is that, though he had the intellectual equipment necessary to have achieved greatness, he was never able to rid himself of the dead weight of tradition.'[70] Taylor also asks and answers an interesting contextual question:

But how was it then that Leibniz' studies led to positive results such as the invention of infinitesimal calculus, whereas those of Kircher merely ended in sterility? The problem is a complex one, since so many of the great scientific figures of the century, such as Kepler and Newton himself, had their mystical side as well. The answer is probably one of degree, depending on the proportion in which the two tendencies, the mystical and the scientific, were present. For one thing, Kepler, Newton, and Leibniz did not have the exaggerated respect for tradition and authority which we find in Kircher, and which made him reject the heliocentric system but accept Hermes Trismegistus. Then, too, the others had a much firmer grasp of the difference between genuine quantitative mathematics, which was the key to scientific progress, and mystical mathematics with its Pythagorean and Cabalistic connotations.[71]

While drawing on Protestant Biblical traditions and exegeses, Wells obviously also drew on the strong visual and symbolic mysticism which characterized Jesuit thinking in Europe in the late sixteenth and seventeenth centuries, and thereby offers in his studies a fascinating set of indices of changing geographical, theological and scientific ideas.

Conclusion

Wells' historical geographies of the Old and New Testaments provide interesting examples of the early use of the term 'historical geography', of the nature of interest in the Holy Land and Biblical geography, and consequently of the use of an extremely heterogeneous range of derived and adapted ideologies for the purpose of reconstructing the landscapes and places pertinent to the understanding of the 'sacred' geography' of the Holy Land. His work is an ideological amalgam or compendium of Biblical exegesis and the synthesis of a wide range of authorities and travellers, ancient and modern. One can see in his attempts to locate and describe important Biblical events and places the influence of many aspects of the thought of his times (in his case mainly conservative), including notions of the age of the earth, the effects of the great Flood, an obsession with reconstructing the geography of Biblical events and places (in some instances from mythological sources which more modern analysts feel defy precise determination) and a sustained strong interest in the evidence of classical and ancient commentators and geographers. The strong influence and authority of the Bible to a scholar such as Wells is a central theme which deserves further exploration, both in a scientific and theological sense, bearing in mind, perhaps G. L. Davies' observation, in his important history of geomorphology, *The earth in decay*, that 'The bibliolatry of the seventeenth- and eighteenth-century scientists can only be understood if we remember firstly, that they were subject to a time-honoured belief in Scriptural infallibility, and secondly, that even in the comparatively toler-ant atmosphere of Restoration England, there were still very strong pressures hindering the growth of a liberal and critical outlook towards the Bible.'[72] In that Wells published in 1723 a paraphrase of part of the Bible as 'An help for the more easy and clear understanding of the Holy Scrip-tures'[73] it might be thought that he was attempting at least to make them more accessible to ordinary people, but in practice wider access to printed copies of the Bible remained limited until the mid nineteenth century, and a more critical use of the Bible as documentary source, geographical and theological, was still a long way off from Wells and those for whom he wrote and to whom he ministered. Nonetheless, the longevity of the various editions of his works (they were translated into French, German and Dutch, and continued to be used, in various editions and revisions into the nineteenth century) show that his interpretations of the landscapes and the places of the Bible were found to be both acceptable to and popular with a large number of people of affluence and influence.

ACKNOWLEDGEMENTS

I am particularly grateful to Professor J. H. Andrews, Dr M. J. Heffernan, Mrs A. Tarver, Dr. D. E. Cosgrove, Dr D. J. Gregory, Dr R. Rehav, Dr S. Stern, and Dr C. Delano Smith for their helpful comments on the manuscript and for assistance with providing important references for the genealogies of maps and geographical ideas.

NOTES

1 Edward Wells, *An historical geography of the New Testament* (Oxford 1708); *An historical geography of the Old Testament*, 3 vols. (Oxford 1710, 1711, 1712)
2 Edward Wells, *A treatise of antient and present geography* (Oxford 1701)

3 E. G. W. Bill, *Education at Christ Church, Oxford, 1660–1800* (Oxford 1988) 232

4 J. Nichols, *The history and antiquities of the county of Leicester* (London 1807), reprint (Wakefield 1971) 149

5 T. Hearne, *Remarks and collections of Thomas Hearne, Suum Cuique. Vol. IX (August 10, 1725–March 26, 1728)*, edited by Rev. H. E. Salter (Oxford 1914, for the Oxford Historical Society) 65, 330

6 T. C. Cooper, Edward Wells, *Dictionary of national biography* (London 1899) IX, 227

7 Edward Wells *A new sett of maps, both of antient and present geography (being forty-one in number, three whereof contain the Bible geography), wherein not only the latitude and longitude of many places are corrected, according to the latest observations; but also the most remarkable differences of antient and present geography may be quickly discerned by a bare inspection or comparing of corres-pondent maps; which seems to be the most natural and easy method to lead young students (for whose use the work is principally intended) unto a competent knowledge of the geographical science* (Oxford 1701)

8 Wells, *Historical geography of the New Testament*, 1

9 *Ibid.*, 3

10 *Ibid.*, 5

11 *Ibid.*, 52b–53b

12 G. Adam Smith, *The historical geography of the Holy Land* (London 1966 reprint of the 25th edn of 1931) 288

13 Wells, *Historical geography of the Old Testament*, I, ii

14 *Ibid.*, 1

15 *Ibid.*, 37

16 *Ibid.*, 93

17 For a recent review of geological events in the Bible, see Y. K. Bentor, Geological events in the Bible, *Terra Nova* 1 (1990) 326–38

18 Wells, *Historical geography of the Old Testament*, II, 11 *seq.*

19 Wells, *Historical geography of the Old Testament*, III, 43 *seq.*

20 R. Greaves, Religion in the university, 1715–1800, in L. S. Sutherland and L. G. Mitchell (eds.), *The history of the University of Oxford, Vol. V. The Eighteenth Century* (Oxford 1986) 243

21 Carl Ritter, *The comparative geography of Palestine and the Sinaitic Peninsula, Vol. II. The Geography of Palestine*, translated and adapted by W. Gage (Edinburgh 1866) 54

22 C. Glacken, *Traces on the Rhodian shore: nature and culture in Western thought from ancient times to the end of the eighteenth century* (Berkeley 1976 edn) 35

23 M. Bowen, *Empiricism and geographical thought: from Francis Bacon to Alexander von Humboldt* (Cambridge 1981) 58

24 Wells, *Historical geography of the New Testament*, 1

25 *Ibid.*

26 J. A. May, *Kant's conception of geography and its relation to recent geographical thought* (Toronto 1970) 58–9

27 J. N. L. Baker, The history of geography at Oxford, in *The history of geography: papers by J. N. L. Baker* (Oxford 1963) 119–29

28 E. W. Gilbert, *British pioneers in geography* (Newton Abbot 1972) esp. Chapter 2, 'Geographie is better than divinity' 44–58
29 J. N. L. Baker, Academic geography in the seventeenth and eighteenth centuries, in *The history of geography*, 17
30 Wells, *Treatise of antient and present geography*, i
31 A. Downes, The bibliographic dinosaurs of Georgian geography, *The Geographical Journal* 37 (1971) 379–80
32 Baker, Academic geography in the seventeenth and eighteenth centuries, 16
33 Downes, The bibliographic dinosaurs of Georgian geography, 382
34 Wells, *A treatise of antient and present geography*, 125
35 Bowen, *Empiricism and geographical thought*, 71–2
36 F. de Dainville, *La Géographie des humanistes* (Paris 1940) 182
37 Baker, The history of geography at Oxford, 120
38 J. A. May, On orientations and reorientations in the history of western geography, in J. D. Wood (ed.) *Rethinking geographical enquiry* (York 1982) 31–74
39 Wells, *A treatise of antient and present geography*, 10
40 W. Warntz, Newton, the Newtonians, and the Geographia Generalis Varenii, *Annals: Association of American Geographers* 79 (1989) 165
41 Bowen, *Empiricism and geographical thought*, 82
42 R. Rubin, Historical geography of Eretz-Israel: survey of the ancient period, in R. Kark (ed.), *The land that became Israel: studies in historical geography* (Newhaven 1989) 23–4
43 Dainville, *La Géographie des humanistes*, 56
44 *Ibid.*, 56
45 *Ibid.*, 47–9
46 *Ibid.*, 56–7
47 Wells, *Historical geography of the Old Testament*, I, i
48 M. T. Hodgen, *Early anthropology in the sixteenth and seventeenth centuries* (Philadelphia 1964)
49 *Ibid.*, 230
50 Wells, *Historical geography of the New Testament*, 93
51 Hodgen, *Early anthropology in the sixteenth and seventeenth centuries*, 242
52 Bowen, *Empiricism and geographical thought*, 94
53 G. A. Williamson, *Josephus: the Jewish war* (Harmondsworth 1959)
54 'W. P. C.', Henry Maundrell, *Dictionary of national biography* (London 1894) 92–3
55 H. Maundrell, *A journey from Aleppo to Jerusalem at Easter, AD 1697* 5th edn (Oxford 1732) 122
56 *Ibid.*, 115
57 *Ibid.*, 114
58 P. M. Holt, The study of Arabic historians in seventeenth-century England: the background and the work of Edward Pococke *Bulletin of the School of Oriental and African Studies* 19 (1957) 444–55
59 Wells, *Historical geography of the New Testament*, second part, ii
60 S. Anderson, *An English consul in Turkey: Paul Rycaut at Smyrna, 1667–1678* (Oxford 1989) 210
61 E. Laor, *Maps of the Holy Land* (New York 1986) 175

62 *Ibid.*, 114

63 Dainville, *La Géographie des humanistes*, 312

64 C. Delano Smith, 'Maps in Bibles in the sixteenth century' *The Map Collector* (1982) 2–14

65 *Ibid.*

66 *Ibid.*; see also C. Delano Smith and E. M. Ingram, *Maps in Bibles in the British Library. Finding List: 1500–1600* (London 1988); C. Delano Smith, *Maps in Bibles in the Newberry Library, Chicago. Finding List 1500–1600* (Ann Arbor 1989); R. Rubin, *Jerusalem through maps and views* (Tel Aviv 1987, Hebrew)

67 Laor, *Maps of the Holy Land*, 165–6; Rubin, *Jerusalem through maps and views*

68 R. Taylor, 'Hermetism and mystical architecture in the Society of Jesus', in R. Wittkower and I. B. Jaffe (eds.), *Baroque art: the Jesuit contribution* (New York 1972) 64

69 *Ibid.*, 66

70 *Ibid.*, 82

71 *Ibid.*, 90–1. See also J. Godwin, *Athanasius Kircher: a Renaissance man and the quest for lost knowledge* (London 1979)

72 G. L. Davies, *The earth in decay* (London 1969) 12

73 E. Wells, *An help for the more easy and clear understanding of the Holy Scriptures* (London 1723)

3

Land–God–man: concepts of land ownership in traditional cultures in Eretz-Israel

RUTH KARK

One observation that emerged from a study of changes in land ownership and land rights in Palestine from the beginning of the nineteenth century up to the establishment of the State of Israel was that an effort should be made to understand the underlying concepts of land and land ownership in different societies throughout the world, and in various periods. Such a comparative study was indeed carried out, focussing on the reciprocal relationship between land, God and man, and providing the basis for a comprehensive and schematic presentation of this relationship.

Also examined in this chapter is the influence of concepts (altering at times in conformation to modernization and secularism) held by traditional societies on thinkers and movements that flourished in eighteenth- and nineteenth-century Europe in the wake of the Industrial Revolution, the French Revolution and sweeping changes in political and agrarian regimes. The thinking of these social philosophers and movements, together with Biblical concepts concerning land ownership and rights, redemption and messianism, influenced the founders of the first nineteenth-century Jewish societies for the settlement of Palestine, members of the Old Yishuv (the Jewish community in Palestine predating the late-nineteenth-century immigration) and the ideologues and activists of the Zionist movement. Preliminary findings are also presented with regard to the impact of such concepts in the years following the establishment of the State of Israel, and after the Six Day War in 1967. Also analysed are the attitudes of particular sectors – such as Israeli Arabs, the Arabs of Judea, Samaria and Gaza and the members of Gush Emunim (the 'Fidelity Bloc').

The comparative analysis of these concepts in different periods and parts of the world relied mainly upon secondary sources – the findings of historical, sociological, anthropological and geographical studies – and at times upon oral testimony by tribe members. Other secondary sources include studies of the writings on agrarian issues by eighteenth-, nine-teenth- and twentieth-century sociologists, utopians, liberals, anarchists and religious movements, some of whose members settled in New World territories such as North America and South Africa. Various literary selections by writers with a social message are also scrutinized.

On the other hand, research on pre-Zionist and Zionist ideology was based upon contemporary sources, including the writings of prominent activists in land-related issues, particularly the settlement of Palestine. Some such discourses were published in books, journals and newspapers, while others are contained in archival documents. The discussion of the post-state period was supplemented by secondary sources such as *belles-lettres*, academic studies and journalistic reportage.

Concepts of land and land ownership in pre-modern cultures

In the mythological imagery of most traditional nations, land is construed as the creation of either God or a mythical hero, as a component of the divine mating of land and the heavens. In most cases, land is perceived as a goddess who mothers all living creatures. An outgrowth of this myth is the complementary belief that man is, literally speaking, a child of the earth. According to Patai, land worship among most nations is expressed in the attempt to mollify and appease the land hurt through hoeing and plough-ing, or to merit its fertility.[1] The widespread custom of laying newborn infants upon the ground stems from the traditional belief that contact with the earth imbues man with strength.

In the societies under discussion, which are based upon ties of kinship, the relationship of an individual or group to the land is paramount, and regardless of whether the society is nomadic, pastoral or agricultural, social relations are largely governed by the norms of interaction with the land. In these cultures, concepts of the land are mythical, religious, symbolic, animistic and sentimental, whereas in civil societies they are instrumental, economic, rational, legal (based upon property rights) and territorial.

In many traditional cultures, whether in historical times or more recently, land – the main source of life – is regarded as belonging to God or the gods. It is perceived as sacred, in contrast with the secular concept of land as an economic resource or geographic object. It follows that human beings are merely tenants or transient holders, in most cases granted the right to use (but not own) this resource by a human agent who is thought to represent God on earth – such as a ruler, leader or clergyman (see Figs. 3.1 and 3.2). It is ultimately God who grants this right to land usage. This generalization as to how land was perceived in traditional cultures seems to hold true for numerous tribal societies, such as the North American Indians, Australian Aborigines and various African tribes – as the follow-ingh selections demonstrate.

Ancestors, tribal leaders

Tribes of Israel, Jubilee
American Indians
Australian Aborigines
African tribes

Kings, emperors, sultans

Church

Tribal shaman
Catholic clergy
Muslim imam
Mormon sect
Shakers

Religious states

Fig. 3.1   Concepts of land ownership in traditional cultures.

According to Stuart Udall, to the Indian mind the land belonged collectively to the people who used it. The notion of private ownership of land, or of land as a commodity to be bought and sold, remained alien to their thinking even after two centuries of acquaintance with the whites:

The land belonged, they said again and again – in the hills of New York, in the Pennsylvania Alleghenies, and in the Ohio Valley – to their ancestors whose bones were buried in it, to the present generation which used it, and to their children who would inherit it. 'The land we live on, our fathers received from God,' said the Iroquois Cornplanter to George Washington in 1790, 'and they have transmitted it to us, for our children, and we cannot part with it.'[2]

Similarly, the chief of one of the large Blackfeet bands, upon being asked by US delegates to sign one of the first land treaties in his region of the Milk River, near the northern border of Montana and the Northwest Territories, responded negatively:

Our land is more valuable than your money. It will last forever. It will not even perish by the flames of fires. As long as the sun shines and the waters flow, this land will be here to give life to men and animals. We cannot sell the lives of men and animals; therefore we cannot sell this land. It was put here for us by the Great Spirit and we cannot sell it because it does not belong to us.[3]

The indigenous semi-nomadic Aborigines in Australia were associated with a particular stretch of territory – the 'estate'. Ownership of land was

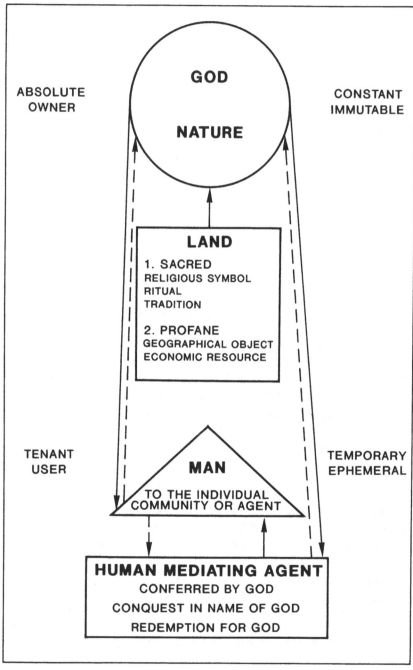

Fig. 3.2   Some examples of human mediating agents.

untransferable; members of the group collectively held land in trust by means of an unwritten charter deriving from the 'Dreaming'. In one of its meanings the Dreaming was the formative or creative period at the dawn of time, when mythic beings shaped the land, gave life to various species and created human life and culture. Although these mythic beings died or were transformed, they lived on eternally in spirit.[4]

A few years ago, a local Aborigine functionary told Bruce Chatwin, who was on a journey through Australia: 'Before the whites came, no one in Australia was landless, since everyone inherited, as his or her private property, a stretch of the Ancestor's song and the stretch of country over which the song passed.'[5] This outlook conflicts with the legalistic concepts of the Western world, reflected in a controversial juridical opinion handed down by Justice Blackburn: 'Although Aborigines did have a system of land tenure it could not be recognised as legal title under Australian law.'[6]

In parts of Africa there existed a hierarchy of powers: from God – the Supreme Being (the Spirit and the Creator) – to the gods and then to the first fathers-ancestors; from them to the more recent ancestors, and finally to the chiefs and elders. In much of East and South Africa, ancestral cults involving the land are clear marks of religious life.[7]

Among the Basuto, Swazi, Ganda, Lozi (Barotse) and Zulu, the land of the tribe-nation is held in trust by the king or chief, who parcels it out to the headmen. The latter, in turn, allocate it to the people. There is no such thing as landed property; instead there are rights to ancestral soil.[8]

The social and religious considerations long predominant in the economic activity and landownership of traditional African societies, together with their inclination to retain the traditional system of land tenure might be viewed as a form of 'agrarian socialism' protecting the right of equal access to land resources. These trends, however, could also play a restrictive role in modern agricultural economies.[9]

Similar perceptions of the land are also embodied in the monotheistic religions: Judaism, Christianity and Islam. An early legal expression of one such concept is found in Leviticus (25:23–4, King James version): 'The land shall not be sold for ever: for the land is mine; for ye are strangers and sojourners with me. And in all the land of your possession ye shall grant a redemption for the land.' A clear distinction is drawn between God's ownership of the land and man's tenure upon it in the capacity of sojourner.

The institution of the Jubilee year stemmed from this differentiation between divine ownership of the land and man's utilization of it. Right of possession was transferred to the people as a collective for a 49-year period, after which the land regained its original status and was reapportioned among the inhabitants. During the Jubilee year land remained fallow, plots were redeemed by their previous holders, and slaves were manumitted. In today's idiom, every sale would be referred to as a sale of leasing rights terminating in the Jubilee year. This practice was meant to prevent monopolies and create greater social and economic equality.[10]

In the opinion of several scholars, Islam was influenced by Jewish ideas regarding the divine ownership of land. Since conquest was a religious act for the Muslims, land and land ownership became endowed with qualities of holiness and tradition. Following conquest, land came under the absolute sovereignty and control of the caliph, as the representative of God on earth, and was regarded as belonging to Muslim society. The state or ruler could parcel it out or lease it in exchange for tax payment or for various services rendered to the central government.[11]

This same principle is espoused in modern times by Seyyed Mahmood Taleqani (1911–79), an ideologue of the fundamentalist Shi'ite regime in Iran: 'Land and natural resources are not the particular property of anyone (neither individual nor society).'[12] He stresses that according to the Kor'an the earth and its resources belong to God, who is their absolute owner.

In the eighteenth and to an even greater extent the nineteenth century – in the wake of the Industrial Revolution, the French Revolution and changes in political, economic, social and religious concepts in Europe – social thinkers, reformers, political and religious movements, nationalistic streams and various groups that settled in the New World (Boors, Shakers, Mormons and so on) began to adopt new ideas with regard to the state, economy, society and religion. Many of these novel concepts touched upon land ownership and agrarian reforms.

While some of these ideas stressed private ownership and the termination of feudal servitude to landowners, others were based upon principles of collectivism, joint ownership of land and an antagonism to land monopolies – tenets that characterized ancient or traditional societies, or the monotheistic religions in their early stages.

Many thinkers and movements can be discussed in this context, such as the physiocrats, Comte de Saint-Simon, François-Emile Babeuf, Georg David Hardegg, utopian and other socialists, liberals, anarchists such as Elisée Reclus, Pëtr Kropotkin, agrarian reformers such as Henry George, the Fabians and so on. Likewise, authors with a social message stressed the importance of eternal forces such as land, nature and God in the lives of humans, whose lives and property are transitory. Among these men of letters were Heinrich von Kleist in the early nineteenth century (who not only set his ideas down in print but also attempted to implement them on a personal basis), as well as writers and poets in the first half of the twentieth century, including Knut Hamsun, Pearl Buck, Robert Frost, John Steinbeck, and Marjorie Kinnan Rawlings.

Religious groups and movements, some of which moved to the New World from the seventeenth century onward, were greatly influenced by

Biblical and early Christian ideas regarding divine ownership of land. Hence they viewed themselves as God's representatives charged with redeeming the land in order to found the kingdom of God on earth. The Boors and Huguenots in South Africa, the Shakers and Mormons in North America and the Templars from Württemberg, Germany, expressed this outlook in both word and deed. Their endeavours merit a separate discussion.

Land and redemption: the Jewish people and Eretz-Israel

The following are examples of salient Jewish historical themes dealing with Eretz-Israel, land rights and land redemption:

1. Eretz-Israel was given by God to the Jewish people as a gift. The Jews' successful conquest of the land resulted not from the use of force, but from the fact that God had promised it to the Jews, and had acted to fulfil his promise.[13]

2. Right of possession remains in the hands of God, and is transferred to the Jewish people as a collective for a period of forty-nine years. The return of the land constitutes an act of redemption and perpetuates the system of divine ownership. Although this process is enacted through technical procedures (redemption, purchase), it is fraught with profound social and religious implications.[14]

3. The Responsa literature expounds in great detail upon the obligation of Jews to settle the land, redeem it through purchase from non-Jews and cultivate it. As an inducement for Jews to discharge their duties in this regard, various concessions were granted, such as allowing land to be cultivated during the Sabbatical year.[15]

4. With time, the concept of redemption took on a wider and more abstract meaning, dissociated from land. Henry Near refers to this as God's collective redemption of the people (entailing rescue, liberation, salvation from Egypt and from the enemies of the people). The term 'redemption' retains this sense in the Talmud, the writings of Sa'adia Gaon, Yehuda Halevi and Maimonides, the Kabbals of the Ari (Rabbi Yitzhak Luria) and the ideologies of various messianic movements. Until the Haskala (a movement in the late eighteenth century to disseminate modern European culture among Jews) and the emergence of Hassidism, redemption was always collective. Hassidism began to favour a personal, or miraculous, redemption. Parallel concepts developed in Christian thought.[16]

From the mid nineteenth century onward, the proponents of settling the land of Israel (Zevi Hirsch Kalischer, Elijah Güttmacher, Judah Ben Solomon Hai Alalai, David Gordon, Hayyim Luria, Simeon Berman) spoke of redemption in a messianic sense, as part of a gradual process destined to unfold in stages. Their beliefs in some ways resembled

millenarian ideas popular in the Christian world at that time.[17] In this
process, redemption (through purchase from gentiles) and settlement of the
land are prerequisites for the redemption of the Jewish people, the coming
of the Messiah and the resurrection of the dead. In two letters from 1863,
Rabbi Zevi Hirsch Kalischer wrote:

As the sickle begins [to cut] the standing corn as part of the efforts to settle
Eretz-Israel, they will eat their fill of bread and this is the beginning of the end of
days ... One wave of immigrants will follow another, and support for this idea can
be found in the holy Zohar [a Kabbalistic work], which states that there will be four
redemptions. The first is redemption of the land itself, planting and raising the crops
of the land, and holy fruit ... The second redemption is the building of the altar,
and [the offering of] sacrifice under a government ... The third is after the wars
between Gog and Magog, whereupon Ben-David [the Messiah, the son of David]
shall come. The fourth is the resurrection of the dead, upon which the passage 'No
eye has seen a God other than thou' shall be fulfilled, speedily in our day, Amen.[18]

Rabbi Judah Alkalai sounded a similar note in 1865: 'And now our
brothers, the people of our redemption! Pay honour to the Lord your God,
and put your heart to the holy goal – to return the Divine Presence to Zion
through the settlement of our holy land.'[19] Rabbi Elijah Güttmacher called
upon people to bestir their hearts 'to the settlement issue, which is a vital
cornerstone for the complete redemption, to spend money on purchasing
land for settlement in Eretz-Israel, to settle the poor members of our nation
there, so they can earn a livelihood from the produce of the land.'[20]
Within the Old Yishuv itself, similar ideas emerged, forming the basis for
the establishment of societies for the purchase, settlement and cultivation
of the land (for example, the Land Cultivation and Redemption Association, founded in 1876, and the Kehillath Jacob Association and the
Nahliel Society at the start of the twentieth century). The platforms of these
societies expressed their desire to integrate messianic and practical forms of
redemption. Their founders were orthodox Jews who aspired not only to
the redemption of the land but also to the performance of divine comand-
ments hinging upon the land, as well as heightened productivity and an
enhancement of the economic situation of the Old Yishuv, whose members
lived on *haluka* (organized charity distribution).[21] The platform of the first
such association attests to this:

We see the rays of dawn beginning to shine gradually, and signs of salvation
drawing near. It seems to us that there has been nothing like this since the day we
were exiled from our land, with respect to the construction of numerous homes and
the purchase of many tracts of land [Motza, Mikve Yisrael and other new neighbourhoods built outside Jerusalem's Old City walls] ... And we have no straighter
path or a more binding obligation than [settling] this place, which is ripe and ready
for our manual labour, so that we eat of its fruit and satisfy the hunger of our small
children with bread, and teach them to revere the Torah, and make them godfear-

ing, and not be shameful, and there will not be jealousy within our borders, and the land shall desire us too, and all our brothers, the children of Israel, and this shall be the first of four redemptions.[22]

These words are reminiscent of Kalischer's ideology.

Certain concepts that developed with regard to land were an amalgam of ideas deriving from contemporary European and American society, Biblical guidelines, the messianic orientation of the 'heralders of Zion' and nationalistic thought. These eclectic concepts originated with the Lovers of Zion movement and gained currency with the rise of the Zionist movement.

Among the general ideological streams that influenced Zionist thought were the following:

1. The physiocratic school, which sprouted in France in the mid-eighteenth century and spread to Eastern Europe. It claimed that only the cultivation of land had a potential for enhancing a society's material worth, and that nature was the sole source of this worth. Society would be improved through productivity, which entailed retraining people for agricultural work. Wealth would grow in proportion to land cultivation, and reach a point allowing for the existence of people on 'sterile' levels of society, such as craftsmen and merchants, who are beneficial and vital socially, but who do not add to the material worth.[23]

The ascent of the physiocratic world view followed upon the waning of mercantilism, which regarded trade as the source of a state's wealth. The physiocrats gave priority to nature and land, and favoured agriculture above all other pursuits.[24]

2. The combined influence of the Puritan revolution in England, the French Revolution, and various schools of thought regarding Russian farming. Zionist ideologues desired to rectify injustices being perpetrated against the common citizen. They sought to grant the average person the right to utilize or own land, while encouraging revolt against foreign rule and its faulty administration, and against the landed nobility. In short, they adhered to the idea of 'land to those who worked it'.

3. Disappointment with the Industrial Revolution.

4. The agrarian reforms whose implementation began in post-Revolutionary France, and later in England, Sweden, Denmark, Spain, Germany, Italy, Ireland, Russia and the United States. The approach to reform and the means of carrying it out differed from one country to the next, but the underlying thrust was to eradicate the vestiges of feudal ownership rights by liberating the farmers from servitude to the land, and turning them into landowners. It is interesting to note that when used in this context, the term 'redemption' recaptured its earlier meaning of paying a 'ransom' for the liberation of the farmer from servitude or from the burden of rental payments for the land.[25]

One of the foremost American thinkers in this realm was Henry George

(1839–97), born in Philadelphia and active in California and New York. His book *Progress and poverty*, written in 1879, deals with the large gap between rich and poor, which was caused by 'the economic rent' (the difference between the productivity of any given piece of land and that of the least productive land in actual use). George referred to this situation as 'exploitation', as it entailed the diversion of increasing income from the many who produced to the few landlords who did not. Rent from land, George said, was unearned because land, indispensable to production, is a free gift to all mankind. The solution he offered was the single tax – a 100 per cent tax on the annual economic rent of land.[26]

George, a religious man although not a church-goer, was influenced by Biblical concepts of land. In an article entitled 'Moses', written in the last year of his life, he depicted the Biblical leader Moses as the greatest of fighters and spiritual mentors on behalf of freedom for the individual and for society. The social legislation of the Torah – especially that concerning the Sabbath, Sabbatical year and Jubilee year – was described as an ideal model for emulation by all generations.

George directly and indirectly influenced numerous social movements both inside and outside the United States. The British Fabians drew upon his ideas, and over the course of the twentieth century attempts were made in a number of places in Canada and the United States to implement his plan for a single tax. Similar attempts were made in Britain during the 1930s, and briefly in Russia right after the 1917 Revolution.

Franz Oppenheimer, the German–Jewish economist and sociologist who laid the philosophical foundations for co-operative settlement in Eretz-Israel at the beginning of the twentieth century, also absorbed George's ideas. George's views had a significant impact on the Zionist movement and the Jewish workers' movement in their early stages. The idea of a Jewish National Fund (JNF) and the demand for nationalization of land were largely inspired by George's writings and preachings, despite the fact that he himself never called for nationalization of land.[27]

5. Shmuel Almog suggests that Russian populism, i.e. the strengthening of farming communes as a means to avoid the inequities of capitalism, should be included in this list of ideologies influencing Zionist thought. The major utopian–populist ideals adopted by the Jews in Palestine were: personal fulfilment, agricultural settlement, manual labour, nationalization of land, co-operation and equality. The atmosphere of liberalism characteristic of the industrial nations in the wake of the Great Depression, which lasted from 1873 to 1896, also permeated Zionist thinking. Two fundamentals of neo-Romanticism that entered Zionist thought are nationalism and the idealization of nature.[28]

The Jewish Haskala's vision of universal redemption was predicated on emancipation and the termination of prejudice on the part of general society toward the Jews. As far as Eretz-Israel was concerned, this blend of Biblical, messianic and nationalistic views led to the adoption of ancient concepts that had undergone secularization, and to the creation of the Jewish JNF. The people, as a collective rather than as individuals, would replace God as landowner. The land was detached from economic considerations of negotiability and exchange, and placed within the context of a nation's tie to its land.[29]

With the secularization of the concepts of land, the practical dimension of redemption – gaining control of land – became prominent. Settlement was necessary in order to establish holding rights. Purchase and possession of a legal title deed were insufficient. Redemption was for the sake of the people, not God; it was a redemption from foreigners and from desolation, attended by land cultivation and general progress. In a summary address to the Eighth Zionist Congress in August 1907, Otto Warburg, a central figure in the Zionist movement, stressed that Zionist activity had to be geared not to the ideal of the redemption of Israel, but to that of practical labour for the sake of 'the redemption of Israel in Eretz-Israel'. He felt obliged to mention 'the unfortunate Uganda affair [the plan to establish a Jewish state in Uganda rather than Palestine] ... and the fact that a number of people dropped out [of the Zionist movement] ... those who had joined Zionism not for the sake of Eretz-Israel but for the sake of the redemption of Israel'.[30] In 1908, Yehoshua Eisenstadt-Barzilai, a leading figure in Palestine, wrote an article stating that 'the essential element of our labours must always be land, the basis for the redemption of Eretz-Israel'.[31] In the opinion of the influential agronomist Yizhak Wilkansky, land purchase through legal procedures was insufficient, and the land had to be redeemed a second time through settlement activity:

Our world hinges on redemption of the land. However, the land can be redeemed in two ways: via legal channels and via settlement.

Through the first way one gains rights on paper, but actual rights can only be attained by means of the second way.

The land of Hauran, acquired through legal channels, is not yet ours, even though the purchase record lies signed and sealed in the archives of the Baron [Rothschild], since it has yet to undergo a second redemption through labour. And not even the moshavot are ours yet, so long as we ourselves have not cultivated their land. The legal concepts are transitory, and the mouth that prohibits things today is the same one that permits things tomorrow. And only 'holding' rights last forever, and the holder is – the labourer.[32]

This type of redemption is accompanied by a private kind, one which derives from the individual's agricultural endeavours, self-labour, self-fulfilment and pioneering. One of the books reflecting this outlook is *The blessing of the earth*, a Nobel Prize-winning novel written in 1917 by Norwegian author Knut Hamsun and translated in 1922 into Hebrew. It

האדמה

יו"ל פעם בחודש ע"י „אחדות העבודה" בעריכת י. ח. ברנר.

⁓⁓⁓

כרך ב'

(ניסן–אב התר"ף: חוברת־זכרון – ניסן התרפ"ט)

בדבר „האדמה"

א.*)

...תכניתה של בטאוננו – כשמהו: ה א ד מ ה.

„האדמה" – זאת אומרת, לא רק־קרקע (חקלאות) ולא רק ארץ (מדיניות),
כי אם עוד דברנוה. ה„אדמה" – היינו: השאיפות לבסיסיות בכל צדדי החיים, חזון
דברים כהוייתם, הכרת יסודי המציאות; ה„אדמה" – היינו המשיכה למקור החיים
והגדול, לסקור ההתחדשות, למקור האמת האנושית:

...וליד האדמה, לאדם העובד, לרגשיו ולרעיוניו, לדרכיו ולחפושי־דרכיו,
לאידיאות ולאידיאלים שלו, לשאיפותיו ולתקוותיו, להשתלמותו ולבקרת־עצמו,
לסלחמתו כער קיומו בעולם ובעד גאולתו – לכל זה ישמשו גליוני ירחוננו.

לא אמצעי לחכסיסי סמלנה; לא סמלנתיות; לאו אבל, כמובן, גם לא כלי־
שרת ליסיסיות עתונית. לא לגדריל ולהאדיר את נוסח המליצות העבריות החדשות,
בכדי להראות למשנאים את כל עשרנו, כי אם בטוי לחיים, לחיינו, לחיר־העבודה
ולמחשבת־העבודה, אנו צריכים להיות בבטאוננו, כל דבר חי, המעיד על יחס אדם
לשאלת חיים זו או אחרת, לחזיון חשוב זה או אחר, באיזו צורה שהיא, צריך
להשמע אצלנו בסלואו, בלי טשטוש. כל סגלת־ספרות, שסני אדם נשקסים מתוכה,
תמצא לה את סקוסה אצלנו, וסקום מרחב – לא מטא־סרום.
הנה התכנית אשר לפני „האדסה".

– – –

tells the story of a farmer who by himself nurtured the land back to health, while overcoming bureaucratic obstacles and the harsh conditions of the Scandinavian wilderness. Isak, the protagonist of the story was,

a tiller of the ground, body and soul; a worker on the land without respite. A ghost risen out of the past to point the future, a man from the earliest days of cultivation, a settler in the wilds, nine hundred years old, and, withal, a man of the day.
... Isak sowing his corn. The evening sunlight falls on the corn that flashes out in an arc from his hand, and falls like a dropping of gold to the ground. Here comes Sivert to the harrowing; after that the roller, and then the harrow again. Forest and field look on. All is majesty and power – a sequence and purpose of things.[33]

It is not coincidental that the journal put out in Tel Aviv by the Labour Party and edited by Hebrew author Josseph Hayyim Brenner was entitled *Ha'adama* (The Land). The first issue appeared in the autumn of 1919, after which it was published on nearly a monthly basis (Fig. 3.3). Prior to the publication of the initial issue, Brenner formulated a credo explaining the selection of the journal's name:

the plan proposed by this mouthpiece of ours is identical to its name; *Ha'adama* [The Land]. What is meant by 'the land' is not only soil and not only political territory, but something else. 'The land' is the aspiration to solidify all aspects of life, a vision of things as they are, a recognition of the elements of reality; 'the land' is the attraction to the source of life and growth, the source of renewal, the source of human truth ... And to the child of the ground, to the labourer, to his feelings and ideas, to his way of life and search for a path, to his perceptions and ideals, and to his desires and to his hopes, to his striving for perfection and to his self-criticism, to his struggle for existence in the world and for his redemption – it is to all this that our journal shall give voice.[34]

A wide variety of articles appeared in this journal, focussing on the utopian socialist ideas of the early Christians and of thinkers during the Middle Ages, as well as the ideals of the French Revolution, the nineteenth-century utopians, religious movements such as the Doukhobors that proposed social reform and land reform, socialist colonies, religious communes, anarchists who rejected the notion of private property and so on. It seems that these ideals, all revolving around the concept of common property – including land – had filtered into the consciousness of the labour movement activists in Palestine at that time.

Following the First World War, when the modern Jewish settlement enterprise in Palestine was resumed, Zionist thinkers, settlement activists and leaders of the Zionist institutions grew more convinced that the redemption of land bore a social (Katznelson) or educational (Bialik) message of national dimensions. The redemption process was also regarded as a pioneering act (S. Shalom) and as a means of obtaining territory for the people (Ussischkin).

Shortly after the end of the First World War, Berl Katznelson, ideologist of the Palestine labour movement, wrote the following:

The national fund is not, in the eyes of the worker, just another settlement institution, but the 'Jewish National Fund' for the People of Israel, a revelation of divine spirit of the nation, a symbol of the idea of complete redemption, the intellectual creativity of the nation that will build and mould the shape of the Hebrew society of the future, the Holy of Holies of Zionism.[35]

Hayyim Nahman Bialik, the leading Hebrew poet of the period, echoed these sentiments in a speech before a convention of teachers in 1929:

The redemption of the land – which is the task of the Jewish National Fund – is a concept likely to develop and be elevated to sublime heights. For truly all the national creativity of all sorts stems from the land, and there is no place for any growth and development of the nation without the primary source of the revival. The land is the starting point not only of education but of the entire life and creative efforts of the nation.[36]

Pioneering on the land finds literary expression in a play by the poet Shin Shalom entitled *Earth and heaven, homeland chapters*. It describes the sacrifices – in some cases the supreme sacrifice – made by members of a settlement called Adama (land):

Five of our comrades. Five rank-and-file heroes. A large harvest of death from among us. For we pay a high price for the transition from slavery to redemption, from life floating on air to a life of ground, a life of land. It is for this that we are pioneers, representatives of the afflictions of our people wandering in the wilderness of foreign nations and representatives of the yearning for a lost world in its distress.

And the father of a fallen member nicknamed 'the mule driver' recounts his reasons for immigrating to Palestine and joining Adama:

I said that my son shall continue in the path of his fathers ... he will complete what they began ... he will strive toward the land of Zion ... In the forest I kindled the flame of Zion within him ... and he grew up ... as a new Jew he grew up ... with the tree and with the sky and with the land ... and he dedicated himself to Zion ... And he waited for a voice ... it is a tradition in our family that one of its sons hears the voice ... the voice of Zion ... the voice of the land ... a sign that the redemption is drawing near.[37]

In parallel, the conviction that physical presence on the land was no less important than the legal right of possession continued to gain ground, and found expression in poetry and literature in which land was portrayed as a woman and the title deed as a marriage contract. Such imagery is found, for example, in a poem called 'Only One' written by Shlomo Tanny in the 1980s:

Like a woman,
No marriage contract can hold her
Without labour, without love.

Even if officially registered,
Even if etched upon your heart –
If you have not slept between the ears of corn,
She shall not belong to you or yours.

A night in the vineyard, the second watch
In the long-furrowed ploughing,
Thinning the corn by moonlight,
The blaze sends off vapours in the noon air,
Blessed twilight dust,
And again the chill of enchanted dawn –
Unwritten land.

Like a woman. Without labour and love
Strangers shall come upon it.
And you have but one.[38]

Besides discharging its original functions as a redeemer of land, the JNF took on the duty of guarding and cultivating the land prior to actual settlement.[39] These developments stemmed from a combination of factors, including ideological changes, growing practicality among the Arab sector of Palestine, and mandatory policy up to 1947 regarding all aspects of land transference.[40]

Menahem Ussischkin, head of the JNF, reassessed the situation in a call to the Jewish people in February 1940, upon the British government's imposition of restrictions on land purchases by Jews:

I am entitled to speak about the sin that the nation of Israel has committed with regard to redemption of the land in our homeland. For the sixty years since the beginning of the Hibbat Zion movement, for the forty-two years since the founding of the Zionist Organization, for the twenty-two years since the Balfour Declaration, few of the nation's leaders have ceased crying out and warning: the most important factor in guaranteeing our future is to redeem and continue redeeming the land of Palestine through all possible ways and means. And what effect has their cry had? Out of twenty-seven million dunams in western Palestine under the British mandate, the entire Jewish nation – the Jewish National Fund, Baron Rothschild, the Palestine Jewish Colonization Association, organizations and individuals – has purchased, over six years, only a million and a half dunams! This amounts to 5% of the land of Palestine. If you take into account the great wealth of Russian Jewry before Bolshevism, of German Jewry before Hitler, of the Jews of America and other countries even today – you will understand the terrible historical sin that the People of Israel have committed over these sixty years ... And we must respond to this not only through protests, but also through concrete action, large-scale enter-prises for the whole world to see – for a ghetto is being created in our land, and we must reply to this by stepping up activity and creating new tools for the redemption of the land, and through this the redemption of the nation.[41]

With the establishment of the State of Israel and the transition to Jewish sovereignty, most of the land was 'redeemed' through purchase and settlement, and as a consequence of the War of Independence. Since the state came to own around 95 per cent of the land, the question arose whether there was still room for those Zionist institutions that engaged in land acquisition, such as the JNF. This question was sharpened with the passage of the Basic Land Law of 1960, and the establishment of the Israel Land Authority, which became responsible for administering state land, absentees' property and JNF land. The Israel Land Authority adopted the renewed practice of leasing land for a 49-year period. David Ben-Gurion, the prime minister of the new state, addressed this issue in one of his first speeches, stressing that the need still existed to conquer the land by fertilizing it and making it bloom, to provide a foundation for agriculture and an infrastructure for raw materials and industry.[42]

An examination of the JNF today, forty years after the founding of the state, reveals that it mainly deals with forestation, preservation of state lands, cultivation of land and acquisition of new territory. The Himanuta Company, founded in 1938, still operates as an important executive arm of the JNF. Shimon Peres, in a speech to the Knesset in 1980 on the topic of immigration to Israel, returned to the more abstract concepts of redemption when expressing his opinion that '[Zionism] is not only territorial redemption but also national redemption and social redemption.'[43]

After the Six Day War in 1967, a new group, small in numbers but extremely important in the modern settlement history of the State of Israel, came into existence. This body, as mentioned before, is called 'Gush Emunim' (the 'Fidelity Bloc'), and is composed of religious Jews motivated by a nationalistic–messianic ideology. Gush Emunim regards the redemption of the land and the settlement of Eretz-Israel – processes in which the Zionist enterprise takes part – as vital stages on the road toward the final redemption of the people. The members of this movement are called upon to take action toward the achievement of that goal.[44] Sociologist Gideon Aran notes that Gush Emunim is a religious group whose *modus operandi* is secular, whose veneer is modern and whose means are political. He regards it as a revival movement of traditional Jewry, carrying the banner of contemporary Zionism. Aran calls Gush Emunim's creed a 'messianization-mystification of Zionism', based upon a 'messianic departmentalization' of issues. Its focus is on practical redemption, which entails settlement and sovereignty in a Greater Israel, with the 'Green Line' serving as the border between the territory in the process of being redeemed (Judea and Samaria) and the remaining land of Israel, which remains unredeemed and free of messianic qualities.[45]

Right-wing ideologist Yisrael Eldad, in his review of the novel aspects of Gush Emunim's ideology, employs imagery surprisingly similar to that popular among the leaders and thinkers of the Jewish community in Palestine prior to the establishment of the state:

Gush Emunim and all those closely associated with it are not frustrated and complex-ridden, for their basis is firm and consistent: we are returning to Eretz-Israel. And just as [in Jewish law] a woman is bought in three ways – money, title deed and sexual intercourse – so Zionism is fulfilled in three ways: money (the Jewish National Fund), title deed (the Balfour Declaration and the international resolutions of 1947) and intercourse (coming onto the land, mounting it through settlement, through construction). This is in adherence with the religion of Moses in the past and the religion of Israel in the present – or in the words of Gush Emunim, 'in accordance with the religion of Moses and Israel', in the present just as in the past. For various reasons too numerous to list here the pioneering settlement movement – even its most idealistic component, the kibbutz movement – waned. Perhaps because of age or fat, perhaps because of statehood, itself a new frame-work, perhaps in accordance with the biological and sociological law of every ideological movement: an exhaustion of its forces.[46]

It appears that Gush Emunin tends to adopt not only modes of operation from the past, but also linguistic patterns. For example, a defence league organized in mid-1989 is called Hashomer ('The Guard' – the name of a pioneering organization in the pre-statehood days).[47]

It would be worthwhile conducting an in-depth study of the diffusion of Zionist ideas concerning land and its redemption – ideas partially forged in the process of creating a national entity – in the land-related ideologies prevalent today among Israeli Arabs, the Palestinian national movement and the Arabs of Judea, Samaria and Gaza. The influence of Jewish nationalism on the ideology of these sectors can be discerned in some of their symbolic events, for instance Land Day (an annual day of protest by Israeli Arabs against expropriation of Arab land, initiated in 1976), and practical actions, such as the planting of olive trees to deter land expropri-ation, and the policy of *Sumud* (i.e. of not abandoning land and settle-ments). This topic is touched upon orally or in various recent publications, by such commentators as Yehoshua Porat, Danny Rubinstein, Anton Shamas and Mati Steinberg.[48] Certain selections from a poem by Mahmud Darwish hint at the mood that prevails among these groups:

Oh, those making their way through the passing words,
Carry your names, and get out,
Take your hours from our time and get out,
Steal what you wish from the blueness of the sea and from the sands of
    memory,

And take for yourselves whatever pictures you want, so that you will
   understand,
What you will never understand,
How a stone of our land builds the ceiling of the sky ...
Oh, those making their way through the passing words,
Pass by like the bitter dust, wherever you want to pass, but
Don't pass among us like flying insects,
Because we have things to do on our land,
We have wheat to grow and to water with the dew of our bodies,
We have what will not satisfy you here:
A stone or partridge,
So take the past, if you wish, to the market of antiquities,
And return the skeleton to the hoopoe if you wish,
On a plate of clay,
For we have what will not satisfy your desire: we have the future,
And we have things to do on our land.[49]

NOTES

1 R. Patai, *Man and land, a study on customs, beliefs and legends among Israel and other nations* vol. II (Jeruṣalem 1943, Hebrew) 67–120

2 Stuart L. Udall, The land wisdom of the Indians, in R. Detwilder *et al.* (eds.), *Environmental decay in its historical context* (Glenview, Ill. 1973) 30–1

3 T. C. McLuhan, *Touch the earth: a self portrait of Indian existence* (New York 1971) 53

4 *Encyclopaedia Britannica*, XIV, 423; N. Peterson and M. Langton (eds.), *Aborigines, land and land rights* (Canberra 1983) 3–12

5 B. Chatwin, *Songlines* (New York 1987) 57

6 Peterson and Langton, *Aborigines*, 3; J. M. Powell, Patrimony of the people: the role of government in land settlement, in R. L. Heathcote (ed.), *The Australian experience* (Melbourne 1988) 14–24

7 E. G. Parrinder, God in African mythology, in J. M. Kitagawa and C. H. Long (eds.), *Myth and symbol: studies in honor of Mirca Eliade* (Chicago 1969) 122–3

8 H. Ashton, *The Basuto* (Oxford 1952) 144–9; M. Gluckman, *The judicial process among the Barotse of Northern Rhodesia*, (Manchester 1955) 306; E. J. Krige, *The social system of the Zulus* (Johannesburg 1936) 176–8; H. Kuper, *An African aristocracy: rank among the Swazis* (London 1947) 44–7; Martin Southwold, The Ganda of Uganda, in J. L. Gibbs (ed.), *People of Africa* (New York 1965) 92–5; Thorrington-Smith, Rosenberg and M. Crystal, *Toward a plan for Kwazulu: a preliminary development plan* vol. I (n.p. 1978) 88–98

9 *Ibid.*

10 Y. Yaacobson, *Wisdom in the Bible: studies on the weekly Torah portions according to exegetic literature* (Tel Aviv 1958, Hebrew) 139–42; *Biblical Encyclopedia*, III, 578–82

11 A. Granovski, *The land regime in Palestine* (Tel Aviv 1949, Hebrew) 9–29

12 M. Taleqani, *Islam and ownership* (Lexington, Ky. 1983) 89

13 W. Brueggemann, *The land, overtures to Biblical theology* (Philadelphia 1977) 47–53

14 Henry Near, Redemption of the soil and of man: pioneering in labour Zionist ideology, 1904–1935, in R. Kark (ed.), *The redemption of the land – ideology and practice* (Jerusalem 1990, Hebrew) 33–47

15 Y. Eisenberg, Land redemption in Rabbinical law and legend (unpubl. manuscript; Hebrew)

16 Near, Redemption of the soil

17 Central Zionist archives, A9/55, letter from Z. H. Kalischer to Rabbi Yehiel Michael Zablodovski, 1863 (Hebrew); Yisrael Kloizner, from the archives of Rabbi Zevi Hirsch Kalischer *Sinai* 2, 23–4 (1939) 385 (Hebrew)

18 *Ibid.* (both references); letter from Kalischer to Rabbi Zablodovski

19 C. Merchavia, *The call to Zion* (Jerusalem 1981, Hebrew) 122

20 *Ibid.*, 123

21 Haim Peles, The attitude of the 19th-century Jewish population to the settlement of Palestine *Zion* 41, 3–4 (1976) 148–62 (Hebrew)

22 A. Druyanov, *Documents on the history of Hibbat-Zion and the settlement of Eretz Israel*, new edition by Shulamit Laskov, vol. I (Tel Aviv 1982, Hebrew) 49–56

23 Matityahu Minc, A long remark in the margins of Drazivan's opinion in 1800, in S. Almog *et al*, *Between Israel and the nations* (Jerusalem 1988, Hebrew) 103–12

24 Shmuel Almog, From 'Brawn Judaism' to the religion of labor *HaZionut* 9 (1984) 137–43 (Hebrew); *idem*, 'The land to its tillers' and conversion of the Fallahin, in M. Stern (ed.), *A nation and its history* (Jerusalem 1984, Hebrew) 165–75; *idem*, Productivism, proletarianism and Hebrew labor, in S. Almog *et al.* (eds.), *Modern Jewish history in flux* (Jerusalem 1988, Hebrew 41–2

25 *Encyclopedia Americana*, international edition, XXII, 561–6

26 *Encyclopaedia Britannica*, XII, 514 and 24, 842–3

27 *Ibid.*; *Hebrew Encyclopedia*, 470–71 (Hebrew); Avraham Halleli, 80 Years of land redemption by the Jewish National Fund *Am Ve'Admato* 52 (1987) 4 (Hebrew)

28 Shmuel Almog, Land redemption in Zionist rhetoric from Hibbat Zion to the end of world war I, in R. Kark (ed.), *The redemption of the land – ideology and practice* (Jerusalem 1990, Hebrew) 13–32; Yaakov Shavit, Hebrews and Phoenicians: a case of an ancient historical image and its usage *Cathedra* 29 (1983) 173–91 (Hebrew)

29 Natan Rotenstreich, Right of possession *Am Ve'Admato* 51 (1986) 5–8 (Hebrew)

30 Y. Thon (ed.), *The Warburg book* (Herzlia 1948, Hebrew) 123–4

31 Yehoshua Eisenstadt-Barzillai, Letters from Palestine *Ha'olam* 25 (24 June 1908) 337

32 Yizhak Wilkansky, *On the road* (Jaffa 1918, Hebrew) 129

33 K. Hamsun, *The blessing of the earth* (New York 1923), part II, 252 (translated from Norwegian)

34 Josseph Hayyim Brenner, With regard to the land *Ha'adama* 2 (1923) 523

(Hebrew); this article was first published in *Kuntras* 6 (1919), shortly before the first edition of *Ha'adama*; H. Halprin, On 'Ha'adama' – a change in formulation, in Y. Levine (ed.), *Research notes on the work and activities of J. H. Brenner* (vol. II (1977. Hebrew) 107–18; see Fig. 3

35 B. Katznelson, *With an eye toward the days to come: the writings of B. Katnelson III* (Tel Aviv 1946, Hebrew) 74
36 Hayyim Nahman Bialik, On land, settlement and education in the spirit of the Jewish National Fund – a digest *Am Ve'Admato* 49 (1984) 11 (Hebrew)
37 Shin Shalom (pen name of Shalom Joseph Shapira), *Land and sky, homeland chapters* (Tel Aviv 1949, Hebrew) 214–23
38 Shlomo Tanny, Only one *Karka* (July 1983) 4–5 (Hebrew)
39 Bezalel Amikam, Arab riots and Jewish settlement policy, 1936–1939, in R. Kark (ed.), *The redemption of the land – ideology and practice* (Jerusalem 1990, Hebrew) 290–7
40 K. Stein, *The land question in Palestine, 1917–1939* (London 1984) 212–22
41 Merchavia, *Call to Zion*, 234–5
42 David Ben-Gurion, Land redemption in Israel *Am Ve'Admato* 51 (June 1986) 2 (reprint of speech given in 1950; Hebrew)
43 Transcript of Knesset discussion from 10 October 1980, in Merchavia, *Call to Zion*, 264
44 G. Aran, From religious Zionism to a Zionist religion: the origins and culture of Gush Emunim, A Messianic movement in modern Israel (unpubl. Ph.D. thesis, Hebrew University of Jerusalem 1987, Hebrew); H. Peles, The dialectic development of the Zionist idea *De'ot* 45 (1976) 333–41 (Hebrew); Yohai Rudik, The 'jewish underground' between 'Gush Emunim' and the 'redemption movement' *Kivunim* 36 (1987) 85–98 (Hebrew); Yossef Shilhav, Interpretation and misinterpretation of Jewish territorialism, in D. Newman (ed.), *The impact of Gush Emunim politics and settlement in the West Bank* (London and Sydney 1985) 111–24
45 Aran, Religious Zionism, xxiii–xxiv, 522–4, 572–6
46 Yisrael Eldad, The innovation brought on by Gush Emunim, *Ha'aretz* (9 February 1984) (Hebrew)
47 Nadav Shragai, Gush Emunim establishes Hashomer [the Guard] organization for prevention of car burnings in Jerusalem *Ha'aretz* (18 August 1989) (Hebrew)
48 D. Rubinstein, *The fig tree embrace* (Jerusalem 1990, Hebrew) 55–61; Mati Steinberg, The PLO vs. Palestinian fundamentalist Islam, in his *From Theory to Practice: fundamentalistic streams in light of the problems of our region* (Davis Institute, Jerusalem 1989, Hebrew) 27–41
49 Tom Segev, The poet's intention – a continuation *Ha'aretz* (25 March 1988) (Hebrew)

4

Religious ideology and landscape formation: the case of the German Templars in Eretz-Israel

YOSSI BEN-ARTZI

The immigration movements of the nineteenth century wrought significant changes in the appearance of the world. Individuals, families, ethnic and religious groups abandoned their native countries in great numbers, either voluntarily or under duress, in the hope of establishing a better life elsewhere. Most of these movements originated in Europe and were directed at the New World and the continents of Africa and Australia. These movements have been studied extensively in terms of their history, causes, character, and influence on the host countries. In relation to historical geography, the principal issue is the role and function of such groups in influencing and shaping the culture and patterns of settlement in the regions in which they established themselves.[1]

A number of small distributary streams detached themselves from the great flows of migration to make their way to that remote part of the southwest Orient known as Palestine. The most publicized aspect of this particular immigration has been that of the Jewish settlement of the country and the crucial part the Jews had in changing the face of Palestine and in the development of the region. But, at the same time that Jews were settling the region, and even earlier, other ethnic and religious groups sought refuge in the Holy Land. Alongside Muslims arriving from territories that had passed from Turkish rule, there came adherents of a variety of Christian groups from the United States, Western Europe and Russia; some were members of established churches and missionary movements while others belonged to dissenting congregations. One of the most fascinating among the non-Jewish groups of settlers that had a significant effect on the landscape of Palestine in the nineteenth century was the Templar community, which originated in Württemberg in south Germany. The Templars have been studied in detail from a historical and political perspective.[2] Nor have historical geographers of the nineteenth century ignored their evident contribution to the development of Palestine in a broad range of social, economic and political spheres. However, no comprehensive account has been presented to date of the unique and significant contribution of the Templars to the settlement landscape in Palestine, especially as regards rural settlements.

The aim of this chapter is to describe the characteristics of the settlement pattern created by German colonies in Palestine during the period in which the Templars established themselves and exerted an influence on their neighbours; and to examine the contribution and role of the settlement on the landscape of Palestine at a time which was most significant in the country's development.

The Templar's relationship to Palestine

The Society of Templars (*Tempelgesellschaft*), which was active in Palestine from 1869, rose in the context of social, political and religious developments in Europe and Germany at the beginning of the nineteenth century, and the Kingdom of Württemberg.[3] The religious beliefs of the Templars and their particular conception of Christianity derive ultimately from the Pietist movement within the established Protestant church, whose history goes back to the seventeenth century. But the creation of the Society itself is associated with the person of Christoph Hoffmann.[4] The son of a family of teachers, clergymen, theologians and men of affairs, Hoffmann was himself a teacher and religious thinker. He developed an articulated and coherent, religio-social conception of the relationship between man and God, of the function of the Christian church in his day, and the nature of society and the state in the conditions of the social and political changes which were taking place in Europe.

Hoffmann and his followers sought to right the social and spiritual ills that corrupted society and the established church in Germany. They hoped to accomplish this end by returning to the simplicity of faith and practice of primitive Christianity, believing that the Millennium was at hand and that fulfilment of the Biblical prophecy of the Messiah's coming was imminent.[5] Hoffmann's group had made the prophets' vision of the rebuilding of Jerusalem a cornerstone of its doctrine, and it was this tenet in its creed which caused the group's expulsion from the established church and the repressive measures taken against it. The rebuilding of Jerusalem was also the article of the Templars' faith underlying their ambition to gather in the City of God. Therefore the Friends of Jerusalem, as they called themselves, formed a comprehensive programme of action, according to which 'Gods' People' (*Gottes Volk*) would collect in Palestine, where through the purity and perfection of their deeds, conducted in the spirit of the first Christians, they would serve as an example to the world and raise up the people of the Land from their corruption in order to prepare for the Second Coming and the building of the Temple of Jerusalem. (It is from the desire to fulfil this last aspiration that the group later took the name of Templars, by which they are now generally known.)[6]

The establishment of the Templar colonies and their geographic distribution

Templar immigration to Palestine and settlement of the country was not a spontaneous affair which took place without guidance. Nevertheless, the initial conception of an organized movement into the country and orderly establishment of communities was to a significant degree not realized according to plan. For years the leaders of the Templar Society hesitated in taking the practical step of establishing themselves in the country. Instead the Templars waited to be joined by masses of recruits, for assistance from international agencies, and for growth of significant support in Germany itself. But while the heads of the Society waited in anticipation of the fulfilment of these things, many of the members lost patience and left for Palestine on their own initiative, before the leadership had issued its call. Some members settled in Beirut, others in Nazareth and Jaffa, and they looked for ways in which they could secure their sustenance.[7] This development, conjoined with the failure of the group of young settlers in Samunie and the declining influence of the Society in its home country, accounted for the decision of the Templar leadership to send the Society's founders, Hoffmann and Hardegg, to Palestine in order to found a Mission station in Nazareth.

When they disembarked in Haifa on 30 October 1868, they were awaited by a handful of families and young people from the community that had preceded them and had rented living quarters in the city.[8] It was at this juncture that the earliest significant decision was made concerning the siting of the first Templar station in Palestine. Nazareth was discarded in favour of Haifa. (Haifa was then a small coastal town in its early phase of development and was beginning to assume some importance in the life of the region.) There were various reasons for this change of plan. Chief among these were the relative backwardness of Nazareth and its remoteness from the coast, and the need of convenient access to the representations of the European powers. Haifa was the more favourable location in respect of both these factors, and had the added advantage of offering conditions to establish a reception centre for absorbing the large numbers of incoming colonists arriving at its harbour, who could then be sent on to settlement locations elsewhere in the country. Quite apart from being a coastal town with anchorage facilities, Haifa was also headquarters for European vice-consuls. It had, moreover, a wholesome climate – certainly compared to that of Nazareth, where so many Templar settlers had been felled by disease, attributed to effects of the unsalutary climatic environment of the area. Finally, there was land available for purchase in the vicinity of Haifa.[9] But while this pilot group of settlers was getting ready to purchase land and the equipment it needed to build the facilities of the community, news arrived from Jaffa of an appealing offer from the missionary Metzler in Jaffa of a 'ready-built colony' that could be occupied immediately. The Templars were presented with the opportunity of simultaneously establishing a second settlement on the coastal plain.[10]

The Jaffa colony was located on a small hill north of the town. The complex contained five large houses suitable for habitation (there were an additional four that had not yet been offered to the Templars); a steam-powered mill, for which a shipload of coal had already been deposited at the port as fuel; a hospital with five beds; and a nineteen-room hotel.[11] The offer had much to recommend it on practical grounds alone, but it also furnished the leadership with a means of solving the problem of the rivalry between the Society's founders, Hoffmann and Hardegg. The two men could simply be separated from each other's sight by locating them far apart in different settlements. The Templars were therefore able to acquire a site for a second colony, and this decision, too, had been arrived at fortuitously rather than by intention – or, in Templar terms, they had been guided by the hand of Providence.[12]

The decision to acquire the Jaffa property proved to be advantageous in every respect. The possession of the steam-powered mill gave the Templars immediate economic support and the hotel served as temporary accommodation for additional settler families. Principally, however, it would allow them to conduct the sort of 'mission by example', which was central to their conception of the Templars' function in the Holy Land. In concrete terms, this meant that they could put to immediate use the hospital and school they had taken over from Metzler, and establish a central educational institution for the community. The only drawback to the Jaffa colony was the absence of any farmland that was adequate for a large settlement. For this reason they cultivated land they leased to the north of the colony from local people and Europeans, such as the Isaacs Model Farm, the Montefiore Garden and Mount Hope.[13] But these were temporary measures, and it was evident to the Templars that they would have to establish a large agricultural settlement of their own. Therefore in the summer of 1871 they acquired land off the road leading to Shechem, near the intersection of the Ayalon and the Yarqon. The way was now clear to establish the third of their colonies, an achievement which the community had accomplished within two years of arriving in the country.[14]

Nor had the Templars neglected to obtain a foothold in Jerusalem, but in this goal, too, the organization had been anticipated by individual members. A number of families had settled in the city from 1870, and had started up various small enterprises. When they joined the general endeavour of the Templar community, the Jerusalemite group began acquiring land, and in 1873 laid the cornerstone for the first house in the colony intended to serve as the spiritual centre for God's People.[15] In the same year other Templar stations were established in regions adjacent to Palestine. It would appear that the Templars took a very broad view of the geographic extent of the Promised Land. A number of families settled in Beirut and Alexandria, where they ran hotels, steam-powered mills and workshops.[16] Within the territory generally accepted as Palestine, a small community was founded at Nazareth in the early 1870s, at the English Orphanage. By 1882 the community in Nazareth numbered 28 persons.[17] In 1873 at Tira, near Haifa, land was leased to build a first sister colony to the town. However, because of depredations by the neighbouring local population, actual settlement of the land was put off and activities were confined only to work. Several families also leased farmland at Ramla and Asqelon, although this did not result in the formation of separate community. Meanwhile a number of internal crises and events within the Templar group put a brake on the pace of immigration and settlement.

The rigours of the climate, disease, epidemics, exposure to attack, civil disorder and a variety of agricultural problems were making it increasingly difficult to absorb new settlers. By 1874 the Templar 'immigrant convoys' (*Wanderzüge*) had ceased to arrive in Jaffa and Haifa, and the Templar settlements directed their energies to entrenching and extending what they had already established (Fig. 4.1).

The stage of establishing a firm foothold in the country, and of preparing the succeeding generation to carry on with the work of the founders, lasted longer than the timetable the Templars had set for the task; and they were unable in the course of this initial phase to found any additional colonies. Only towards the end of the 1890s, after many had left and pressure mounted from the second generation, did the Templar leadership turn its attention to mobilizing the resources for setting up new colonies. By now, however, the country was covered with Jewish settlements established throughout its various districts between 1882 and 1898. Land and building permits were no longer as easily acquired, even for German subjects, so that it was only with considerable effort that the Templars were able to purchase land. The visit of Kaiser Wilhelm II in Palestine in the autumn of 1898 gave new impetus to Templar activity at this time, for it bestowed on their movement the status of a German national enterprise, in addition to its self-appointed role as a Christian mission. This official recognition and the public exposure the movement enjoyed as a consequence was useful in helping the Templars collect the money required to purchase land in the vicinity of established colonies for the use of their children and grand-children.[18]

The factors taken into consideration by the Templars at this time regarding location were entirely geographical: all that concerned them was

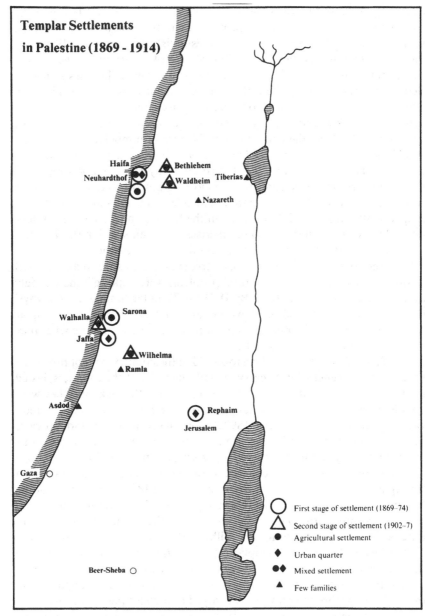

Fig. 4.1   The distribution of Templar settlements in Palestine, 1869–1914.

that the land should be close to the old colony and near urban settlements and major markets. They no longer sought out locations that were of religious significance, nor did they seek to penetrate into the country's interior in order to present themselves to the inhabitants as exemplary patterns of Christian piety worthy of imitation. The settlers in Jaffa and Sarona looked to the Sharon Plain; those in Haifa searched in the Jezreel Valley. Then, in 1902, the cornerstone was laid for Wilhelma, located less than 10 kilometres from Jaffa-Sarona;[19] and in 1906–7 the cornerstone was laid for the 'twin colonies' of Bethlehem Galilee and Waldheim, at the foot of the hills of Nazareth, only 18 kilometres from Haifa (see Fig. 4.1).[20] Later a fresh attempt was made to settle in Tira, but ended in failure because of a blood feud with the inhabitants of a nearby Arab village.[21]

This, then, was the process that determined the geographic distribution of the German colonies which was in force until the outbreak of the First World War. It should be pointed out that at the time these colonies were negligible in number compared to the approximately 55 Jewish settlements and farms. All the same, despite being so few, the German colonies succeeded in maintaining themselves on a qualitative level which did not decline with the passage of years, and even managed to preserve their power and influence.

To summarise, the factors which were taken into account when locating a colony were contingent on political, personal and local considerations; the *site factors*, on the other hand, were universally applicable. If land was sought with the intention of establishing a colony, the site had to be suitable for building houses and farmsteads and farmyards, in accordance with requirements common to all rural regions: priority was given to elevated sites that would protect the colony and allow the surrounding neighbourhood to be kept under watch; or to dry points in low-lying areas; or to centrally placed sites that allowed convenient access to nearby fields.

Settlement patterns

The Templars chose a clustered settlement pattern as opposed to a dispersed pattern of separate farms and hamlets. In this they differed from most European settlement movements to America, Australia and Africa. The settlement movement conditions in the country, the character of Turkish rule and the need to defend themselves against incursions made it inadvisable for the Templars to disperse as independent settlers. Their situation differed from that of colonists elsewhere, say in East Africa, who had immigrated under the protection of a colonial government that had the power to enforce its rule and had taken possession of land that they put at the disposal of incoming settlers. On the other hand, the settlers who came to Palestine – the Germans in the first instance, followed by the Jews – were obliged to obtain land at their own expense, and to depend on the protection of consuls and armed naval vessels to secure their property and personal rights. These circumstances compelled them to gather into clustered settlements. Indeed the few attempts at establishing individual farms had ended in failure, even in tragedy.[22]

From the very onset Templar settlements were designed on a regular pattern of buildings and squares laid out on generous lines and on the basis of a prepared plan – although, contrary to what might be expected from a community settling under the direction of a central authority, the German colonies did not follow a homogeneous plan laid down for all.[23] All of the colonies were built in accordance with a defined plan which covered streets and the allocation of land use for building, workshops and enterprises, agriculture and communal spaces and facilities. The building area (*Bauplatz*) was clustered, as we noted earlier, and within its confines plots of varying size were allocated to houses, farmyards and common facilities. The size of building plots and the location of structures within the allotted spaces was dictated by the plan, whereas the form of the house was determined by the individual settler alone. Generally speaking, the principle of absolute equality existed over the size of building plot, and only the acreage of the field depended on the amount of money a settler had invested for its acquisition. The building plots were generously large by the standards of the country, and in the range of 2.2–4.5 *dunam* per family. The frontage measured between 25 and 40 metres across, and the depth was dictated by the features and character of the site. An area that extended behind the house was set aside for a garden and a variety of plants. In this way a standard settlement pattern was established which was at once spacious and convenient for the erection of farmstead facilities, and where areas were also allotted to communal buildings. The latter were usually located at the centre of the settlement. In Haifa a small, rather confined plot was set aside for the Communal House (*Gemeindehaus*) at what had been the original core of the colony; as the colony grew its relative area within the settlement declined – though it still retained its centrality of function, if not of location. In Sarona, Wilhelma and the more recent colonies of the Lower Galilee, such areas were located at the geographic centre of the settlement, at the intersection of the main streets.

In Haifa, the plan for allotment of land use in an agricultural settlement was broadened to accommodate non-agricultural activities as well as adapting to suit the local topography.[24] This colony accepted for membership viticulturists, farmers, craftsmen and a variety of tradesmen and service personnel; so that land was allotted for a broad range of economic activities. The area was traversed by a continuous straight road that stretched from the seashore to the foot of the Carmel, and divided the area as a whole into eastern and western halves. This allowed the buildings to be arranged in a way that permitted breezes to move freely through the colony from the west, thus significantly reducing the discomfort brought on by the heat of the summer months. At its lower end, near the sea, the main street was intersected by the trunk road that led from Haifa to Jaffa (the Jaffa Road); a major road junction was thus established that furnished, as it were, two entrances to the colony from two directions – one from Haifa and the west, the other from the sea. The nodal area was reserved for services and enterprises for accommodating travellers, hotels, and larger businesses. Nearby, on the main road, areas were given over to light manufacture and the crafts. It was here that a soap factory, a slaughterhouse, a carriage service-station and hotels were rapidly established, as well as one of the largest wholesale trading establishments in the country. The Carmel slopes – which reminded emigrants from the Neckar Valley of the hillside vineyards in Germany, and therefore appeared to them well suited for the same purpose – provided areas that were used for the cultivation of vines. There was a residential street intersecting the main street at its southern upper end, with the vineyards above; in the middle was a feature resembling a broad avenue, along which land was allotted to the remainder of the settlers, consisting of farmers and craftsmen, whose fields were located directly behind the building plots (Fig. 4.2). Three types of land-use allotment were thus established that suited the economic activities and topography of the region. The fact that this layout had been designed not by experienced planners but by the settlers themselves, and at so early a stage in the pioneering development of the country, makes the achievement all the more remarkable. The planning of this group was on a truly ambitious scale, for in laying down from the very outset two streets parallel to the main street it was anticipating the future development of the colony along these roads, where land would have to be allocated for a variety of activities.

The settlement forms of the Templar colonies were developed as well, and followed a defined geometric plan, although in this, too, there was no single pattern imposed on all of the settlements. Here again one might have expected that a small community, governed by a centralized administrative body and settled in an alien land, would have taken the line of least resistance and relied on a single, received form of settlement. But in fact each colony selected a form that seemed best suited to its particular conditions and requirements. The form developed in Haifa was that of the 'street village', in its modern sense, although from the start two parallel streets were added to the design.

In Jaffa the Templars inherited the 'cruciform' ground plan, consisting of two short streets intersecting at right angles; and at Sarona the 'gridiron' pattern was used, although not in its purest form.[25] Initially two main streets were built that intersected at right angles. However, because of the

Fig. 4.2   General view of the German colony in Haifa, 1890–1891.

Fig. 4.3   Wilhelma: main street and settlers.

relatively large number of colonists and the elongated elliptical shape of the littoral sandstone ridge on which the colony was sited, two streets parallel to the original ones were added. The relationship was not symmetric but it nonetheless formed a gridiron pattern. At Wilhelma the 'double-cross' form was used: a single main street was laid down, north to south, for a distance of 1,400 metres, with two streets intersecting it perpendicularly, one of which was longer than the other.[26] Bethlehem adopted the 'single-flank street' form, in which only the northwest side of the street was allotted to buildings and farmsteads, with the intention of eventually developing the southern flank of the colony. In neighbouring Wilhelma the full cruciform pattern was adopted of two perpendicularly intersecting streets.

The settlement patterns and forms of the Templar colonies therefore shared the characteristic features of regularity, spaciousness, and a well-defined system of allotment of land use (Fig. 4.3). These conjoined with geometric-ally laid out settlement forms. Taken together, these features are the manifest expressions of advance planning in accordance with clearly estab-lished rules and a clear conception of the external shape and character of the settlement. In all of these respects Templar colonies were diametrically opposed to the haphazard and random appearance of rural settlements in the country prior to the arrival of the German settlers which were the result of long centuries of development. The Templars can thus be regarded as pioneers in this sphere, and their colonies were the first planned rural settlements to have been built in Palestine in the modern era.

Houses, farmstead facilities, and communal buildings

The manner of building of Templar colonies formed an immediately striking feature of the landscape of Palestine because of the evident contrast these settlements presented to the traditionally built towns and villages of the country. During the earliest phase of their settlement the Templars introduced a great many innovations in building style and methods of construction. These attracted the attention of observers to such an extent that people came great distances in order to see the colonies. However, the Templar settlement landscape in its familiar form was not created at a single stroke. Because of the limited means at their disposal and their initial ignorance of local conditions, the Templars constructed their facilities in a style all their own, in which they applied their particular building methods and system of planning while at the same time adopting local approaches to building in the way of materials, shape of roof and the use of stone.[27] Instead of building in wood, which was the common building material used at home, they employed sandstone, which was widely used for the purpose locally. Flat roofs rather than gabled roofs were used initially; and the arch, a characteristic architectural motif in the Middle East, was also incorporated into Templar buildings. In this they may have sought to accommodate themselves to local conditions, on the assumption that these features suited the climate, because they evolved from a long regional tradition of construction. But they soon learned that flat roofs tended to leak, a drawback which apart from the discomfort it caused, also exposed those living beneath to illness and disease. Within a short time, therefore, the Templars placed tiled roofs both on their new buildings and on those they had already provided with flat ones. And when their colonies began to prosper, they enlarged their homes and started building houses of two or three storeys, with basements and attics. These houses were impressively large by the standards of the country at the time, and were built of a mixture of materials (Fig. 4.4). The main body of the house was constructed of limestone or littoral sandstone, and to ensure proper construction experienced craftsmen were employed in their own communities. For the ceiling and superimposed framework tiles and wood imported from Trieste and Salonika were used, which the Templars were able to bring into the country through the firm of Briesch, who was also a member of the community.[28] As regards the roof of the house, they introduced many innovative features. The entire interior space which the roof enclosed was used as a work or living area, by the expedient of the gabled form; with the narrow lower sides made of stone and furnished with

Fig. 4.4   A typical family of Templar settlers and their house.

windows, and the broader sloped sections covered with tiles. In this way they obtained an interior space which was sufficiently tall and spacious for maximum use to be made of it. Shaped stone gutters ran along the tiled slopes, and these gave a stylish look to the façade and gable. Over the years the interiors of the houses expanded to accommodate two storeys, excluding the attic. The ground floor generally comprised a kitchen and a large living space, and the upper floor contained the bedrooms. In most houses both storeys were used entirely for human habitation. However, the Galilean colonies had a different arrangement, which was arrived at for budgetary and other reasons.[29] In these settlements houses were built with two storeys. The ground storey was used to store hay and keep animals; the upper storey, which had a separate entrance from the outside, was used by the family as living accommodation. Originally this building was put up at the back of the farmyard, and only later, when the family had established itself sufficiently, was a separate house put up at the front end of the plot, and the earlier building turned into an ample hay storage and animal shelter.

Apart from the house, separate farmyard facilities were constructed as well. In farms in the home country, the house and facilities (hay store, manger and storage) formed a single unit. In Templar colonies the facilities were located in the farmyard away from the living quarters. The Templars

Fig. 4.5   A farmhouse in the Templar colony in Haifa.

followed the procedure of erecting farmyard facilities for housing carriages and wagons, farm animals, hay and crops. Some of the structures were workshops: carpentry shops, smithies and even small-scale factories – such as the soap factory in Haifa, which was operated in a farmyard.[30] The number of such farmyard structures was quite large relative to those customarily found in local villages; and they resembled the houses in point of building style, being covered by a tile roof (Fig. 4.5). Some of these were large, even luxurious in comparison with houses in neighbouring towns and Arab villages.

Various decorative and functional elements went into stamping the character of the houses. They generally had a large number of glazed windows, each with a set of twin wood shutters. Small balconies canti-levered out from the façades, and were supported by wood corbels and protected by tile-covered awnings. The entrances were framed by large stone lintels with carved inscriptions of quotations from prayer books, Scripture and Templar writings.

Construction was carried out by using local materials and methods. Local labour was used as well, although under the guidance and close supervision of Templar craftsmen. Men experienced in the arts of construction had been included in the very first immigration convoys of Templar colonists, and candidates in this category were given explicit preference over farmers, viticulturists and those in the service trades. Christian Beilharz, for example, was sent because of his expertise in the building crafts, most particularly that of plastering.[31] For the same reason preference was also given to Schumacher and Oldorf from America, and to Gottlieb Schumacher of Wangen.[32]

The leaders gave considerable thought to the problem of suitable building construction for local conditions, and to that of economic opportunities for Templar settlers. Initially it was thought that local material and labour should be employed in putting up the buildings of the colonies. Arab workmen appeared to be experienced in such matters, dexterous in their work, familiar with the materials of building and able to endure the heat.[33] But the Templar experts who supervised their work were also instructed to learn local building techniques themselves. It took them a very short time to master these to the point that they were sought out by the local population to carry out jobs for them.

On the whole it can be said that the Templars created a style of building which was unique to their colonies. At first glance Templar structures appear to conform to the general pattern just described of houses and farmstead facilities. But beyond the general impression of homogeneity deriving from the elements of stone walls and tile roofs, the architectural variety among individual houses is perhaps more remarkable than the immediate impression of stylistic unity. This difference is the result of the settlers having themselves built and designed the structures they used to suit their own requirements, without any intervention whatsoever from the movement's authorities. It was entirely the taste, intentions and financial capacity of the individual settler, not of the Society, that determined the design. This characteristic of Templar building is especially striking when we compare it to the situation in the Jewish settlements built under the auspices of Baron Edmond de Rothschild and of the Jewish Colonization Association (JCA). In the latter, a single model was established for living quarters and farm facilities which was then imposed on all sites selected for settlement, without regard for the character of the landscape.[34]

The only structures in the Templar colonies that were built at public expense were the communal buildings. In Jaffa such structures were already *in situ* as part of the property which was purchased. In Haifa and Sarona, however, common facilities had to be built by the colonists. Communal life was an essential principle of Templar ideology, and the Templars were therefore unstinting in using their resources to build the Communal House of the colony. This was the centre at which the settlers conducted their religious and secular affairs. Churches were disallowed by Templar religious convictions, and it was the Communal House that fulfilled the functions of a church in their colonies. In Haifa the Communal House was the first structure that was raised when the construction of the colony was undertaken. There the building was called the Model House

(*Müsterhaus*), in token of the Templars' ambition to uplift the country by serving as an example to its inhabitants.[35] Among its functions, the Model House initially also served as a school, and was therefore enlarged in 1890. Only in 1902 was enough money collected for building a separate school-house.[36] The situation developed along similar lines at Sarona, as well. The Jerusalem Templars were the only community in Palestine who from the first built a Communal House together with a building to house an educational institution (*Lizeum*) which was intended to become the crowning achievement of the Templar educational enterprise in the country.[37] All these buildings stood out by virtue of both their ambitious scale and their simplicity of form, which was bare of showy ornament on the outside and within. Templar communal houses tended to be absorbed into the complex of the rest of the colony's buildings, from which they were not marked off by courts or garden precincts. In this respect, too, the Templars deviated from prevailing practice in Palestine among both Arabs and Jews. Arab villages usually had no communal buildings, and in the case of both Arabs and Jews the number of public buildings was never commensurate with the size of the settlement.[38]

The Templars thus created a highly original approach to building that arose out of the convergence of ideology, the needs of the time and the financial resources at their command. Their desire to furnish a practical moral example to the world at large produced, in the domain of architecture, amply proportioned houses and farm buildings into which a great many innovations were incorporated, while their economic circumstances led to a choice of building material and techniques and an architectural style that lay within their financial means. The result of this concatenation of factors is unique, and the Templar buildings which still survive impress us by their soundness and quality. Extant Templar buildings are still serviceable and remain in use for a variety of functions that serve the needs of our own day.

Templar influence on the settlement landscape

From the foregoing description of the character of settlements, some notion can be obtained of the extent of the German colonists' influence on the development of the Palestine landscape. If only because of their great novelty, the presence of this handful of colonies on the region's landscape not only attracted the attention of the native population in their vicinity and of the casual wayfarer, but became a cynosure that drew visitors from afar who came for the express purpose of witnessing the miracle that was assuming shape before their very eyes. The Templar activities in the spheres of trade, transportation, agriculture and politics have been examined by many scholars; and the political and historical context of the movement has been studied at length and in detail by Alex Carmel.[39] However, our concern here is confined to the issue of the physical aspect of Templar presence in the country and the influence it exerted on the emergence of the modern settlement landscape of the region. This question can be examined in the context of the population that was active in shaping the country's landscape. This population comprised the Turkish government, the Arab inhabitants, missionary groups, European societies and enterprises which were then active in the country, and finally the Jewish settlers (who were to become a major factor in their own right, in contributing to the creation of the country's modern landscape).

The Templars influenced the population among which they settled in a great variety of ways. Although the native inhabitants of Jaffa had witnessed the erection of modern European buildings prior to the arrival of the Templars, these consisted only of individual houses or religious institutions which had only recently been put up. The construction and development of the Templar colonies at Haifa and Sarona, however, were of an altogether different order whose significance was beyond anything previously experienced. To the local population they seemed like living wonders brought into existence by conjurers. Many among the native population copied elements of the German buildings in their own. These were primarily people who had left the precincts of the old city to move into newly established neighbourhoods. To them, the Templar structures represented examples of modern architecture, the features of whose style they borrowed for their own buildings: the tiled roofs, cantilevered balconies, new building materials and (most significantly) the ample interior spaces which were so rarely encountered in the cramped architectural environment of the old urban settlements. The Templars also influenced the life of the city as a whole, and this did not consist merely in their contributions to the development of trade and commerce, and in the importation of modern consumer goods from abroad. Their impact extended to the areas of internal communications and transport, civic order, hygiene, and attitudes and values concerning the natural environment.

The effect of Templar presence was especially evident in Haifa, thus the streets of that city, which were originally too narrow to allow the passage of horse-drawn vehicles, were widened to some extent; and the city's gates were gradually torn down to ease the movement of traffic to outlying neighbourhoods. A visiting Templar spoke with pride of the improvement of communications within the city and along the roads leading to town which had resulted from the use of carriages by the German colonists.[40] Two new carriageways that the Templars themselves had built resulted in improvements for the city and its environs which were impossible to ignore: the road to Nazareth, which furnished a source of livelihood for coachmen and guides who plied the route between Haifa and that sacred city; and the road to Akko, which provided a line of convenient communication with the district capital. The Templars even laid down a carriageway on the slopes of the Carmel that linked the city with the mountain that dominated it.

In the summer of 1891 a teacher of the Russian Mission in Nazareth took his students on a tour and pointed out the German colony to his pupils, whom he exhorted to learn from this example of cleanliness and order about how human beings should live – in clean and spacious homes, rather than in a room shared with beasts of the field.[41]

Laurence Oliphant, who had direct experience of the influence of the German colony on the life of the city makes this point a number of times in his writings:

Meanwhile, the influence of three hundred industrious, simple, honest farmers and artisans has already made its mark upon the surrounding Arab population, who have adopted their improved methods of agriculture, and whose own industries have received stimulus which bids fair to make Haifa one of the most prosperous towns on the coast. Already, since the advent of the Germans, the native population has largely increased.[42]

None of this escaped the notice of local government authorities. The governors of Akko and Haifa were often guests at the German colony; and in Jaffa and Jerusalem the German settlers established close ties with local officials. A short time after the arrival of the Templars, the professional and skilled personnel of the community were asked by government authorities and by European organizations and various missions in the country to undertake public works and perform other services. Thus in 1871, Jakob Schumacher was asked to enlarge the pier at Haifa and deepen the harbour at Akko, and to design a bridge over the Kishon – each of these requests being made by the Governor of Akko.[43] The Italian and English Missions asked him to design and oversee the construction of their institutions in Nazareth; in doing so he was also able to provide employment for young Germans arriving in Palestine. The German settler, Oldorf, received a concession to build the Haifa–Jenin–Shechem Road.[44] And in Jaffa, Theodor Sandel became a much sought-after engineer by the city's inhabitants; he received a concession to improve the road from Jaffa to Jerusalem.[45] Later, it was the members of the Beilharz family who carried out construction and road work in the north, and whose greatest accomplishment was the construction of bridges, stations, depots and water towers along the Hejaz railway.[46]

In December 1885 the local government appointed Gottlieb Schumacher to the office of engineer for the Akko district, and within a short time he submitted a number of proposals for the 'uplifting' of Palestine: a port at Haifa, a bridge spanning the Kishon, a railway line to Damascus, and refurbishing government buildings in the cities of the north of the country.[47] Only a fraction of these schemes was ever carried out; but Schumacher established himself as a key figure in the planning and development of the northern district for a period of decades. He was later joined by the engineer Voigt and the Beilharz brothers, who initiated and carried out a variety of development schemes. Thus the Templars, who had earlier regarded themselves as too inexperienced to handle local building materials by themselves, became the teachers and guides of the local population and Turkish government authorities in everything connected with the planning, construction and development that was to determine and shape the modern landscape of the country.

The Templars exerted an equally important influence upon the Jewish settlers in the country. The achievements of the German colonists were a source of inspiration to the Jews, and the Templar successes greatly encouraged them in their own practical endeavours. They often used the example of the Templars to buttress their claim that native Europeans could establish themselves and prosper in this part of the world.

The German settlers served as practical models for comparison and imitation in the spheres of agricultural methods, transportation, hotel facilities and manufacture – in regard to each of which areas they were the pioneers. Many advocates of Jewish settlement in Palestine seized on the example of Templar institutions in order to compare these efforts with their own and to learn from them. So, for example, Y. M. Pines, as early as 1882, pointed to the Templar colonies in order to refute, point by point, all of the arguments that were being put forward concerning the supposedly insurmountable difficulties that people from Europe would encounter in attempting to colonize Palestine.[48]

Arthur Ruppin often compared the achievements of the Templars and their approach to settlement with the colonization efforts of the Zionist Organization. Thus he regarded the milk farm established at Ben-Shemen as being completely dependent on the examples of Sarona and Wilhelma. In considering the questions of the education of women and agricultural work, he looked to the Templar colonies and concluded that only when Jewish women would learn the work of milking cows would Jews become real farmers.[49]

Hirshberg, on the other hand, was impressed by the independence with which the Templars conducted their affairs, which they were able to do without 'support'. In the practical domain Hirshberg found there was much to be learned from the Templars about land deals, the dairy industry, wine production, the marketing of grapes and their way of life as farmers ('the Germans of Sarona, even the richest among them, feel no shame in bringing in the manure by themselves with their own cart').[50] And if the Jewish settlers and their leadership regarded the German colonists with admiration, they could not help but be aware of the physical layout and features of Templar colonies. Shimon Berman was already impressed in

1871 by the accomplishments of the German settlers in Haifa and regarded their colony as clear proof of the feasibility of his own scheme for settling Jews in the country.[51]

The appearance of the German colonies did not fail to impress Jewish travellers in Palestine. The new and entirely modern settlement pattern embodied by the Templar colonies contrasted starkly with the patterns of urban and rural Arab settlements. But the German settlements also had transformed the landscape of the country by the deeds of their settlers, who were guided in their actions by ideological aspirations. Their planned and regularly laid-out settlement patterns, with homes surrounded by farmyards and well-tended gardens, caught the eye of every traveller, visitor and resident in the country. Many described Templar settlements as forming a 'European landscape'. This is what Brill has to say of his impressions of Jaffa and Sarona:

Here, too, in Jaffa, the German (Christians) from Württemberg have breathed new life into it. This colony, which was set up in the vicinity of Jaffa and is visible from it, enhances the beauty of the city. Anyone coming to visit the colony feels as though he were in Europe. And the second colony, which is an hour's walk removed from the city, is also built in European fashion.[52]

The Templars conformed to the 'European' idea that the Jewish settlers had in mind when they came to establish their own colonies. Little wonder then, that so many advocates of Jewish colonization should have regarded the Templar settlements as a model to be learned from. However, the relationship of the Jews to the Templars was not confined only to the sphere of ideas. The Templars became *actively* involved in Jewish enterprise in more immediate ways, even at the initial phase of Jewish settlement. The first maps of Petah Tiqwa and Eqron were drawn by Sandel.[53] The settlers of Rishon le-Zion sought the advice of Christoph Hoffmann concerning where they should sink a well and locate their settlement on the sandy land they had purchased.[54] The settlement maps of Zikhron-Yaakov were made by the Haifa Templars, Schumacher and Voigt.[55] And when the dismantled prefabricated wood shacks were received by the settlers of that colony, the task of putting them together was done by the Haar family of the Haifa Templars, who were the only ones in the country that knew how to do it.[56] German building contractors of the Haifa Templar community did the work of planning and building the settlements of Bet Shlomo and Shaffiya; and the firms of Dück in Haifa and Breisch in Jaffa were the suppliers of these colonies for many years. Once again, when a winepress was established at Rishon le-Zion, it was Gottlieb Schumacher who was engaged for the job.[57] The claim has also been made that the designs of the settlements of Hadera, Rehovot, Petah Tiqwa, Mishmar ha-Yarden and others as well, were influenced by German models.

The influence of the Templars as regards the physical layout of Jewish settlements was therefore strong both on the theoretical level of furnishing an ideal model, and on the practical level of providing the professional expertise for the actual building of the settlements.[58]

Conclusion

The settlement of the Templars in Palestine was unique in both its historical and religious aspects, compared to the other European immigration movements of the nineteenth century, most of which were spurred by economic motives. Unlike the latter, the Templars had immigrated out of religious conviction and the aspiration of creating a Christian mission that was not concerned with turning the inhabitants of Palestine away from their own faith, but rather with setting an example for them of the Christian society which conformed to the true ideals of that religion. This example was to be set by them through their family lives as expressed in a living community and in their day-to-day interactions with their fellow man in trade, at work and in industry. The expression of this ideal in respect to the landscape of Palestine was through the creation of a model colony built on modern lines that differed from the physical pattern of the traditional Arab village in the region. All of these spheres of action conjoined in accomplishing the goal of 'uplifting the Orient'.

The Templars were few in number and their colonies but a handful. The extent of their influence was, however, inversely related to their quantitative presence in Palestine. They also differed on an important point from other Christian missions in the country. Unlike the latter, whose efforts to convert the local populace tended to exacerbate ethnic tensions and to encourage political conflict, the Templars succeeded in becoming an object of admiration and an example which the population at large sought to emulate. Their many innovations in a variety of economic and social spheres within their own colonies aroused respect in both the Arab population and its Turkish rulers, and this circumstance gave them a key role in the country's development. It was only after the arrival of Jewish settlers in increasing numbers and the dispersion of their colonies throughout Palestine that the Templars were supplanted by the Jews as a major force in the country's development. This process, however, took time, and the Templars maintained their position of importance in the life of the country for a period of three decades.

NOTES

1 See for example, A. Lemon and N. Pollock, *Studies in overseas settlement and population* (London and New York 1980)
2 The main research was done by A. Carmel, *Die Siedlungen der württemberg-ischen Templar in Palastina 1868–1918* (Stuttgart 1973); a recent publication is P. Sauer, *Uns Rief das Heilige Land* (Stuttgart 1985)
3 On the establishment of the movement and its development until 1898, see Fr Lange, *Geschichte des Tempels* (Jerusalem and Stuttgart 1899) 1–121; on its place among Pietism H. Lehmann, *Pietismus und weltliche Ordnung in Württemberg* (Stuttgart 1969) 240–4
4 About Hoffmann and his personality, *Neue deutsche Biographie* 2, 392–3; C. Hoffmann, *Mein Weg nach Jerusalem* (Jerusalem 1881) 1884; this auto-biography was translated in part as *Jerusalem journey, the autobiography of Christopher Hoffmann 1815–1885* (Stuttgart 1969)
5 Lange, *Geschichte*, 1–4; Lehmann, *Pietismus*, 210–17; an unpublished manu-script, D. Lange, Die Templer (1950) 30–40
6 Carmel, *Siedlungender Templer*, 7–14; Lange, *Geschichte*, 60–118; C. Hoffmann, *Orient und Occident* (Stuttgart 1875) 14–18
7 *Die Warte des Tempels* (1869) 28, 105. *Die Warte des Tempels* (hereafter *WT*) is the organ of the Templars, it was first published in 1845, and continues to be published to the present day
8 *WT* (1968) 50, 197
9 *WT* (1869) 4, 13–15, describes the town as the 'Gate to Palestine'
10 He was a 'Pilgermission' missionary in Basle, who purchased the property of the American Colony. In 1869 he left for Russia and offered it to the Templars for purchase
11 An exact list of the purchase in Jaffa: *WT* (1869) 15, 57
12 In *WT* (1969) 16, 61, Hoffmann describes the Jaffa deal as 'the finger of God', guiding in the proper direction
13 *WT* (1870) 1, 3; Kark, *Jaffa – a city in evolution 1799–1817* (Jerusalem 1984, Hebrew)
14 *WT* (1871) 37, 145; The name 'Ebenezer' was at first suggested, but Sarona was preferred
15 The cornerstone for the first building in Rephaim Colony had been laid on 25 April 1873
16 *WT* (1876) 1, 2, a report on the Society
17 On the creation of a community in Nazareth, *WT* (1882) 51 8–10
18 Carmel, *Siedlungen der Templer*, 157–74 explains the impact of the visit on the development of the settlements
19 *WT* (1902) 50, 249; on its development see K. Imberger, *Die deutsche landwirt-schaftlichen Kolonien in Palastina* (Ohringen 1938) 37–40
20 On Bethlehem, *WT* (1906) 41, 329–30; on Waldheim, *WT* (1907) 46, 353, see also Sauer, *Heilige Land*, 121–4; Imberger *Deutsche Kolonien*, 42–5
21 On Tira, Imberger, *Deutsche Kolonien*, 45–6; Carmel, *Siedlungen der Templer*, 190–6

22 The most famous one is the attempt of Mrs Minor and her supporters to settle in 'Mount Hope', near Jaffa, which ended in the rape and murder of the group

23 The following discussion is based mainly on fieldwork, carried out in all the settlements between 1985 and 1989

24 Hardegg describes his opinion in a letter to the committee in Germany: *WT* (1869) 13, 49–50; a later analysis made by C. Paulus is to be found in *WT* (1874) 39, 153–4

25 *WT* (1871) 39, 153

26 The streets in Wilhelma are very long in comparison to those of any other village in Palestine. The main street was 1,400 metres long, the intersecting streets, 518 metres and 560 metres, and their width 18 metres

27 The committee in Germany selected the candidates for immigration according to their ability to contribute to building: *WT* (1869) 31, 122–3

28 On timber imported to Palestine: *WT* (1874) 42, 168

29 Imberger, *Deutsche Kolonien*, 42

30 *WT* (1885) 10, 5–8

31 *WT* (1870) 21, 81–3

32 *WT* (1869) 38, 149; Gottlieb Schumacher (1837–87), was a builder from Wangen near Stuttgart. He is not to be confused with the famous architect, archaeologist and topographer with the same name (1857–1925)

33 *WT* (1869) 31, 122–3

34 On the Jewish settlements and their planning, Y. Ben-Artzi, *Jewish Moshava settlements in Eretz-Israel 1882–1914* (Jerusalem, 1988) 290–1 Hebrew

35 The cornerstone was laid on 23 September 1869. It was known as a *Müsterhaus* (Prototype): *WT* (1869) 42, 165

36 *WT* (1902) 46, 363

37 *WT* (1884) 1, 3; Lange, *Geschichte*, 732–4, 851–3

38 Ben-Artzi, *Moshava settlements*, 152–79

39 Carmel, *Siedlungen der Templer*, 224–94; Sauer, *Heilige Land*, 125–31

40 *WT* (1888) 25, 193–4

41 *WT* (1891) 28, 217, quoting a Russian teacher, who guided his Arab pupils

42 L. Oliphant, *Haifa – or life in modern Palestine* (Edinburgh 1887) 20

43 *WT* (1871) 23, 89

44 *WT* (1876) 21, 83

45 Carmel, *Siedlungen der Templer*, 230

46 On their history see P. Sauer, *Beilharz Chronik* (Ulm 1975); and recently in English P. Sauer, *The story of the Beilharz family* (Sydney 1988)

47 *WT* (1886) 35, 36–7

48 Y. M. Pines, *Binyan ha-Aretz* (*Building the land*) 2 (Jerusalem 1932, Hebrew) 145–55

49 A. Ruppin, *Ha-Hityashevut ha-Haqlait shel ha-Histadrut ha-Zionit* (The agricultural settlement of the Zionist Organization) (Tel Aviv 1925, Hebrew) 14–15

50 A. Z. Hirshberg, *Mispat ha-Yeshuv ha-Hadash be-Eretz-Israel* (The Jewish settlement in Palestine) (Wilna 1900, Hebrew) 45, 47, 83–5

51 S. Berman, *A journey to the Holy-Land – 1870* (Jerusalem 1980, translated from the Yiddish source), 94–6

52 Y. Brill, *Yesud Hama'ala* (photocopy edition, Jerusalem 1978, Hebrew) 196

53 *WT* (1884) 14, 3, on land-surveying Petah Tiqwa by T. Sandel

54 Z. D. Levontin, *Le-Eretz Avoteinu* (Our Homeland) I, 44–5 (Hebrew)

55 The map of Zikhron-Yaaqov made by G. Schumacher was published in A. Samsonov, *Zikhron-Yaaqov – Parashat Divrei Yameha* (The history of Zikhron-Yaaqov) (Tel Aviv 1943, Hebrew)

56 *WT* (1885), 27, 1–5 reports in details about this history of the colony and its building

57 D. Yodelovitch, *Rishon-le-Zion* (Rishon le-Zion 1941, Hebrew) 152

58 A. Hashimshoni, Architecture, in B. Tamuz (ed.) *Art in Israel* (Tel Aviv 1963, Hebrew) 221 (Hebrew); on the links between Jewish and Templar settlements, Carmel, *Siedlungen der Templer*, 259–94; Sauer, *Heilige Land*, 131–45; D. Trietsch, Deutsche und judische Kolonien in Palastina *Kolonial Rundschau* (1915) 336–55; L. Saad, Die deutschen und judischen Kolonien in Palastina *Kolonial Rundschau* (1918) 160–71; H. Seibt, *Moderne Kolonisation in Palastina* (Jerusalem 1938) which includes two volumes, one dealing with the Templar colonies and the other with the Jewish

5

Planned temple towns and Brahmin villages as spatial expressions of the ritual politics of medieval kingdoms in South India

HANS-JÜRGEN NITZ

In spoken English and German there is a difference in the meaning of 'ideology' respective to *Ideologie* which is of some importance for this chapter because the term is used much more as an 'ordinary' one than as a well-defined scientific concept. At least for this contribution to the volume a clarification seems useful.

The term *Ideologie* in spoken German has a narrower, generally pejorative meaning. Implicitly the *Ideologie* of a section of society or of a group is seen as a hidden tool for its self-interest and is viewed as not being in accordance with the 'true' interests of society as a whole. *Ideologien* are viewed as inexperienced views of the world and even as dangerous, and a threat to society. Religious or political beliefs of certain groups are suspected of being *ideologisch* (ideological).

This critical, pejorative conception of the term *Ideologie* dates back to German philosophers of the nineteenth century, including that of Karl Marx. In present sociological and political discourse this concept is maintained in the 'Critical Theory' of the Frankfurt school. *Ideologie* in this restricted meaning especially refers to cases where certain elements of a religion or, more generally, of a belief system are used for the purpose of an efficient legitimization of rule or government, to serve the self-interest of a ruling minority or class against the whole society.[1] According to Habermas it is this kind of ideology which was characteristic of the ruling class of early civilizations: 'legitimation of power by the cosmological interpretation of the world'. It is in this specific meaning that *Ideologie* (written as 'ideology') is used here because it best serves the interpretation of the case.

The much more general, unspecific use of the term in English which refers for example to any religion as an ideology[2] would, in the view of this author, make it useful for social sciences such as human geography to apply one or the other concept of 'ideology' as offered by sociology/philosophy and in this way sharpen the tool of analysis and interpretation for a project such as *Ideology and landscape* and to make the respective studies easier to compare. The Swiss geographers Gallusser and Meier offered a proposal with their concept of *Geisteshaltungen*, which they translated by the English term 'belief systems'.[3] As Sitwell proposed, an orientation to the conception of ideology in the sociology of knowledge, as understood in K. Mannheim, *Ideology and utopia* would be a rewarding approach.[4]

Ideology and ritual policy

The Chola dynasty in medieval South India offers an instructive example especially for the practice of instrumentalizing religion as a strategy by intentionally building and carving symbols of ideological meaning into the landscape: royal temples, towns, villages and even geometrical patterns of survey lines, field lanes and irrigation canals. Similar patterns in the same ideological context have been observed and interpreted by Tichy for Meso and South America.[5] Spatial patterns based on cosmological conceptions are widespread phenomena in urban and rural landscapes of East and Southeast Asia, and here again connections can be drawn to the ideologies of the historical ruling powers.[6]

   In the past empires with a large spatial extension created through subjugation of neighbouring principalities always faced the problem of maintaining political control over the growing masses of their subjects, in the peripheral newly attached regions as well as in densely crowded core regions. Generally large armies had to be employed for this purpose. In many early states the standard of administrative control was poor. Facing such problems medieval Indian kingdoms additionally, or even in the first place, employed the ideological strategy: they invested in the socio-ideological sector and managed to effectively control the provinces of their empire through a 'ritual policy'.[7] They bound their subjects to the Crown through the political implementation of the Hindu religion which was instrumental in legitimizing royal power as the only authority to rule each and every part of the empire and its people as subjects. This was effected by establishing an intimate religious linkage between the emperor and an important deity which was declared the personal god of the king and of his empire as well. It was the task of obedient Brahmin priests to convince the subjects of the indissoluble identity of the divine and the secular authority, through the Hindu temple ritual. Naturally, it was part of the ritual policy of the ruler to heavily support the Brahmins in terms of wealth (income, property) and of raising their social rank and thus to transform this caste into a political class closely connected to the ruling dynasty. The number of Brahmins was increased by thousands, even through invited immigration from as far as the Ganges plain around Banaras. In several royal temples images of the rulers have been found which indicate the deification of the king who was conceived as a personification of the deity. Through the indoctrination of the pious subjects by the Brahmins 'the imperial authority was regularly and systematically integrated into and equated with the divine authority'.[8]

It was part of the implementation of this ideological policy to make the subjects perceive with all their senses that the divine cosmic order really was the basis, the divine superstructure of the imperial rule. This policy was achieved by marking the secular landscape as the lifeworld of the people not only with divine symbols – especially temples of the royal deity in the political centres as well as in the main villages – but by literally drawing lines upon the land which should project the cosmic order onto the ground which they worked day by day, and upon the layout of the villages which they lived in.

This will be shown in detail by taking as an example the ritual policy of the Chola kings of South India who conquered and ruled a vast empire from the late ninth century to about AD 1300. The early kings of the Chola dynasty chose Lord Shiva as the personal deity of the royal family. It was especially King Rajaraja (literally 'king of kings', AD 985–1014) as the most prominent of the Chola monarchs and the third in the known chronology of the dynasty, who invested in the royal patronage of worship of Lord Shiva as 'a method adopted by an ambitious ruler to enhance his very uncertain power'.[9] Shivaism as a version of Hinduism was converted into an ideology with the political intention of legitimizing the rule of the Chola dynasty over many regions and peoples with different cultures and traditions. With the willing co-operation of the priest caste of Brahmins, Hinduism in the Chola empire was transformed into a canonized religion; the numerous goddesses of the traditional folk religions of various regions were reduced to the status of companions of Lord Shiva in his various personifications and in this way were integrated into the state religion. Rajaraja claimed the identity of Lord Shiva and added the title *deva* (divine) to his name.

The newly constructed Shiva temples were places to demonstrate to the regional nobilities, as well as to the common man, the ritually sanctioned supreme sovereignty of the king as a personal mandate of 'his' Lord Shiva. This was one of the main elements of the ritual strategy to secure the loyalty of the subjects to the emperor without the need for applying expensive physical force. To implement this strategy, about thirty impressive and huge Shiva temples were built all over the provinces of the empire as centres of pilgrimage for their hinterlands.[10] Each imperial Shiva temple was manned with hundreds of Brahmin priests and temple servants along with their families. To house them, as well as additional service castes, towns had to be built around the temples.

Four steps were taken by Rajaraja and his successors to implement this concept of imperial Shiva temples:

1 At the royal capital of Tanjavur, the largest Shiva temple of the empire was constructed. It was even the largest temple of the Hindu world with a gigantic pyramidal tower of 57 metres.

2 About thirty new temple towns with Shiva temples at their centres were founded in various parts of the empire. Their distribution depended on the density of population. The largest concentration of people and respectively of imperial temples was to be found in the delta of the main river, the Cauvery, the 'paddy bowl' of the empire. Tanjavur, the capital, was located at the head of this delta.

3 In a later phase of Chola rule (twelfth century), when it became obvious that local deities of pagan sects attracted strong adoration from the regional rural peoples for which the rituals of the large imperial Shiva temples proved to be much too esoteric, these pagan deities were also used for the purpose of the royal ritual policy: for them, too, the kings built magnificent temples to demonstrate to their subjects the royal affection towards these deities. The royal Brahmins gradually managed to incorporate into the Shivaitic pantheon the local deities in order to again attract and confirm the affection of the local people for the ruler.[11]

4 A last strategic step was to establish a close local network of Brahmin centres below the level of temple towns. This lower tier consisted of Brahmin villages – so-called *brahmadeya*s – with a Shiva temple which was to form the ritual focus for the surrounding peasant villages. In the densely settled Cauvery delta, the Cholamangalam province, there existed about 250 *brahmadeya*s among a total of approximately 1,300 villages; on average one *brahmadeya* served four peasant villages.[12]

In order to combine ritual and rule, the emperor followed certain practices: (1) He regularly visited the large temples at the occasion of the annual temple festivals when thousands of pilgrims assembled to worship Lord Shiva or the respective main deity of the temple.[13] In this way the intimate link of the ruler to the deity could be demonstrated to the subjects. In connection with this event the king held the imperial diet within the precincts of the temple, in a large pillared hall, the *mandapam*, the 'Hall of a Thousand Pillars', frequently named *rajasabha*, the royal congregation hall. It was through this institution of royal temples and *rajasabha*s that the rulers finally covered their whole empire with a network of direct royal influence, ritual and political at the same time. Kulke refers to similar practices of European medieval rulers who also used to rule by travelling through their empires and to hold diets not only at the royal palaces but also at the main royal church centres – the sees and the royal Benedictine monasteries. (2) Regional vassal princes and high-ranking royal officers were obliged to participate in certain ritual services during temple festivals and to give grants to the temple. In this way they were intimately bound to the royal ritual. (3) Temple servants of the main imperial Shiva temples were drawn from the provinces of the empire. Their service was seen as an honour for their home village, which kept strong ties to the temple and its royal patron by its obligation to supply rice for the respective temple servant. This institution as well as the pilgrims from the villages served as transmitters of the messages of the ritual legitimation of royal power emerging from the temple towns.

These were the strategies to cover the whole empire with a 'ritual superstructure' of royal temples and a 'ritual network' of pilgrimages connecting them to each and every village.[14] This kind of ritual policy was not only applied by the Chola kings but by all the dynasties which succeeded them, especially the kings of the Vijayanagar empire, and by the rulers of other South Indian territories, for example, Orissa.[15]

In order to establish and to maintain this ritual superstructure with tens of thousands of Brahmins, temple servants and specialized craftsmen along with their families who all had to reside in temple settlements, where large numbers of residential, administrative and, of course, ritual buildings had to be constructed, a stable economic base was needed. It consisted of a considerable agricultural surplus production which was extracted as revenue from the peasant villages.[16] This surplus production was secured by several innovative activities of the Crown. Important were the intro-duction of plough cultivation in paddy fields and the construction of highly effective canal irrigation networks.[17] The latter could most successfully be implemented in the delta regions where extensive swamps were brought under cultivation during the reign of the Chola kings.[18] It was the already-mentioned king Rajaraja 'the Great' who through the establishment of an effective irrigation department started to convert the swampy tracts into a well-irrigated and densely settled paddy region.[19] On similar lines, the other deltas and river valleys were converted into paddy bowls, though it is not yet clear to which extent these activities were undertaken by the Cholas or by other dynasties.

From these highly productive agricultural core regions large amounts of revenue could be drawn which enabled the king to support the ritual superstructure. In addition and even more importantly royal grants were made, of whole villages and even of agricultural districts, to imperial temples. These generous royal gifts stimulated additional deeds from the nobility and from independent peasant communities.[20] These resources permitted the continuous increase of the magnificence of the temples and temple treasures which consequently reinforced their ritual splendour and their impression on the subjects, especially at the occasion of temple festivals with thousands of pilgrims from the hinterland villages.

Ritual spatial structures in the cultural landscape

These ritual concepts and the activities to implement them resulted in some quite remarkable spatial structures in the landscape. Three tiers of settlements are described and discussed: the capital of Tanjavur with the royal palace and the imperial temple; the temple towns of the provinces; and finally the *brahmadeya* villages.

The Chola capital of Tanjavur

A capital did already exist at Tanjavur at the head of the Cauvery delta when the Cholas overthrew the Pallava dynasty about AD 900, of which they had been regional vassals. A first demonstrative act of ritual policy was carried out by King Rajaraja the Great; he allowed the building of the first large Shiva temple – named Brihadeshwara – in the southwest of the capital (Fig. 5.1). On its huge temple tower of 57 metres the crowning dome rests on a single block of granite weighing 8 tons (well over 8 tonnes). An earthen rampart had to be constructed to move the block to the top of the tower. Thousands of labourers must have been employed. As the first royal temple built to demonstrate the close connection of the ritual centre to the king, it was called Rajarajeshvara temple after him. The temple complex is surrounded by a moat and wall. Through a gate it is connected to the capital.

The layout of the town (Fig. 5.1) displays regular as well as some irregular features, the latter in the northern part with a haphazard pattern of lanes around some temples and temple tanks seeming to date to pre-Chola times. The curving bazaar street to the east of this area could have been a main axis of the earlier town. A regular feature is the large rectangle of four wide streets (about 760 × about 820 metres). It is subdivided into three smaller rectangles of which the eastern one contains the palace and a large open market place to the north. The other two rectangles and the areas along the southern, western and northern main streets are residential. The most regular structures are to be observed along the western and southern main streets where many narrow lanes branch off at right angles. They indicate parallel house plots the longest of which are to be found along the western street. They are occupied by Brahmins as are those along the northern street. The houses lining the southern street belong to peasants, temple servants and to merchants. The regular shape of this large ring road and the residential pattern with the main Brahmin quarter in the western street which leads to the Rajarajeshwara temple seems to indicate that this layout was intentionally created in connection with the construction of the large royal Shiva temple which employed about 300 priests and several hundreds of temple servants. The hypothesis of a replanning of the layout of the

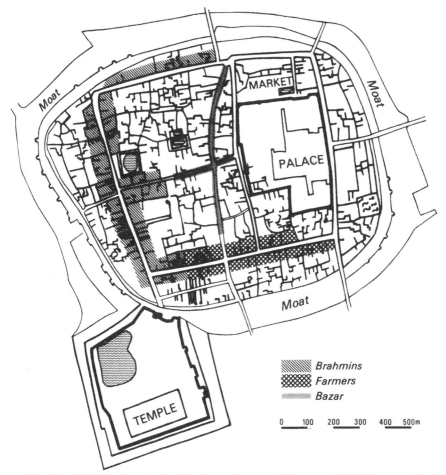

Fig. 5.1 The Chola capital of Tanjavur: layout of the town, 1906. Topographical map, Survey Office, Madras, 1906. Original scale 10 in. to 1 mile.

Chola capital under King Rajaraja the Great is supported by the ritual function of the ring road – it serves as the processional street on the occasion of royal temple festivals. As generally in temple towns of the Chola model (see Figs. 5.2 and 5.3), on such a very wide ring road (40 feet or about 13 metres) the huge wooden temple car, shaped as a temple on wheels, carries the tiny idol of Lord Shiva, and smaller cars contain the idols of his companion and of other important deities. At the respective temple festivals devoted to their glorification they are moved from their shrines onto the temple cars accompanied by a group of Brahmins who on the slow drive around the processional street – the car is drawn by

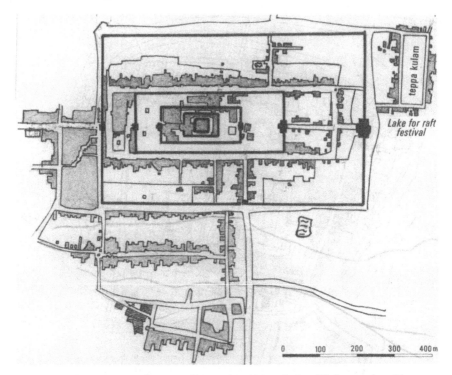

Fig. 5.2   Tirunannayanellur: small temple town of the Chola empire. Topographical map, Survey Office, Madras, undated copy. Original scale 10 in. to 1 mile.

hundreds of male pilgrims – receive the pious gifts of the devotees and return them after their symbolic consecretion to the deity. In Tanjavur the ring road is laid out in such a way that, through the procession of Lord Shiva, the capital and the royal palace become part of the ritual and thus are sanctified. All the inhabitants of the town join the procession, as do numerous pilgrims from the province of Cholamangalam and from other parts of the empire. The circumambulation of town and palace united the divine and the mundane lords – Shiva and the king – with their subjects. Obviously this must have been one of the most effective actions of ritual policy for legitimizing the political position of the ruler as a divinely sanctioned one.

The royal temple towns of the provinces

As has already been explained, temple towns were built around Shiva temples and later on around temples of other deities, too. The main purpose of the town around the temple was to accommodate the Brahmins and the additional staff. So we find large temple towns around the important Shiva temples and smaller towns around temples of lesser-ranking deities. An example of the latter group is presented by Tirunannayanellur (Fig. 5.2), named after the main deity of its central temple: Tiru (Lord) Annai, the elephant god, deity of wisdom, in Hindustan known under the name of Ganesh. (The suffix –ur means 'settlement'.)

The planned layout is perfectly rectangular and oriented to the cardinal points, with a slight deviation. There are five concentric walls and open spaces in between, so-called *prakrama*s. The four interior *prakrama*s enclose the main temple with the shrine of Lord Annai, additional temples of other deities including the companion goddess, and buildings of the temple administration. The third *prakrama* contains the temple pond and the spectacular 'Hall of a Thousand Pillars', the royal assembly hall the function of which has been described earlier. The wide open space of the fourth *prakrama* permits the worshippers to congregate. The widest outer *prakrama* contains the Brahmins' and the temple servants' settlement, which constitutes the true temple town. The quarters beyond the outer wall are later additions for lower-ranking castes. Their straight streets seem to have been adapted to the rectangularity of the temple town. This is especially true for the small settlement around the large temple tank in the northeast. The main entrance to the town and its temple is in the east which conforms to a strict rule; the idol of the main deity looks east. Here we find the three largest of six pyramidal gate towers, the so-called *gopuram*s. The axial access street from the east – the Sannathi (street) – is a regular feature of every temple complex, and ranks highest as residential location for Brahmins.[21] The wide ring road of the outer *prakrama* also serves the double function of a residential and processional street, and as such forms part of the sacred space.

Viewing this pattern as a 'ritual townscape' three observations ask for explanation. How have we, in the ritual context, to interpret (1) the strict rectangularity, (2) the concentric arrangement of the *prakrama*s with the shrine of the main deity in the core, and (3) the orientation to the cardinal points?

Even more striking is this kind of arrangement in the larger temple towns like Shrirangam (Fig. 5.3) which has been studied in detail by the German architectural anthropologist Pieper who offers some convincing explanations of the patterns in question. The town is located about 50 kilometres west of the capital on a narrow island between the Cauvery and the Coleroon Rivers.[22] The Cauvery is the 'Southern Ganges', the important ritual river of South India. The two *ghat*s, the place of the ritual bath of the pilgrims on the bank of the Cauvery in the south and the place of cremation of the corpses on the bank of the Coleroon in the north (Fig. 5.3b) are located precisely on the southern and the northern corner of the diamond-

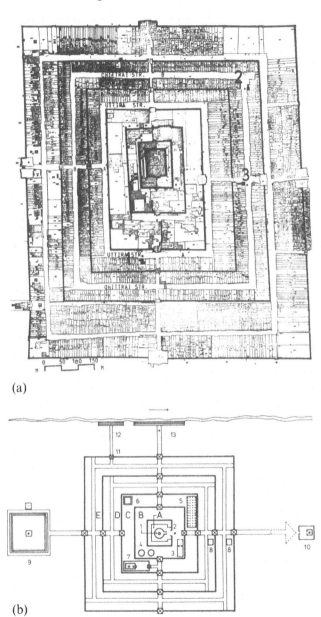

(a)

(b)

Fig. 5.3   (a) Shrirangam: large temple town of the Chola empire; (b) Schematic layout of a South Indian temple town (*after* K. Fischer, M. Jansen and J. Pieper, *Architektur des indischen Subkontinents* (Darmstadt 1987) Figs. 33 and 34).

shaped river island.[23] The central Shiva temple of the town is enclosed by seven concentric walls of which four *prakrama*s constitute the temple area, again with the royal Hall of a Thousand Pillars in the corner of the third *prakrama*. The outer three *prakrama*s contain the residential and service areas. Each of their three wide ring roads serves as a processional street. Four *gopuram*s top the main gates of the access streets in the south, east and north and three in the west.

Pieper argues that the idea behind this concentric pattern of temple towns is derived from the mythological equation of temple and temple town with the Hindu cosmos.[24] According to the Brahmin Vedic cosmology, a circular central continent with the mountain of the gods, Mount Meru, in its centre is concentrically surrounded by six ring-continents and seven ring-oceans.[25] Beyond the outermost ocean a ring of rocky mountains confines the cosmic world. On top of Mount Meru the town of the Creator of the World, Lord Brahman, is located, surrounded by the towns of the eight World Guardians. The fabric of the temple town with its strict concentric arrangement is conceived as an image of this Hindu cosmos, though not as a perfect copy because circularity is replaced by rectangularity. However, the concentric pattern has been implemented, in the case of Shrirangam even with the complete set of seven rings, although they were not created in one step, but by subsequent enlargements of the town carried out under several pious kings. The temple complex in the core is considered to represent the cosmic Mount Meru with the town of the Creator of the World, and the *prakrama*s with their concentric walls, temple precincts and processional streets respectively represent the ring-continents and the ring-oceans. The concept of concentricity around the temple core is the standard rule to be followed even in smaller temple complexes, such as Tirunannayanellur (Fig. 5.2). 'Through the equation of

---

Legend

A, B, C	Inner *pakramas* of the temple
D, E	Residential areas
1	*Mulasthana* (cellar) with idol of main deity
2	Flagpole
3	*Vahana Mandapa* (treasury and pavilion for wooden temple cars of the main deity)
4	Office
5	Hall of a Thousand Pillars (*rajasabha*, i.e. royal congregation hall)
6	Temple pond
7	Shrine of the companion goddess to the main deity
8	*Ratha Mandapa* (two-storeyed pavilion from where the cult idol of the main deity is moved on the temple car (*ratha*))
9	*Teppakulam* (pond for the 'raft festival')
10	Temple of a 'related' deity
11	'Gate of corpses'
12	Cremation *ghat* (terrace steps) along sacred river
13	Ritual bathing *ghat*

Published by Thiru S. GANESAN, B. A., PL.I.S., Deputy Director of Survey and Land Records,
Madras, on behalf of the Government of Tamil Nadu, 1971 (Elev. No. 1881)

No. 102.
DEVARAYAMPETTAI
PAPANASAM TALUK
THANJAVUR DISTRICT
Scale 1 : 10000

No. 101
RAJAGIRI

No. 6
PANDARAVADAI

No. 99
TIRUVAIYATTUKKUDI
(BIT I)

No. 98
MEL SEMMANGUDI

No. 97
PONMANJANALLUR
(BIT I)

No. 5
RAGHUNATHAPURAM

No. 103
VADAKKAMANGUDI

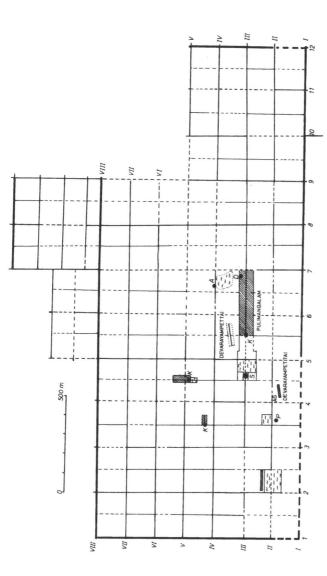

Fig. 5.4 (a) Cadastral map of the township of Devarajampettai (Tanjavur District), surveyed 1923; (b) Reconstruction of the medieval field-survey pattern of Devarajampettai.

the temple and the temple town with the Hindu cosmic order, divine order was projected upon the respective centre of the political authority which the temple town represented.'[26] This was no doubt one of the most important spatial concepts of ritual policy to legitimize political authority through the means of religion, which had been 'instrumented' as the ideology of the ruling class.

Some other geometric features of this temple and settlement pattern seem to be based on mythology, too. The orientation to the cardinal points quite obviously is conceived as an adjustment to the holy spatial structure of the cosmos. Slight deviations from the meridian which can be observed in most of the temples might possibly be explained by the location of the point of sunrise at the date of the foundation of the temple. Tichy was able to prove this for Meso-American ceremonial centres, but for South India this question has not yet been studied.[27]

Another aspect of regularity relates to the observation that width and length of the temple walls and town walls are laid out in plain numerical ratios. The four temple walls of Shrirangam have an identical ratio of 2 : 3, and the walls of the sequent *prakrama*s are laid out as 3 : 4, 6 : 7 and 5 : 6. There can be no doubt that these plain ratios were deliberately chosen. Are they again to be seen in the cosmological context as another expression of divine order? I have observed plain ratios in several other temple towns and they obviously present a general rule. Hence it is of no surprise that also the absolute extent of the sections of the walls is based on an exact measurement with the application of the medieval rod (*kol*) which holds 16 spans, equal to 8 cubits (3.50 metres). Eight and sixteen are holy numbers in Hindu mythology.

Cosmic order at the basis of medieval field survey and village layout

Strict rules of geometric regularity can also be observed in the survey lines of village fields and in the layout of rural settlements in the canal-irrigated delta regions, the colonization of which was started under King Rajaraja the Great at the same time as the first royal temple towns were built. The main features will be demonstrated by a sample village, Pulimangalam (Fig. 5.4), which by its place-name suffix – *mangalam* – can easily be identified as a Brahmin village, a *brahmadeya*. The Shiva temple at the western end of the straight street village is a further feature of *brahmadeya*s.

The field pattern exhibits clear features of regularity (Fig. 5.4a); parallel rows of rectangular plots, especially well preserved in the north, are arranged along survey lines running east to west at equal distances. In the south, around the main village, the pattern seems to have been partially destroyed, but relics of former regularity can easily be identified. Even the street village with its east–westerly extension fits into this pattern of parallel lines. The east–west lines as well as the main lines north–south are courses of irrigation canals. Due to the slow current of the irrigation water, the field canals over time tend to adopt a meandering course if they are not properly controlled.

I have tried to reconstruct the original historical layout of the field survey pattern (Fig. 5.4b). The continuous lines are those which still exist (compare Fig. 5.4a), the broken lines are reconstructions. The field pattern can be described as a kind of chequerboard, with a slight deviation from the cardinal directions. This is quite a strange arrangement of canals in a river delta where the natural flow of water tends to form a dendritic, anastomosing pattern. Here it has been transformed into a completely artificial geometric network of drainage and irrigation lines. The explanation may be in the intention of providing an even distribution of irrigation, though the partial destruction of the pattern and return to natural courses at least seems to indicate a certain degree of disfunctionality of such a schematic arrangement of canals. In my view, this strict rectangular design with its orientation to the cardinal points in principle conforms so much to the concept of rectangularity and orientation of the temple towns that it could as well, and even better, be interpreted as an application of cosmological ideas; to bring divine order to the land. As Pieper has observed at Suchindram, another temple town near the southern tip of the Indian peninsula, orientation and basic measurement of the temple and of the layout of the fields conform with each other.[28]

But there were, no doubt, at the same time also practical purposes behind this order, not only in respect to even distribution of water. It permitted subdivision of the land into portions of exactly the same size, for land allocation and for revenue purposes. My metrological analysis, the details of which have been published elsewhere, has revealed that the original rectangular standard plot measures 40 rods from north to south, limited by two parallel canals and field lanes which together hold two rods in width.[29] The breadth of the standard plot is 32 rods (2 × 16, the holy number!). So each plot of 40 by 32 rods contains 1,280 square rods. This is exactly half the size of the basic medieval land-share unit, the so-called *Veli*, which held 2,560 *kuli* the term for the square rod. The number of 2,560 contains the holy number of 16 even twice: 16 × 16 × 10. *Veli*s of this size were the standard units of land distribution to the Brahmins. Thus two plots make up one *veli* – 64 × 40 rods (3.32 hectares). The whole township contains 86 *veli*s, including public spaces. The land share of Brahmins in the Middle Ages was one or two *veli*s.[30] If we calculate this sample *brahmadeya* with the 2-*veli* share we would arrive at about forty Brahmin families who were settled here in the Middle Ages.

The majority of Brahmins were resident in the main village called Pulimangalam. The revenue township is named Devaraya-pettai. The name of Devaraya most certainly refers to rulers of the Vijayanagar dynasty which during the fourteenth century conquered the whole of South India from the Dekkan and succeeded the Cholas. Two Devarayas, father and son, reigned during the early fifteenth century. We suppose that one of them established a second, new Brahmin hamlet – an *agraharam* (another term for a Brahmin settlement) – in the south of the township, and we may assume that it was an intentional religio-political action in the context of establishing his rule by settling 'his' Brahmin communities in the newly conquered core province of the former Chola empire.

Another section of the population of the township is the ritually unclean agricultural labourers who have to live separately in their own small hamlets, to the northwest of the *brahmadeya*, a pattern of ideologically based social and spatial segregation which is practised to the present day. The labourers have to cultivate the fields of the Brahmins who for ritual reasons are not permitted to do manual work. All the settlements are arranged in a strict linear form. The main Brahmin village, Pulimangalam, extends over exactly 96 rods, three basic units of 32 rods. A temple of Kali, the black mother goddess, marks the eastern end of the residential area. Kali is another name for Parvati, the wife of Lord Shiva. His temple is located at the opposite side of the village beyond a public square and the adjoining temple pond, exactly in linear continuation of the east–west line which forms the axis of the whole settlement. The idol of Shiva looks to the village which thus has the same position as the Sannathi street at the east front of the larger temples in the provinces. There can be no doubt that the conception of the layout of this lowest tier of ritual settlements – all *brahmadeya*s of the Cauvery region follow the same pattern – has to be placed within the Hindu cosmological ideology which projected divine order upon the cultural landscape of the Chola empire.

The application of this kind of ideological strategy as an effective instrument of rule was used by other dynasties, too. We have already mentioned the Vijayanagars. Another impressive example is offered by the kings of Orissa further north, who as early as the tenth century started to apply this type of ritual policy which they continued well into the seventeenth century.[31] In establishing Brahmin villages, again with a linear layout oriented to the cardinal points, around Puri, their main imperial temple centre, they, too, strictly observed the holy number sixteen: 16 *sasan* (royal Brahmin) villages, 32 *karbar* villages (inhabited by priests of lower rank) and 64 *para* villages of peasants serving the 16 royal *sasan*s,[32] details of their regular layout have been described by Nitz.[33]

The rule of kings has passed into history. The political ideological message of the symbols placed by them in the landscape is no longer understood. Their metaphoric meaning is lost or reduced to the religious content provided by Hinduism. It is a charge for historians, architectural anthropologists and historical geographers to rediscover the historical 'metaphorical topology'[34] and to contribute to its interpretation in order to understand past societies and their cultural landscapes.

NOTES

1 J. Habermas, *Technik und Wissenschaft als 'Ideologie'*, edition Suhrkamp 287 (Frankfurt 1968, quoted after the 1971 edition), esp. 68
2 D. E. Sopher, Geography and religions *Progress in Human Geography* 5 (1981) 510–24
3 W. Gallusser and V. Meier, Unterwegs zu einer 'Geographie der Geisteshaltung'? *Geographia Religionum* 2 (1986) 31–53
4 O. F. G. Sitwell, The cultural landscape as myth: a study in metaphorical topology (unpub. paper presented to the 25th Congress of the International Geographical Union, Paris 1984); K. Mannheim, *Ideology and utopia* (New York 1938)
5 F. Tichy, The axial directions of Mesoamerican ceremonial centres on 17° north of west and their associations to calendar and cosmovision, in F. Tichy (ed.), *Space and time in the cosmovision of Mesoamerica*. Proceedings of a symposium, XLIIth International Congress of Americanists, Vancouver, Canada, 1979. Lateinamerikanische Studien 10 (Munich 1982) 63–83
6 R. Heine-Geldern, Weltbild und Bauformen in Südostasien *Wiener Beiträge zur Kunst- und Kultgeschichte Asiens IV* (1930) 28–77
7 H. Kulke, Tempelstädte und Ritualpolitik – Indische Regionalreiche, in *Stadt und Ritual* 2nd edn (Darmstadt and London 1978) 68–73
8 H. G. Böhle, Bewässerung und Gesellschaft im Cauvery Delta (Südindien) *Geographische Zeitschrift Beihefte* 57 (1981)
9 G. W. Spencer, Religious networks and royal influence in eleventh-century South India *Journal of the Economic and Social History of the Orient* 12 (1969) 42–56, and quoted by Kulke, Tempelstädte
10 Kulke, Tempelstädte
11 *Ibid.*, 72
12 Y. Subbarayalu, *The political geography of the Chola country* (Department of Archaeology, Government of Tamilnadu, Madras 1973)
13 Kulke, Tempelstädte, 72
14 *Ibid.*
15 *Ibid.*, and Legitimation and town planning in the feudatory states of Central Orissa, in H. Kulke *et al.* (eds.), *Städe in Südostasien* (Wiesbaden 1982) 17–37. Also G. Pfeffer and Puris Sasandorfer, Basis einer regionalen Elite (unpubl. habilitation thesis, Heidelberg 1976)
16 B. Stein, The economic function of a medieval South Indian temple *Journal of Asian Studies* 19 (1959/60) 163–76
17 S. Krishnamurti, *A study on the cultural developments in the Chola period* (Annamalainager 1966) 101
18 K. S. Gopalakrishnan, Cauvery delta: a study in rural settlements (unpubl.

124   Hans-Jürgen Nitz

Ph.D. thesis, Banaras University, Varanasi 1972), and in Böhle, Bewässerung und Gesellschaft im Cauvery Delta, 30 and Fig. 7

19 Krishnamurti, *Cultural Developments*
20 Stein, South Indian temple
21 H. F. Hirt, The Dravidian temple complex: a South Indian cultural dominant *Bombay Geographical Magazine* 8/9 (1961) 95–101
22 J. Pieper, *Die anglo-indische Station oder die Kolonisierung des Götterberges* Antiquitates Orientales, series B, vol. 1 (Bonn 1977)
23 J. Pieper, Ritual movement and architectural space: structural analysis of the spatial system of Srirangam in South India, in *Stadt und Ritual. Beiträge eines internationalen Symposiums zur Stadtbaugeschicte Süd- und Ostasiens 2–4th Juni 1977* (Darmstadt 1977) 82–91
24 Pieper, *Die anglo-indische Station oder die Kolonisierung des Götterberges*
25 Heine-Geldern, Weltbild und Bauformen
26 H. G. Böhle, Bewässerung und Gesellschaft im Cauvery Delta
27 Tichy, Mesoamerican ceremonial centres
28 J. Pieper, The spatial structure of Suchindram, in J. Pieper (ed.), *Ritual space in India: studies in architectural anthropology* Art and Archaeology Research Papers (London 1980) 65–80
29 H. J. Nitz, Order in land organisation: historical spatial planning in rural areas in the medieval kingdoms of South India, in V. S. Datye *et al.* (eds.) *Exploration in the tropics* (Pune 1987) 258–79
30 K. Gough, Caste in a Tanjore village, in E. R. Leach (ed.) *Aspects of caste in South India, Ceylon and Northwest Pakistan* (Cambridge 1960) 11–60
31 Pfeffer and Sasendorfer, Basis einer regionalen Elite
32 *Ibid.*, 423
33 Nitz, Land organization, 273–4
34 Sitwell, Cultural landscape

6

Territorial strategies applied to captive peoples

D. W. MEINIG

Ten years ago at the Cambridge meeting of the IGU Historical Geography Symposium I described a basic framework for 'the geographical analysis of imperial expansion' as a generic phenomenon.[1] I now shift my focus to consider some of the types of geopolitical territories to be found within one of the major imperial states of the modern world.[2]

Imperialism has always been with us. 'It is clear', writes Philip Mason, 'that dominance by one group of people over another is something of which no race or period has a monopoly; it is as old as the Pharaohs and springs from passions that are common to all men.'[3] To be clearly *imperial* in character that dominance must be territorially defined wherein the subordinate people seek to maintain their separate cultural identity and claim to a specific territory within some larger encompassing geopolitical structure. The specific patterns of such relationships take many forms. I shall confine my remarks to those that seem most pertinent to nineteenth-century America, but something of the relevance of these to many other empires and to the current world scene will, I think, be apparent. Empires are created in various ways. My use of the term 'captive peoples' refers to whole societies that found themselves bound into an alien polity by diplomatic or military decisions made in distant capitals without their participation.

The internal imperial character of the United States has received far less attention than it deserves because its scholars as well as its leaders and public have been reluctant to recognize it as such. The reality of empire has been obscured under one of the thickest covers of American exceptionalism. From its beginning the United States has always been treated as a nation and a federation, and often as a set of sections or regions, but American imperialism has normally been applied only to expansions beyond the continent into the Caribbean and Pacific beginning in the later nineteenth century. This unwillingness of American leaders to see their country as an empire has been basic and formal from the first. Although in

1789 nearly half of the sovereign territory of the United States was unceded land occupied by Indians the Constitution made no reference to that fact nor any provision regarding expansion, cession, annexation or accommodation of such peoples within the body of the Union. Despite continuous problems in dealing with the Indians, and despite subsequent annexations of enormous expanses of territory containing substantial numbers of distinct peoples, the Constitution was never amended to specify any provisions whatever for the incorporation of such territories and peoples into the overall framework of the state (although there were occasional proposals to do so). Thus, in formal legal terms the United States is not regarded as an empire, and powerful political and ideological forces have long been effective in denying it to be one.

The reality of empire becomes clear through comparative work. Once one looks around the world and recognizes some of the common patterns of such structures, it is not difficult to recognize similar features within the United States. Simply using the appropriate terminology of empire for such features begins to put American history into a quite different perspective. Let us begin with the simplest minimal form of imperial relationships, the *protectorate*. Here the subordination of a people and their territory is formally recognized, or implied by treaty, but they are left largely undisturbed. Such treaties typically declare the mutual respect and friendship of the two parties, and the promise of protection from external enemies in return for acknowledgement of the exclusive political supremacy of the imperial power. Initial American treaties with Indian nations (following long-standing European practice) usually took some such form. The American state insisted on its ultimate sovereignty over all its territory and the metaphor of the Great Father was formally established in the first major treaty, in 1795, but these newly adopted 'children' were left to shift for themselves so long as they did not disturb the main American household. In American law Indians were at first defined as 'resident foreign nations' and regarded as an anomaly in the body politic. In many cases such treaties purposely left territorial definitions vague, but there were also numerous proposals and formal attempts to set up large explicitly defined areas to accommodate many different Indian nations within a single protectorate, as in the attempts to set aside much of the upper Great Lakes country as a Northwest Indian Territory in the 1820s.

It is of course notorious that such treaty relationships were inherently unstable and temporary, because the United States never regarded them as anything but a first step in negotiations for drastic cessions by the Indians of all or most of their lands, and because the United States was unable, or unwilling, to commit the military resources necessary to provide such protection. An important exception was the case of the Pueblo Indians in New Mexico, who lived in compact villages supported by small tracts of

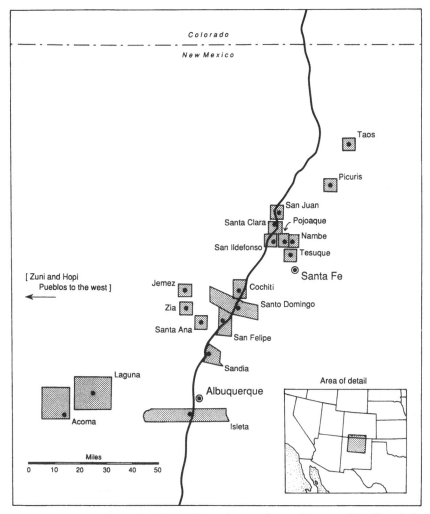

Fig. 6.1   Reservations of Pueblo Indians along and near the upper Rio Grande in New Mexico, 1886.

intensively cultivated lands in a region far removed from American settler expansions (Fig. 6.1). With too little land to covet, the relatively peaceful Pueblo towns were regarded as islands of stability amidst the resentful Hispanic residents and the dangerous nomadic Indians of the region. By quickly confirming them in at least the most intensively occupied of their arable lands, the Americans were simply reaffirming Mexican and Spanish imperial practice. Thus the Pueblo Indians today are unusual in having centuries of continuity in the same place.

A more formalized and coherent form of indirect rule is that of the *indigenous state* wherein the imperial power recognizes a local ruler and assigns him to administer his people and territory as a surrogate of the imperial ruler, under the eye of a resident agent. Local laws, courts, customs and language are left undisturbed so long as these do not contradict any important interest of the empire. Such 'native states' are perhaps more likely to be the product of an alliance rather than direct conquest, as in the case of many of the most famous modern examples, the 'princely states' of British India. The nearest the United States came to having this kind of unit was in the strenuous efforts (with substantial public and congressional support) to recognize a special Cherokee state in the southern Appalachians in the 1830s (Fig. 6.2). The Cherokees had conspicuously made extensive efforts to adapt themselves to the American socio-political system and they were to be accorded a delegate in Congress but retain internal autonomy as a distinct Indian polity. As is well known, the US Supreme Court ruling that the Cherokees could not be expelled from their homeland was ignored by the executive and legislative branches and by the popular will, and the end result was the Trail of Tears and exile in Oklahoma rather than a substantial Cherokee enclave. We might note parenthetically that the largest and most populous Indian reservation in the United States today, that of the Navaho in the Southwest, with about 150,000 residents, has in recent decades taken on a good many marks of an indigenous state. Although still under imperial veto power, the Tribal Council, seated in its modern hogan at Window Rock, makes all local decisions, maintains its own police force and law courts and runs its own newspaper, radio station and variety of facilities, using Navaho as well as English.

In another nineteenth-century episode, had the United States in 1847 granted formal protection (rather than simply sending arms) or annexed Yucatan (as was requested by its creole leadership who feared annihilation in a Mayan revolt and as was supported by the American President James Polk, who had just gobbled up half of Mexico), that detached peninsula might well have evolved into some sort of indigenous state. Indeed, the special commonwealth status of Puerto Rico today makes it some degree a variation of this form.

However, Puerto Rico at first, and New Mexico half a century earlier, were clear examples of a more common imperial type: *direct rule*, wherein the imperial state places a governor and set of high officials in charge of the conquered province. These assigned agents may seek the co-operation of local elites and try to disturb local customs as little as possible, but the basic law, courts and language of empire are imposed and there will probably be strong pressures for further conformity to the dominant society. Hispanic New Mexico was in some ways analogous to French Lower Canada

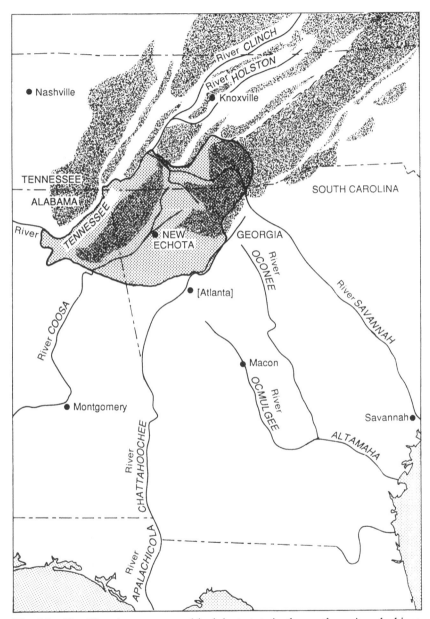

Fig. 6.2   The Cherokee reserve and incipient state in the southern Appalachians, c. 1830. New Echota was the capital town.

(Quebec): a conquered province, a people of foreign tongue and culture jealously guarding their integrity with the new encompassing structure, but nevertheless feeling relentless pressures for change, governing themselves in some degree, but penetrated by an influential minority from the imperial society whose political influence and resources allow them to gain ever greater command of the provincial economy. There was, of course, nothing akin to the broader dualism of the Laurentian geopolitical situation. Although at the time of its annexation New Mexico had more than enough citizens to qualify for statehood, it was not admitted into the federation for more than sixty years because of fears on both sides: many Anglos were contemptuous of the Hispanos as unfit for republican government; many Hispanos feared that statehood would endanger their cultural integrity. New Mexico did enter the Union as a bi-cultural state with certain guarantees accorded the Hispanic population (somewhat akin to the Quebec Act); by then Anglo immigration was steadily eroding the Hispanic position.

The Territory of Utah, which under Governor Brigham Young functioned in many ways more as the *de facto* Mormon State of Deseret than a routine territory of the central government, was another variation of this kind of imperial holding. The tensions and conflict arising from Mormon pressures toward maintenance of an 'indigenous state' and federal insistence on a normal 'direct-rule' territory underlay the crisis of 1857 which saw the dispatch of an American army to subdue and occupy the state. With the mining rushes, railway penetration, and federal marshals pursuing church leaders, the Mormons increasingly felt themselves to be a captive people. Only after the forced cultural conformity of banning the religious practice of polygamy were they deemed eligible to enter the Union.

A favoured American imperial technique in the treatment of annexed Euro-American populations was what may be termed *minoration*, by which the acquired area is so flooded with immigrants from the imperial state as to make the captured society a minority without actual displacement or diminution. When French Louisiana was purchased, the treaty stated that the inhabitants 'shall be incorporated in the Union of the United States, and admitted as soon as possible, according to the principles of the Federal Constitution, to the enjoyment of the rights, advantages and immunities of citizens of the United States; and in the meantime they shall be maintained and protected in the free enjoyment of their liberty, property, and the religion which they profess'. But Jefferson envisioned neither statehood nor cultural integrity for the existing French community. He set a northern boundary for a Territory of Orleans so as to enclose about as much country above the French area along the Mississippi as was already settled below and set about to populate it with Anglo-American soldier-settlers as quickly as possible (Fig. 6.3). And he had in mind for the interim firm rule

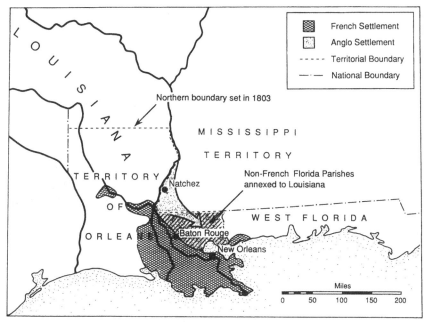

Fig. 6.3 Louisiana and the framework for minoration. The Territory of Orleans plus the Florida parishes were combined to form the State of Louisiana.

under an American governor and the imposition of American law, courts, and language. It didn't work out quite as he planned. Through the addition of a part of West Florida (the Florida parishes) and voluntary immigrations the Anglo-Americans were soon equal in number to the French, but the latter successfully retained their language rights and civil law code. New Orleans, growing rapidly, was soon formally subdivided into three municipalities (under a single mayor) with distinct American and West Indian sectors on either side of the old French grid – a common pattern for imperial entrepots and capitals.

In the case of Texas, the Anglo influx came well before annexation so that the residual Mexican population of Bexar (San Antonio) and the Rio Grande were already a small minority when taken into the United States. In Alta California the influx of the Gold Rush came suddenly, undesignedly, and uncontrollably immediately after annexation, overwhelming everything in its path. There were repeated movements, led by the California Hispanos, to create a separate territory for southern California but none succeeded. Despite official intentions, New Mexico did not become another Louisiana for a long time because there were too few resources within its spacious frame to lure any considerable number of Anglo settlers.

When weaker 'tribal' peoples are annexed, and their lands are coveted by the stronger expanding people, it is of course common imperial strategy to reduce them to a small reserve of the least desirable lands and to make them dependent upon allocations of essential supplies and support from the imperial power. Because they can no longer sustain themselves in their age-old ways of life, such captive peoples are placed under enormous pressures to change themselves into a people more closely conforming to the dominant patterns of the conquering power. Ostensibly such culture change may be intended to release such people from dependency and enable them to become full participants in national life; realistically 'reduction' almost inevitably means not only a drastic depletion of territory, but of numbers and cultural vitality as well. This polity of *reduction and dependency* was applied to most of the Indians in the American West. The development of the reservation system coincided with a shift in definition of Indians from 'resident foreign nations' to 'domestic dependent nations,' and a shift in the bureau of imperial management from the War Department to the Department of the Interior. A federal agent was assigned to each tribe as the legal liaison officer with the American state. Superficially this pattern might appear as similar to an indigenous state; the critical difference lay in the drastic degree of dependency and areal confinement, and the absence of any recognized tribal government.

Prior to this development, however, the United States had applied an even more drastic policy. By 1818 the balance of power between imperial state and indigenous peoples had shifted decisively, and in 1830 the United States undertook the truly radical programme of expelling all Indians then living east of the Mississippi, and resettling them in a large territory in the western plains where they would undergo programmed acculturation to turn them into appropriate imperial subjects, and perhaps, eventually, citizens. A broad belt of country between the Platte River and the Red River in the South (the boundary with Mexican, and soon the Republic of, Texas) was set apart as the 'Indian' or 'Western Territory' and subdivided to accommodate the separate Indian nations. There was a necessary division into two basic categories: 'indigenous village tribes' – those already living there – and 'emigrant tribes' – those removed from their eastern lands; to the west were the 'Plains tribes' of 'wild Indians' – nomadic groups without formal treaty relations with the United States (Fig. 6,4). The Superintendent of this new territory resided in St Louis; Fort Leavenworth and Fort Gibson stood guard at the main entryways up the Missouri and the Arkansas; a resident agent was assigned to each tribe, as were varied sets of specialists to provide technical assistance (such as farmers, millers and blacksmiths) and promote cultural change (such as teachers and missionaries). The geopolitical future of this complicated new protectorate was unclear and heatedly debated in Congress. Formal territorial status, a capital town, a general Indian council, an elected delegate to Congress and

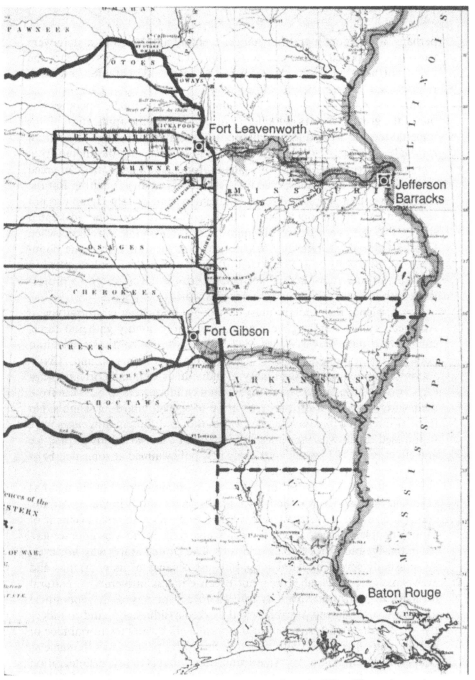

Fig. 6.4 The Indian Territory, as shown on a portion of a 'Map illustrating a plan for the defences of the Western and North-Western Frontier', published by the War Department in 1837. Main bases, routes, and tribal reservation boundaries have been enhanced for clarity.

perhaps, after enough culture change, eventual admission as a state were variously proposed, but none were approved. Within twenty years the whole northern half of the protectorate had been collapsed into a few small reserves and thrown open to white settlers.

This vast relocation project, involving more than 100,000 people, was one of the great examples in modern world history of the application of this imperial technique, a kind of geographical social engineering. The official rationale defended the policy as the only hope of saving the Indian from complete destruction. The obvious main propellents were American greed for Indian lands and widespread indifference to Indian extinction. But the idea of uprooting people and resettling them in some distant region was not restricted in application to such tribal peoples. It would be advocated simply for the convenience of empire, as, for example, in 1853, when Charles Conrad, Secretary of War, urged that the entire Hispanic population of New Mexico (some 60,000–70,000) be expelled and resettled in some better and more accessible region because it cost more to protect them with the army than the territory was worth.

The insistence that Indians must be brought into closer conformity with American culture under government supervision rapidly gathered force, culminating in the Dawes Act of 1877. Indians were to be transformed into good Christian farmers, residing on individual allotments of land. 'Civilization' was premised on the ability of individuals to handle private property. To speed the process of change Indian children were to be taken from their parents and sent off the reservation to special boarding schools. No form of internal tribal government was authorized, Indian languages were suppressed. Such was the general policy (with vast variations in practice and effect) until 1934 when an entirely new policy aimed at some degree of tribal autonomy was initiated.

Now, all of this cries out for much closer examination, but enough has been said to illustrate that the United States was certainly in fact an empire, and made use of a variety of territorial strategies in the management of captive peoples. None of these geopolitical arrangements was unprecedented in the age-old history of imperialisms. The United States was, however, an unusually severe imperial state, not just because of its enormous and ever-expanding material power, but because it was intolerant of cultural diversity in territorial form. It refused to accept any such geopolitical identity as an enduring possibility and worked assiduously – and confidently – to efface any assertive cultural separateness by any territorial bloc of people. The general goal with reference to captive peoples was the same as that applied to immigrants: 'Americanization', that is to say, disintegration of social identities and rapid assimilation of the resultant individuals. In this sense, we may say that American nationalism ruled out the acceptability of empire. In such a view Indian reservations are an embarrassing residue, a blot on the ideal landscape, for they represent the failure of a century of American acculturation to efface tribal allegiance and identity. To probably most Americans they remain an enigma: why, after all this time, do American Indians insist on remaining *Indian*, despite the obviously very severe social costs of doing so?

To a lesser degree (for there was never the same calculated programme for acculturation) the persistence of a French Louisiana, and of the several Hispanic societies in the Southwest, also represent failures of the nineteenth-century design and expectation for the United States. It was widely discussed and commonly assumed that North America was destined to be an 'Anglo-Saxon' continent. The power and capacity of American assimilation was considered to be almost unlimited. During the 1840s and 1850s a succession of presidents had worked hard to take over much more of Mexico than was actually annexed in 1848 – additional areas containing more than 600,000 Mexicans. In the 1850s an American president and various leaders spoke confidently of annexing and 'Americanizing' Cuba – which had a million people at the time, including 600,000 blacks. (Here we may parenthetically note that no one had set forth this vision of an Anglo-Saxon North American more eloquently than Lord Durham in his prescription for the Canadas, wherein he explained why the French must 'abandon their vain hopes of nationality' and acquiesce to integration and assimilation into a progressive English-speaking society.)

Today not only is that overweening national confidence much eroded, but there are many American voices arguing the positive blessings of ethno-cultural diversity, although one may presume that relatively few in the United States would advocate any more overt geopolitical definition of such a thing. As we might expect, Americans take pride in diversity only so long as it poses no challenge to their dominant socio-political order. The United States is in fact becoming more diverse in important ways, rather than less. New immigrations have further blurred and altered proportions in the broad Anglo-Hispanic borderland and new political consciousness continues to alter relationships between dominant and subordinate peoples. I think it is important for Americans to understand much more clearly than they do that there has always been an imperial dimension to American history. Because empires are special kinds of geopolitical structures they invite and require geographical analysis. Here, without doubt, the field of geography could have a major impact on important matters. For there is more involved here than a reassessment of history, there is the enduring topic of how a diversity of peoples can live together and share this earth. That, of course, has always been the positive challenge of empire: how to create an encompassing geopolitical system that can accommodate cultural diversity in the form of discrete territorially delimited societies on a sustained and not destructively subordinate level.

The modern world has been so imbued with nationalism that it may be difficult to recognize anything good about imperialism, but surely any close look at the history of the modern world will show that nationalism has by its very nature been the most relentless and often ruthless enemy of cultural diversity – think of the plight of 'minorities' in dozens of nations; by comparison the older ramshackle despotic empires seem almost benign shelters. The challenge of how to create political frameworks that can accommodate human variety in such territorial terms will ever be with us. It is simply not possible that the modern dictum of 'self-determination' can be translated into full political independence for every aspiring people. By getting underneath the polemics of imperialism and looking closely at the actual workings of empire in its many guises in many different places, geographers might be of some help with this chronic human problem.

NOTES

1 D. W. Meinig, Geographical analysis of imperial expansion, in A. R. H. Baker and M. Billinge (eds.) *Period and place: research methods in historical geography* (Cambridge 1982) 71–8
2 This paper is undocumented because it is a distillation from an extensive presentation of the topic in *Continental America* (New Haven, Conn. 1992), and it is not practicable to provide here an adequate list of sources
3 Philip Mason, *Patterns of dominance* (Oxford 1971) 337

7

Ideology and landscape of settler colonialism in Virginia and Dutch South Africa: a comparative analysis

LEONARD GUELKE

The Europeans who settled Virginia and South Africa created new colonial communities and landscapes. These developments are not seen as the outcome of forces external to the agents of colonialism, but rather as the intelligent application and development of European ideas and assumptions to the circumstances that European settlers encountered in the lands they settled. There is no concern here with looking for laws or generalizations. Rather the concern is to illuminate the nature of colonialism in two cases by comparing the essential logic of development that produced some striking historical similarities in them. Each case is seen as a coherent historical development capable of being understood as a case in its own right.

In making these comparisons, I do so from a base of knowledge which is grounded in many years of research on Dutch settlement in South Africa. The Virginia material is based on readings in the extensive historical and geographical secondary sources on that region. Colonial Virginia and South Africa, notwithstanding differences in crops, environments and peoples had much in common, both being creations of European expansion overseas. The colonizers in both places were unrepresentative, male-dominated fragments of Europe, occupied lands inhabited by non-European peoples, developed plantation agriculture using extensive methods of cultivation and slave labour, and created expanding settlements based upon the exploitation of frontier resources.

The relationship of settlers to indigenous peoples

The English who settled Virginia in 1607 and the Dutch and other Europeans who settled South Africa in 1652 had no historical claims to the lands they occupied. In both areas there were thriving societies of indigenous peoples exploiting the land and its resources.[1] This hard fact European colonizers transformed from a potential problem into an opportunity. In their statements the promoters of colonization stressed the propagation of the Christian gospel and conversion of local peoples as major objectives along with strategic and economic advantages to be derived from colonial plantings.[2]

The idea that Europeans were justified in occupying the lands of other peoples provided that they at the same time undertook to convert them to their own faith was a manifesto of European Christian supremacy. When the rhetoric is stripped down to its essentials European colonizers were proposing what could only become a Christian version of a holy war against the local population should it resist the encroachments of the invaders.

The first small settlements along the James and in Table Valley were in no position to make war against the local people. John Smith sought to establish goods relations with the Indian population in his efforts to establish a viable European settlement.[3] The diplomatic efforts were aided by the marriage of John Rolfe and Pocohontas, which held out the prospect of a peaceful union of two peoples. In South Africa the explorer Jan van Meerhof married a Khoikhoi woman, Eva, amid fanfare and ceremony and with the blessing of the Commander.[4]

These two celebrated inter-racial marriages were not followed by others. The prospects of a blending of cultures foundered because of European assumptions of religious and cultural superiority, and relations between the invaders and native peoples soon deteriorated into open hostility. In Virginia the Indians came close to destroying the English settlement in 1622.[5] The Indian attack was not unprovoked, but it provided the justification for English retaliation, and a debate ensued as to whether the Indian population should be exterminated or enslaved.[6]

In South Africa the Khoikhoi also resorted to armed resistance, but to no avail. The first Dutch–Khoikhoi war ended in 1660 with the Dutch appropriating former Khoikhoi lands in the Cape Peninsula, but not without Khoikhoi objections:

They strongly insisted that we had been appropriating more and more of their land, which had been theirs all these centuries, and on which they had been accustomed to let their cattle graze etc. They asked if they would be allowed to do such a thing supposing they went to Holland, and they added: 'It would be of little consequence if you people stayed here at the fort, but you come right into the interior and select the best land for yourselves, without even asking whether we mind or whether it will cause us any inconvenience.' They therefore strongly urged that they should again be given free access to this land for that purpose. At first we argued against this, saying that there was not enough grass for their cattle as well as ours, to which they replied: 'Have we then no reason to prevent you from getting cattle, since if you have a large number, you will take up all our grazing grounds with them? As for the claim that the land is not big enough for us both, who should rather in justice give way, the rightful owner or the foreign intruder?' ... eventually they had to be told that they had now lost the land as the result of the war and had no alternative but to admit that it was no longer theirs, the more so because they could not be induced to restore the stolen cattle which they had unlawfully taken from us without any reason. Their land had thus fallen to us in a defensive war won by the sword as it were, and we intended to keep it.[7]

In South Africa the Khoikhoi continued to resist European settlement, but European colonists did not consider exterminating or enslaving them. The Dutch East India Company valued the Khoikhoi as trading partners and the settlement of European farmers was not incompatible with continued Khoikhoi occupancy of the land. As the European farming population grew the Khoikhoi were gradually squeezed out and then were reduced by degrees to client labourers on European farms.[8]

The growth of both colonies created large frontier areas and individual settlers took out land on the edges of settlement. In Virginia the cost of frontier settlement was often high as white settlers alienated the land from the Indian population. Indians could not easily be enslaved and defended their territory with tenacity. In South Africa white colonists secured springs in a strategic occupation of the land and gradually gained control. Only the hunting and gathering San, known to Europeans as Bushmen, provided violent resistance and their fate was similar to that of the Virginia Indians.

The violence of the frontier and the apprehensions of white colonists about 'weak' native policies sparked rebellions in both Virginia and South Africa. In Virginia Nathaniel Bacon became in 1676 the standard bearer of frontier settlers who saw every Indian as enemy and wanted their complete extinction.[9] In South Africa one Estienne Barbier briefly gained support of frontier colonists in 1737 by accusing the Dutch East India Company of favouring Khoikhoi interests over those of the colonists.[10] Both rebellions ended with the deaths of their leaders. Bacon died of natural causes. The VOC caught and executed Barbier, quartered his body and displayed it on posts in the countryside as a warning to all.[11]

In occupying new lands in Virginia and South Africa, European colonists lived with violence. The lands Europeans wrested from the original inhabitants were taken at a cost in blood and lives. They were not free lands. Yet it was a cost many settlers were prepared to accept, because they had no attractive alternative for making a living.

Land tenure in the colonies

In both colonies a brief period of Company-directed farming gave way to individual ownership of land and a significant degree of free enterprise. Although there were a few remnants of feudalism in such things as quit-rents and the VOC controlled markets for commodities, Virginia and

South Africa were imbued with a spirit of bourgeois materialism and commercialism. In acquiring ownership and control of land, settlers gained immense control over their own fortunes. As an owner of land, a settler had power and freedom to manage his or occasionally her affairs. The individual not the state would be the beneficiary of good management or good luck and suffer the consequences of failure. In both colonies, and notwithstanding the autocratic rule of the VOC, there was scope for individual enterprise.

The private control of land works well in situations where land is scarce and valuable. Private control spurs investment and improvement. Communal control works well where land is abundant and is well suited to shifting cultivation or nomadic pastoralism. When Europeans settled new lands they applied their Old World institutions, devised for intensive agriculture, to situations in which extensive land use made more sense. The development of extensive agriculture within a framework of private ownership and control was something of a mismatch, because it restricted mobility and put continuous pressure on land, at a time when the resources to improve it were lacking. The system worked but not nearly as well as nomadic pastoralism or shifting cultivation and with negative consequences on land quality measured in terms of fertility and pasture quality.

There was clearly no possibility that European colonists would seek to emulate the ways of the Indians or Khoikhoi, given their ideological commitments to private ownership and control of land. The problem early settlers faced was to develop a workable agriculture within the concepts they took for granted. The resulting patterns of settlement were not called forth by the dictates of either the Virginia or South African landscape: they were the deliberate creations of European colonists logically developing their basic ideas in solving the problems the application of these ideas to new situations had created.

The introduction of private ownership and the establishment of a free market in land spurred development. Individuals in both Virginia and South Africa devised ways of making a profit from their lands and developed efficient ways to exploit them. The rules of competition favoured the successful and in both colonies a small percentage of individuals acquired large estates.[12] The large landholders developed self-sufficient plantations employing scores or hundreds of slaves and indentured workers.

The early settlers based their prosperity on staple crops. In Virginia it was tobacco; at the Cape wheat and the vine. These crops demanded time and effort, but the settlers devised ways of using unimproved lands to pasture stock. They fenced their fields, not their livestock, which foraged for themselves in the forest and the veld.[13] The use of unimproved land became critical to economic survival, and settlers eagerly pushed inland in search of it. The land had been induced to yield its abundance freely and without preparation, but the yield was modest and settlers were ever in search of new tracts of land, both for livestock and fields exhausted after a few years of cultivation.[14]

The large estate holder was able to use vast tracts of land to sustain crop production and, at the same time, maintain livestock close by on unimproved land. The less well-off also benefited from ownership and control of land, but were unable to realize the economies of scale of large estates. Many of the landless were able to acquire land on the frontier and use its abundant resources, in ways that gave them a modest subsistence. They were able to hunt and their animals could graze the forest or the veld. They could produce crops with little effort and maintain themselves with tenuous links to market. The frontier settlers of Virginia and South Africa carried the spirit of bourgeois materialism and free enterprise into the interior and drew upon it in deriving a frontier mode of living in which they had a large manner of self-reliance.[15]

Yet in a new country the population responded to the mutual needs of all. In both Virginia and South Africa residents treated strangers with hospitality. In 1705 Robert Beverley wrote about hospitality in Virginia:

> The Inhabitants are very Courteous to Travellers ... A Stranger has no more to do, but to inquire upon the Road, where any gentleman or good House-keeper Lives and Here he may depend upon being received with Hospitality ... And the poor Planter, who has but one Bed, will very often sit up, or lie upon a Form or Couch all Night, to make room for a weary Traveller ... If there happen to be a ... that either out of Curt ... or Ill-nature, won't comply with this generous Custom, he has a mark of Infancy set upon him and is abhorred by all.[16]

O. F. Mentzel, a German resident of the Cape in the 1730s, made similar observations:

> I must make a little clearer the natural disposition of the inhabitants towards hospitality. Firstly, as there are no inns in the country districts, where one may stop and obtain service for payment, it would not only be somewhat inhuman to refuse a meal or a night's rest to a traveller, but any one refusing would be paid in his own coin; for everyone has to travel at some time and cannot take a shelter with him. Secondly and particularly, the country folk consider it more as a pleasure than an honour if a city dweller, whether a servant of the Company, or a free burgher, calls on them and stays for some time.[17]

Slavery in Virginia and South Africa

The English and Dutch considered slavery to be part of the natural order in colonial settings, but not at home. If European Christians had some scruples about enslaving their own people, they had none about enslaving non-Europeans and non-Christians in their overseas ventures. The institution of slavery was sustained by the same ideology of European Christian supremacy which made it possible for Europeans to conquer and occupy the lands of others. In establishing slavery in many of their colonies European colonizers created a double standard of behaviour for their citizens, and at the same time introduced a gulf between the standards of home life and colonial life.

In Virginia and South Africa the way Europeans thought about work was deeply affected by the idea of slavery. In South Africa it seems probable that colonial expectations of an easier life were present among the very first settlers, who failed to work as hard as they were expected. The first Cape Commander, Jan van Riebeeck, was calling for the importation of slaves within a month of his arrival at the Cape.[18] The Colony acquired slaves shortly thereafter, although it was only in the early eighteenth century that slavery as an institution became fully entrenched. The Cape slaves were drawn mainly from Asia because under the Dutch monopoly system the West India Company had the monopoly of the Atlantic Ocean trade, and the East India Company that of the Indian Ocean.

The Cape attitude to slave labour is clearly revealed in official answers to a question from the Dutch East India Company directors to the Cape Political Council about whether the Cape could be developed without slaves. All but one of the councillors were convinced of the need for slave labour. Their replies written in 1717 included the following comments:

It is more fitting that slaves rather than Europeans should be used, especially in the daily menial tasks ... Daily work performed by slaves could not be demanded of a European in this climate ... The fact that they (European immigrants) have left their country makes them think that they should lead an easier life than at home ... I cannot understand who has dared to trouble our Masters with such useless suggestions ... carefully consider in what way the work is done, throughout the whole of India, all the colonies, the West Indies, etc.[19]

When Commissioner Van Imhoff visited the Cape in 1743, he comments on European work habits: 'Having imported slaves every common or ordinary European becomes a gentleman and prefers to be served than to serve.'[20]

The earliest settlers of Virginia were also notoriously unprepared to work, spending time 'bowling in the streets of Jamestown'.[21] Whether the aversion to work was based on their gentlemanly origins or expectations of a better life in the colonies is a moot point. The fact is that Virginians turned to slave labour when it became profitable to do so and slave labour replaced European indentured labour as the foundation of the Colonies' prosperity. In 1736 William Byrd commented on the consequences of Virginia's dependence on African slave labour in terms similar to Van

Imhoff's remarks on South Africa: 'I am sensible of many bad consequences of multiplying these Ethiopians amongst us. They blow up the pride, and ruin the Industry of our White People, who seeing a Rank of poor creatures below them, detest work for fear it should make them like slaves.'[22]

The entrenchment of slavery created distinctively colonial societies in both Virginia and South Africa. The European community saw itself as separate and superior to the slaves and indigenous people, and emphasized its Europeanness in defining itself. The slave economy, however, generated few jobs for European settlers and encouraged them to seek independent livelihoods. Many smaller planters and their children had difficulties making a living in the settled areas and gravitated to the frontier where it was easier to establish an independent livelihood.[23]

Socio-economic background of settlers

The people who settled Virginia and South Africa were not representative fragments of English or Dutch society. Although most Virginian settlers came from various parts of England, they included many more poor, and landless people than the population as a whole. The fragments were not balanced from a gender point of view, with men greatly outnumbering women.[24] In a male-ruled society it was easier for men to move than it was for women, and throughout Virginia's colonial period there were more men than women settlers. In South Africa the settlers were drawn from Holland, France (French Huguenots) and Germany. The settlers were generally poor and many were poverty stricken. The vast majority of people who made their way to South Africa were males, and there was as in Virginia an enormous imbalance between the sexes.

The large number of male settlers was in keen competition for the small number of eligible European females. In the seventeenth century Chesapeake women married in their late teens and produced, on average, nine children.[25] The men were about five years older than their first wives. The picture in South Africa is similar but more extreme. In the early eighteenth century men married at an average age of thirty. The women they married were much younger, being on average only eighteen years old.[26] The South African colonial family was comparable to that of the Chesapeake averaging seven to eight children.[27]

Peter Kolbe, a European visitor, displayed his male chauvinism when he commented on the large size of Cape families. He wrote:

The European Women in the Cape-Colonies are generally modest, but no Flinchers from Conjugal Delights. They are excellent breeders. In most houses in the Colonies are seen from six to a dozen children and upwards; brave Lads and Lasses with Limbs and Countenance strongly declarative of the ardour with which they were begotten.[28]

In both colonies the age of marriage for women increased as the ratio between settler men and women became more balanced. In South Africa the men who moved to frontier areas married on average two years earlier than those colonists who remained where they were. In 1731 frontier men were on average twenty-five years old compared to nineteen years old for women.[29]

The colonial male-dominated social order worked to the disadvantage of all women, who were more vulnerable to domination and sexual exploitation. In the early years of settlement men applied enormous pressure on women to marry young and have large families. Only women who had passed their childbearing years were able to assert their independence by not remarrying within a few years of the death of their husbands. These widows were able to handle their own affairs and acquired considerable influence.

Conclusion

The settlements created by Europeans in Virginia and South Africa involved far more than the replacement and subordination of indigenous peoples. These European colonies incorporated values of white supremacy and bourgeois individualism that shaped the way people lived and used the environment. Individual European settlers acquired rights to control and use land as they saw fit in pursuit of their own interests. They devised ways of extracting a living from the lands they settled which differed both from the ways of Europe and those of the original inhabitants. The settlers adopted extensive methods of land use, because such methods were more efficient and less time consuming than those in use in Europe. In pursuing private gain the settlers scattered themselves thinly across the lands they occupied, creating distinctive colonial landscapes and ways of life.

The expansion of European settlement involved the destruction of indigenous societies and the annihilation of their values. The new landscapes of Virginia and South Africa were landscapes of European conquest, in which Europeans and their slaves replaced most of the original inhabitants. The Europeans brought much of their culture with them but also devised ways of using indigenous culture and native plants and animals for their own purposes.

The revival of slavery in European colonies made them fundamentally different from European homelands. An institution that was dead or moribund in Western Europe became the foundation of colonial societies in Virginia and South Africa. The ability to acquire and use slaves promoted rapid stratification of the white community. The wealthy gained power and more wealth, the poor were denied the opportunities a free market in labour would have given them. Many of them sought to establish their independence on the frontier, using its abundant resources.

The Virginian and South African settlers were conscious that they lived in remote regions of the world on the fringes of civilization in the wilderness of North America or the wilds of Africa. They were also aware of the potential that existed for their communities to be absorbed or diluted by the non-European indigenous peoples or slaves among whom they lived. In this situation they emphasized their Europeanness, and took steps to ensure that home views of conduct were maintained. These communities conceived themselves as home communities away from home rather than strictly colonial ones.[30]

The landscapes of Virginia and South Africa differed greatly in terms of people, plants and animals, but they both bore the imprint of European colonial expansion in which individual settlers gained power through access to land and labour. They fashioned a rough and ready landscape, but it was a landscape suited to their needs. The settlers were not made on the frontier, but rather made it. The freedom of the frontier was premised on the unfolding of settler freedom already implicit in an ideology which conferred power on certain individuals to acquire land and slaves.

In both colonies wealth was unevenly distributed, and the settlers adopted or adapted a variety of strategies to make a living in accordance with their resources. A few settlers accumulated wealth and power by inheritance, marriage and good management. Others succeeded in making a reasonable living on a more modest scale. The competition for resources was keen, and many settlers moved onto the frontier where they lived in rough comfort. Many male settlers failed to achieve an independent living and lived miserable poverty-stricken lives as bachelors in a society that had few opportunities for unskilled white men.

Male settlers outnumbered female settlers in Virginia and South Africa. This male predominance intensified European patriarchy and women settlers married at very early ages and had large families. The pressure on women to marry young and remarry if widowed put their financial lives in the hands of their husbands, who made important decisions for them. A few older widows did not remarry and gained independence for themselves.

The settler colonial ways of life in Virginia and South Africa reflected a synthesis of European assumptions and values. They incorporated traditional ideas on the organization of agriculture and social roles but included specific visions of colonial life that separated the colony and homeland. The colonial vision included the idea of slavery and the expropriation of the lands and rights of native peoples. In both Virginia and South Africa there emerged distinct settler colonial ways of life that separated both colonies from Europe and united them as colonial undertakings based on shared bourgeois assumptions and ideas of European cultural and racial supremacy.

NOTES

1 W. F. Craven, *White, red, and black: the seventeenth-century Virginian* (Charlottesville 1971) 39–72; R. Elphick, *Khoikhoi and the founding of white South Africa* (Johannesburg 1985) 71–89

2 H. C. V. Leibbrandt, *Précis of the archives of the Cape of Good Hope: letters and documents received, 1649–62* (Cape Town 1896–9) 1–17

3 E. S. Morgan, *American slavery–American freedom: the ordeal of colonial Virginia* (New York 1975) 75–9

4 R. Elphick and R. Shell, Intergroup relations: Khoikhoi, settlers, slaves and free blacks, 1652–1795, in R. Elphick and H. Giliomee (eds.), *The shaping of South African society, 1652–1840* (Middletown 1989) 185

5 W. M. Billings, *The Old Dominion in the seventeenth century: a documentary history of Virginia, 1606–1689* (Chapel Hill 1975) 220–4

6 Morgan, *American slavery*, 98–100

7 *Journal of Jan van Riebeeck, 1651–1662*, H. B. Thom (ed.) (Cape Town 1954) 111, 195–6

8 R. Elphick, The Khoisan to 1828, in R. Elphick and H. Giliomee (eds.), *The shaping of South African society, 1652–1840* (Middletown 1989) 28–34

9 W. M. Billings, J. E. Selby and T. W. Tate, *Colonial Virginia* (White Plains 1986) 77–96

10 G. Schutte, Company and colonists at the Cape, 1652–1795, in R. Elphick and H. Giliomee (eds.), *The shaping of South African society, 1652–1840* (Middletown 1989) 308–9

11 *Ibid.*

12 Billings, Selby and Tate, *Colonial Virginia*, 97–138. L. Guelke and R. Shell, An early colonial landed gentry: land and wealth in the Cape Colony 1682–1731 *Journal of Historical Geography* 9 (1983) 265–86

13 L. C. Gray, *History of agriculture in the southern United States* (Washington, DC 1932) 200–12; H. Giliomee, Processes in the development of the Southern African frontier, in H. Lamar and L. Thompson (eds.), *The frontier in history: North America and Southern Africa compared* (New Haven and London 1981) 76–119

14 P. J. van der Merwe, *Die trekboer in die geskiedenis van die Kaapkoloni, 1657–1842* (Cape Town 1938)

15 R. D. Mitchell, *Commercialism and frontier: perspectives on the early Shenandoah Valley* (Charlottesville, Va. 1977) 230–40; L. Guelke, Freehold farmers and frontier settlers, 1657–1780, in R. Elphick and H. Giliomee (eds.), *The shaping of South African society, 1652–1840* (Middletown 1989) 66–108

16 R. Beverley, *The history and present state of Virginia (1705)* (Chapel Hill 1947) 312–13

17 O. F. Mentzel, *A geographical–topographical description of the Cape of Good Hope* (Cape Town 1944) III, 199

18 W. Blommaert, Het invoeren van de slavernij aan de Kaap, *Archives Year Book of South African History* 1 (1938) 19

19 *Reports of Chavonnes and his council, and of Van Imhoff, on the Cape* (Cape Town 1918) 83

20 *Ibid.*, 137
21 Morgan, *American slavery* 44–70
22 Quoted in Gray, *History of agriculture*, 350
23 *Ibid.*, 85–106; Guelke, Freehold farmers and frontier settlers, 84–93
24 Billings, Selby and Tate, *Colonial Virginia*, 52–9
25 A. Kulikoff, *Tobacco and slaves: the development of southern cultures in the Chesapeake, 1680–1800* (Chapel Hill and London 1986), 54–61
26 L. Guelke, The anatomy of a colonial settler population: Cape Colony 1657–1750 *International Journal of African Historical Studies* 21 (1988) 463
27 *Ibid.*, 467
28 P. Kolbe, *The present state of the Cape of Good Hope* (London 1731) II, 340
29 Guelke, Freehold farmers and frontier settlers, 463
30 L. Guelke, The origin of white supremacy in South Africa: an interpretation *Social Dynamics* 15 (1989) 40–5

8

Municipal sanitary ideology and the control of the urban environment in colonial Singapore

BRENDA S. A. YEOH

Power relations in a colonial city

Scholars who have enquired into the conceptual distinctiveness of the colonial city have often emphasized, first, its racial, cultural, social and religious pluralism and secondly, the concentration of social, economic and political power in the hands of the colonizers, often an ethnically distinct group.[1] David Simon, for example, considers the stark asymmetry of power between colonizers and the colonized to be by far the most powerful independent variable influencing both social processes and urban spatial structure.[2] Such a view has led some scholars to further assume, whether implicitly or explicitly, that the colonial town or city is either wholly or largely a creation of its colonial masters. In Simon's 'schema of colonial urban development', it is envisaged that once the subjugation of the colonized is achieved, metropolitan colonial power, exercised by its agents and institutions, becomes the most profound, and practically the singular, influence on the shaping of colonial urban space.[3] In a similar vein, T. G. McGee portrays turn-of-the-century Singapore as a planned colonial city which, despite a predominantly Chinese population, 'remained a city planned by Europeans', with a residential distribution which 'continued to reflect the intentions of the European rulers to an amazing degree'.[4] The focus here is on the uncontested supremacy of the European colonizers in undeviatingly carrying through their own plans and intentions to fashion a city after their own image. Such a scheme ignores the impact of indigenous or immigrant agencies, whether separately or in interaction with colonial power, in shaping colonial urban society and space. Their lack of formal authority and economic resources is often equated with minimal impact on the urban environment. Such a view stems from a narrow conception of the nature of power relations within colonial societies.

An alternative view of colonial power relations is offered here. Power, following Anthony Giddens, means 'transformative capacity', that is, 'the capability to intervene in a given set of events so as in some ways to alter them'.[5] Colonial power is often mediated through institutional means of control and it is through the application of power as a constant if undramatic force 'running silently through the repetition of institutionalized practices'[6] that the supremacy of the colonialists becomes entrenched. In colonial Singapore, these institutions of control included the executive and legislative councils, the police, the Chinese protectorate and the municipal commission. These provided the necessary structural apparatus through which colonial power penetrated to encounter and control the basic continuities of Asiatic life. By the end of the nineteenth century, as a result of the proliferation of legislative measures and institutions, the scope of colonial control had broadened to encompass several aspects of daily urban activity including the management and use of the environment.

The growing scope of institutional power accruing to the colonialists, however, does not imply a degeneration of the colonized into a state of subjugation and powerlessness. Giddens contends that '[n]o matter how great the scope and intensity of control superordinates possess, since their power presumes the active compliance of others, those others can bring to bear strategies of their own, and apply specific types of sanctions'.[7] It is hence necessary to take into account the strategic conduct[8] of the colonized people who must be seen as knowledgeable and skilled agents with some awareness of the struggle for control, not just passive recipients of colonial rule, or as the authorities often called them, 'ignorant and prejudiced' people.

In colonial Singapore, the Chinese who came to the Nanyang brought with them an entire array of organizations such as clan and dialect associations and secret societies which provided convenient institutional focal points for the consolidation of power and the organization of counter-strategies to confront whatever means of control imposed by the colonialists. These counter-strategies included rioting, holding demonstrations or going on strike as means of expressing grievances. However, more common than active protest, and in the long run less costly in terms of effort and sacrifice, were the 'passive' means of countering and inflecting colonial control. The Asiatic communities could adopt an outward attitude of apparent acquiescence, but in reality disregard the measures imposed by colonial power.[9] As the administrative machinery could not run faultlessly, 'openings' in the system could be exploited to thwart the execution of particular policies and even to influence the decisions and activities of those who apparently held power over the masses.

This chapter focusses on one aspect of the wider negotiation of power between colonialists and the colonized in Singapore in the late nineteenth and early twentieth centuries. It is concerned with municipal efforts at disease and sanitary control on the one hand and, on the other, the response of the non-European urban classes in countering control and asserting their own counter-view with regard to the management of the urban environment.

Sanitary ideology and municipal perceptions of disease and Asiatic domestic practices

By the middle of the nineteenth century in Britain, the campaign for the sanitized environment had imbibed a visionary expectancy. Advocates of sanitary reform envisaged that by improving sewage disposal, house ventilation and water supply, epidemic diseases could be vanquished, mortality rates lowered and life expectancy improved. Towards the end of the nineteenth century, the sanitarians' vision seemed to have come to pass as overall mortality rates began to fall.[10] The triumph of sanitary control over the scourge of diseases in Britain held out similar hope of improving health conditions in her far-flung colonies. Advances made in the new sciences of bacteriology and parasitology imbued medical men working in colonial settings with a confidence that epidemic disease could be controlled and conquered through the application of Western scientific knowledge and reason.

At the heart of the 'science' of sanitation lay the contention that by managing the environment and restructuring space using scientific principles, it was possible to banish disease and improve health. Sanitary experts and medical men were, in Foucault's phrase, amongst the first 'specialists of space'.[11] The confidence of sanitary experts in their 'science' is epitomized by the Malayan malariologist Dr Malcolm Watson's assurance that the day would come when, by precisely manipulating physical and chemical environments using scientific principles, one would be able to 'play with species of *Anopheles*, say to some "Go" and to others "Come", and to abolish malaria with great ease, perhaps at hardly any expense'.[12]

In colonial Singapore, sanitary control was considered the 'mainspring of municipal action'.[13] Indeed, at the turn of the century, the overwhelming concern for a sanitized environment provided much of the ideological impetus which led to the passing of a ramifying web of municipal by-laws to regulate the daily routines and habits of the urban population.

Mortality rates in Singapore

At the heart of the colonialist's concern with sanitation lay the fundamental issues of demography: that is, population, morbidity and mortality. Mortality rates in turn-of-the-century Singapore were unquestionably high among the immigrant and indigenous communities. Between 1893 and 1910, crude death rates ranged between 32.9 and 51.1 per thousand, with an

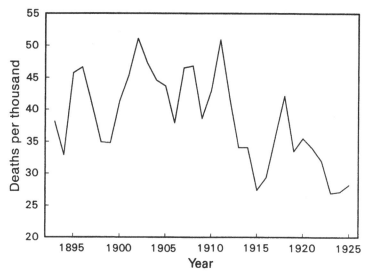

Fig. 8.1   Annual mortality rates of the total municipal population of Singapore, 1893–1925. *Source:* compiled from *Administrative Report of the Singapore Municipality, Health Officer's Report* 1893–1925.

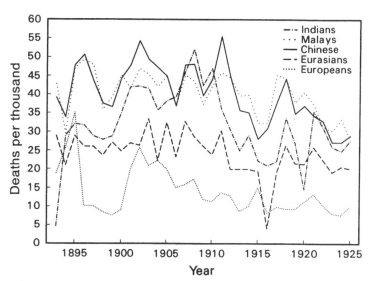

Fig. 8.2   Annual mortality rates of principal races in the municipality of Singapore, 1893–1925. *Source:* see Fig. 8.1.

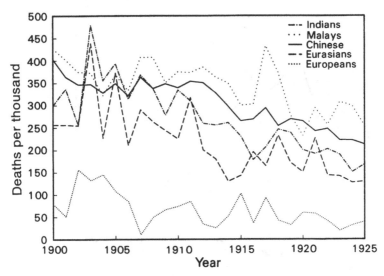

Fig. 8.3   Infant mortality rates of principal races in the municipality of Singapore, 1900–1925. *Source:* as Fig. 8.1, 1900–25.

average of 42.2 per thousand. Between 1911 and 1925, whilst the average had fallen to 32.3 per thousand, high death rates continued to prevail in 1911, 1912 and 1918 (Fig. 8.1). Average death rates varied dramatically among races, particularly between the Europeans (14.0 per thousand) on the one hand and the Malays, Chinese and Indians (40.1, 39.9 and 32.7 per thousand respectively) on the other (Fig. 8.2). The disparity was even greater if infant mortality rates of different races were compared (Fig. 8.3). Between 1900 and 1925, the average infant mortality rate for Europeans was 65.7 per thousand, far below those of the other racial groups – Eurasians (218.9), Indians (257.1), Chinese (305.4) and Malays (343.7).

The principal diseases which consistently felled the greatest numbers throughout the late nineteenth and early twentieth centuries were tubercular diseases, beriberi and malarial or remittent fevers (Fig. 8.4).[14] Another group of diseases, far less dominant in numerical terms but of particular interest to the municipal health department as a result of their capacity to assume epidemic proportions within a short space of time were cholera, enteric fever, dysentery, diarrhoea and smallpox (Fig. 8.5).[15] Both groups of diseases were collectively known as zymotic diseases and, according to contemporary epidemiological theory, were regarded to have sprung from filthy habits and insanitary environments. Outbreaks of zymotic diseases were viewed as ominous warnings of further retribution for sanitary neglect and 'incitement to proceed energetically with sanitary reform'.[16] At the turn of the century, even a disease like beriberi, later proved to be a

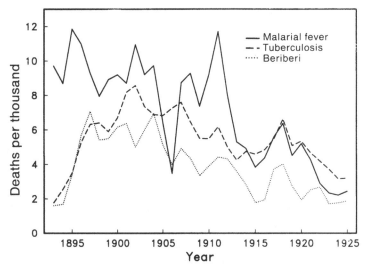

Fig. 8.4 Mortality rates from selected principal diseases in Singapore, 1893–1925.
Source: see Fig. 8.1.

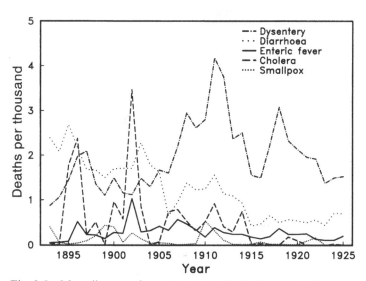

Fig. 8.5 Mortality rates from selected epidemic diseases in Singapore, 1893–1925.
Source: see Fig. 8.1.

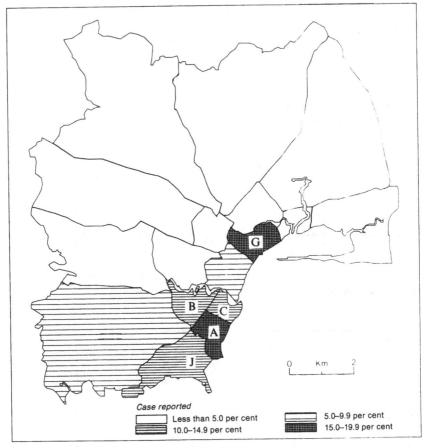

Fig. 8.6   Percentage distribution of reported cholera cases in Singapore, 1900–1904. *Source:* as Fig. 8.1, 1900–4.

deficiency disease,[17] was perceived to be inextricably conflated with insanitary environments. Medical authorities in Singapore contended that the disease could not be attributed to any want of variety or deficiency of diet but was caused by a 'malarious poison' or a 'marsh miasm', described to be 'favoured ... by damp, overcrowding of buildings, and insufficient ventilation'.[18] It was widely claimed that the prevalence of the disease was 'in proportion as the surroundings [were] insanitary'.[19]

Municipal attitudes to disease and Asiatic domestic practices

Typical of the nineteenth-century division of causes of diseases into predisposing conditions and efficient causes, it was believed that suitable moisture and temperature conditions of the soil combined with organic contamination provided conditions predisposed to an outbreak of zymotic diseases.[20] The 'latent poison' was then either washed into surface wells or expelled into the atmosphere with the next rise of groundwater during the wet season. It was then conveyed to human victims through a variety of means, the chief medium being contaminated well-water used for drinking, bathing, rinsing and cleansing household utensils.

From the municipal perspective, the association of a 'filth' disease like cholera with the Asiatic populace, particularly the Chinese, was evident from the spatial statistics collated by the municipal health officer. Fig. 8.6 shows the percentage distribution of reported cholera cases[21] within the municipality for 1900–4, a period with serious outbreaks of cholera. This indicates that the districts which tended to be most severely affected by cholera outbreaks were mainly district A (the southern portion of Chinatown), B (the area around Pearl's Hill), C (the northern portion of Chinatown), G (the area around Kampong Glam) and J (the dock area).[22] The impression that cholera tended to prevail in predominantly Chinese areas was further confirmed by the fact that 84.5 per cent of all notified cases of cholera[23] between 1895 and 1925 were Chinese, mainly males of the coolie class.[24]

From the perspective of the municipal health officer, the prevalence of cholera among the Chinese was not unexpected. Asiatic domestic practices appeared to embody all that could possibly exacerbate the spread of the disease. Carelessness regarding the removal of excreta and household refuse, allowing filth to accumulate in the house, using polluted well-water and the overcrowding of cubicles were perceived as incorrigible Asiatic habits,[25] providing both the bacteria (contained in excrementitious filth) as well as the medium (polluted water or vitiated air) for zymotic diseases to prevail. The Asiatic partiality for water drawn from wells as opposed to municipal piped water was continually singled out as one of the main culprits responsible for the spread of diseases like cholera, enteric fever and dysentery.[26] Even if well-water was not used for drinking purposes, it was felt that the dangers to health were by no means diminished:

[I]t is well-known that it is the universal practice among the Chinese, and probably among other Asiatics to wash out their mouths at the same time as they bathe and the danger from such a practice is little less, if any, than if the same water were drunk. The water, however, if not drunk is used for 'domestic purposes' . . . The idea of 'cleansing' cooking and eating utensils with dilute sewage is sufficiently disgusting, but the danger does not end here . . . Is it not probable that polluted water, containing swarms of microorganisms when thrown on the floors of houses and drying should afford a medium for the propagation of disease, even without the necessity for drinking it?[27]

It was, however, the Asiatic management of nightsoil which aroused the strongest municipal revulsion. Four-fifths of the privies in the municipality

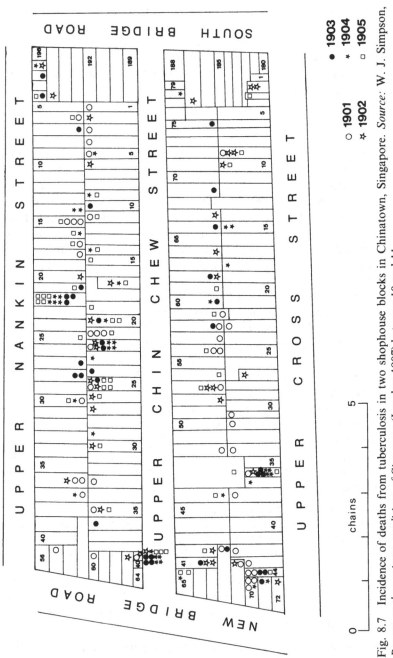

Fig. 8.7 Incidence of deaths from tuberculosis in two shophouse blocks in Chinatown, Singapore. *Source:* W. J. Simpson, *Report on the sanitary conditions of Singapore* (London 1907) between 10 and 11.

'had no catchment apparatus whatsoever, other than the bare earth (with or without a hole dug in it), or a large cesspool built of brick, the bottom being frequently unpaved and the mortar quite porous'.[28] A survey of the conditions of latrines found in two streets revealed that the 60 houses surveyed along the Chinese street had an average of 19 persons to each house and 23.7 persons to each latrine. Quantitatively, the Malay street appeared more respectable, with 6.4 persons per house and 6.8 persons per latrine.[29] However, the large majority of latrines in both streets contained stinking, decomposed, maggoty faeces, ranging in depth from a few inches in some cases to a few feet in others.[30] These large accumulations of excrement, often 'one mass of writhing maggots',[31] were removed only when required for agricultural purposes or when overflowing, which was infrequent, given the continual loss by soakage or leakage into the unpaved floor of the cesspools.

Contamination, filth and a dangerous disregard for dirt were, from the municipal perspective, symptomatic of Asiatic domestic practices, which were hence perceived as highly inimical to health and instrumental to the propagation of disease.

By the first decade of the twentieth century, increasing concern for the prevalence of tuberculosis (Fig. 8.4) drew municipal attention to another aspect of Asiatic urban life: the spatial arrangement of the Asiatic house and building block. Professor W. J. Simpson, an expert in tropical hygiene commissioned by the Governor to investigate high death rates in Singapore in 1906, identified the incidence and recrudescence of tubercular diseases with a particular type of house-form commonly found throughout the city: the Chinese-built shophouse.[32] Fig. 8.7 records the *in situ* number of deaths from tuberculosis between 1901 and 1905 in two blocks of shophouses typical of those found in the heart of Chinatown, illustrating the high prevalence of the disease and its tendency to recur. Professor Simpson concluded that the potential advantages deriving from Raffles' foresight in town planning[33] were increasingly nullified by the peculiar traits of Chinese house design and construction.[34] The Chinese, according to Professor Simpson, preferred to build horizontally rather than vertically: 'Enlargement of the [Chinese] house is effected by erecting another building behind ... separated by a courtyard. There may be a series of buildings each with courtyards approached by passages leading one from the other.'[35] Two separate houses placed back to back in a rather narrow block, might, by the processes of horizontal fusion and vertical additions, culminate in a four-storey tenement house (Fig. 8.8). This arrangement was condemned as destructive to health and the lighting and ventilating of houses as well as obstructive to efficient scavenging and drainage for several reasons.

First, such a style of house construction excessively covered the house-

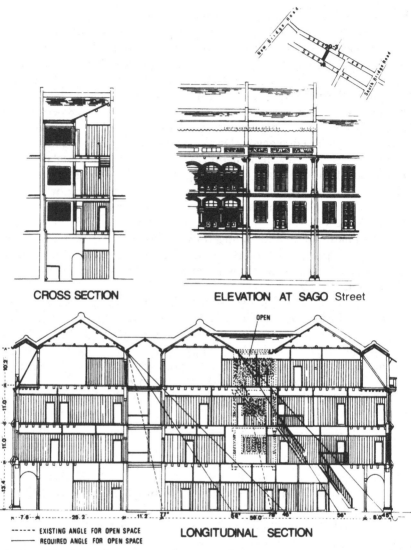

Fig. 8.8   20–23 Sago Street and 18 Sago Lane: four-storeyed tenements enlarged by horizontal and vertical accretion. *Source:* as Fig. 8.7, between 14 and 15.

site of a building and 'shut out from the interior an adequate and necessary supply of sunlight and fresh air'.[36] In Fig. 8.8, the existing angles for open space (77–78°) fell far short of the recommended angles (45°, or at most 56°). Secondly, in the case of tenement houses, defective house design was further aggravated by the subdivision of each floor into a large number of cubicle rooms. Since the houses had no lateral windows, these cubicles were

Fig. 8.9 Typical mosaic of back-to-back and abutting houses in Chinatown, Singapore. *Source:* as Fig. 8.7, between 26 and 27.

'windowless rooms, dark and cheerless, receiving neither light nor air direct from the outside'.[37] Furthermore, the cubicles were filthy, overcrowded with inhabitants and cluttered with goods. For the Chinese coolies who rented out these cubicles, spaces for work, sleep and storage were undifferentiated. The sick breathed the same air and shared the same space as the healthy, and all activities occurred amidst 'immense quantities of rubbish and filth'.[38] Thirdly, in the traditional Chinese quarter, both narrow and wide blocks were packed with buildings, which, in the narrow blocks formed back-to-back houses, and, in the wider blocks, created congeries of buildings abutting one upon another as in a piece of mosaic work (Fig. 8.9). Such an arrangement intensified the obstruction to the admission of light and the free circulation of air already hindered by the design of the individual houses themselves. In some of these blocks, the amount of open space to the covered-over area was only 6–7 per cent, instead of 33 per cent which was considered the required health minimum.[39]

At the crux of municipal understanding of tuberculosis lay the belief that a panacea could be found for the disease by ensuring that each individual was assured an adequate supply of pure air and light. Such a view is evident not only in Professor Simpson's recommendations but also in the Principal Civil Medical Officer Dr D. K. McDowell's propositions that '[w]herever the amount of air to the individual is increased when deficient, the death rate from tubercle of lung ha[d] been greatly reduced', and that '[i]f all inhabited rooms were properly lighted the health of the community would improve materially'.[40] Municipal sanitary reforms were hence motivated by the quest for better health through the banishment of overcrowding, darkness, stale air, filth and clutter from Asiatic tenements.

From the vantage point of the municipal authorities, the entire catalogue of Asiatic and especially Chinese domestic practices and habits signified all those features which seemed to set the Chinese irrevocably apart as a most insanitary race characterized by 'incurably filthy and disorderly habits'.[41] By concentrating attention on the Asiatic environment as the cause of disease and the focus of public-health intervention, the municipal health department eschewed arguments which related mortality and disease to poverty and economic deprivation, or which questioned the colonial socio-economic structure. The characterization of the Asiatics as 'incurably filthy' also obviated the colonial government's responsibility to provide sufficient financial outlay for costly sanitary facilities in the city, for it was argued that even if the government were willing to make the 'financial sacrifice', 'oriental stubbornness would in all probability continue to cling on to old habits and customs, and hang on to dirt'.[42]

Municipal strategies of disease and sanitary control

Imbued with an unquestioned confidence in the supremacy of Western sanitary science on the one hand, and an antipathy for Asiatic domestic practices on the other, the municipal health department contrived various sanitary strategies to improve the health of the city. These included first, replacing Asiatic methods of water supply and sewage disposal which had developed under an era of *laissez-faire* with municipally controlled systems; secondly, reordering the built-form to conform with idealized sanitary standards through back-lane and area reconstruction schemes; and thirdly, coercing the Asiatics to reform their habits and management of the environment through a system of surveillance.

As municipal waterworks, sewage disposal schemes, back-lane and reconstruction schemes invariably drew heavily on the municipal purse, the authorities relied heavily on surveillance to secure the basic sanitary conditions of the town. The process of surveillance involved the accumulation of information pertaining to the Asiatic population and the supervision of various aspects of Asiatic life to ensure the enforcement of municipal by-laws. In practice, municipal surveillance took several forms, including the collection of birth and death statistics, the compulsory notification of 'dangerous infectious diseases' and the division of the municipality into 'sanitary districts', each under the charge of an inspector armed with powers to carry out inspections of Asiatic premises, post notices, serve court summons and remove persons suspected to be suffering from infectious diseases. Substantial sections of the municipal ordinance of 1896 and subsequent amendments were devoted to the control and policing of various aspects of Asiatic domestic practices including scavenging, the removal of nightsoil and household filth, the organization of space and the ratio of people to space within buildings, the maintenance of latrines and the prevention of the spread of infectious diseases. Transgressing municipal requirements in any of these areas would incur prosecutions and fines. Satisfactory regulation of the environment hence depended on effective surveillance to break down the barriers which kept the physical and immoral disorder of the Asiatic classes 'secluded from superior inspection and common observation'.[43] The success of surveillance as a strategy of control in turn depended on whether it was able to work in a pervasive and meticulous manner, or in Foucault's terms, on its ability to infiltrate, reorder and colonize.[44]

Manpower assigned to the enforcement of municipal sanitary law was, however, severely limited. In 1894, for a town well over 150,000 inhabitants, the sanitary staff only comprised one health officer, six or seven inspectors and a similar number of peons and detectives.[45] The municipal commission contributed annually towards the maintenance of the police force and in return was entitled to assistance from the six hundred odd police officers employed within the municipality.[46] However, in the enforcement of 'sanitary order and decency',[47] the efficacy of the police officers fell far short of anticipation. The municipal president, Alex Gentle complained that sanitary offences were habitually committed in the sight of the police and went unnoticed, hence perpetuating the impression among the Asiatics that sanitary offences were 'no impropriety at all, or practically condoned'.[48] That the police force was notoriously inefficient in enforcing municipal law might have made the system of surveillance less intrusive, but it also made it more arbitrary. In the next section, I turn to the response of the Asiatic plebeian classes to municipal attempts to impose such a system of sanitary surveillance.

The Asiatic plebeian response to municipal sanitary control

Municipal purposes, and in particular the sanitary cause, were little appreciated by the Asiatics as means of improving civic life or mitigating the scourge of disease. In terms of its objectives, legislative framework and institutional style, the municipal commission, patterned after the municipal authorities of British towns and cities, was a form of control alien to the majority of the Asiatics. For the Chinese, civic authority and control were more firmly invested in the myriad clan and dialect organizations to which they belonged. The disparity between municipal and Asiatic views of sanitary reform could not have been more plainly manifest then when Alex Gentle expressed forlorn hopes that the municipal ordinance would 'convince the ignorant people ... that sanitary rules [were] not the outcome of fitful zeal, or worse still, attempts at extortion on the part of a few subordinate officers [of the municipality], but the settled purposes of the Government of the Colony, not to be evaded or resisted, but to be consistently enforced for the public good'.[49]

In elucidating the Asiatic response to municipal sanitary control, I will first argue that the plebeian response was not of 'ignorance, apathy and superstition' but a rational reaction based on different perceptions and priorities derived from alternative systems of medicine and health care. Ethnic medical systems provide a set of contextual parameters within which ethnic communities interpret matters of health, disease, death and its relation to the wider environment. Among the immigrant Chinese, the strength of their own medical traditions and practices was a crucial resource drawn upon by the Chinese in responding to the imposition of municipal sanitary control.

Traditional Chinese health and medical provision

By the turn of the century, various elements deriving from the traditional organizational structure of Chinese medicine were well established in Singapore in order to serve the burgeoning Chinese urban population. The colonial government generally tolerated the establishment of non-Western medical systems as it obviated the responsibility of welfare provision among ethnic communities such as the Chinese. The structure of the Chinese medical system in Singapore comprised first, at the institutionalized level, a number of charitable medical organizations set up by wealthy Chinese merchants to provide free medical advice and treatment by Chinese physicians for indigent persons. The pioneer among these organizations was the Thong Chai Yee Say (later known as the Thong Chai Medical Institution) established in 1867 along North Canal Road.[50] According to Li bai gai, a prominent Chinese physician who first served from 1901 to 1905 in the Thong Chai Medical Institution, he alone treated a total of 6,125 patients within three months when he returned to the Institution in 1922.[51] A second institution, the Sian Chay Ee Siah was established in 1901 along Victoria Street to minister to the needs of the burgeoning Chinese population in the Rochore area.[52] In 1910, the first in-patient hospital providing Chinese medical care, the Kwong Wai Shiu Hospital, was opened by Cantonese.[53] Another important element in the structure of Chinese medicine in colonial Singapore was the traditional Chinese medicine-shop. Between 1870 and 1928, there were at least fifty-eight Chinese pharmacies established in various parts of Singapore by the Hakkas, the principal dialect group associated with the retailing of Chinese medical supplies.[54] Besides dispensing Chinese herbs, drugs and medicaments, pharmacies also provided readily available medical advice on a wide range of ailments. The care of the sick and aged among the Chinese was also based on the clan organization. Several clan associations ran sick-receiving houses which provided free food, shelter and medical care for the diseased and chronically ill. Finally, at the most basic level, the Chinese medical system which emerged in colonial Singapore comprised freelance physicians or *chung-i* who operated from clan associations, temples, the market place or their own homes.

In the absence of a municipal system of medical care and poor relief, the Chinese medical system filled a much-felt void in the lives of the Chinese plebeian classes. As a medical system based on different conceptions of disease, it also placed a different construction on the relation between the maintenance of health and the management of the environment. Whilst Western sanitary science advocated the removal of filth, the disinfection and ventilation of houses and the isolation of the sick as essential preventive measures in stemming the tide of disease, Chinese medical theory did not necessarily imbue these measures with similar significance. Traditional Chinese pathological theory distinguished between internal and external causes of disease. Illness was considered to result from a lack of moderation, either in the external sphere (such as excessive dryness, damp or cold) or in the internal sphere (such as excessive joy, anger, melancholy or trauma). These imbalances could then interpreted in terms of disharmony between the *yin* and the *yang*, the primogenial forces which govern the universe.[55] Diagnosis of diseases was based on detailed descriptions of symptoms and prescribed medications were drawn from a rich pharmacopoeia. The Chinese physician Li bai gai for example, classified common ailments in Singapore into fourteen categories, each characterized by meticulously observed symptoms.[56] Unlike Western sanitary science which regarded disease as the product of germs which were 'enemies' to be conquered by 'scientific' means, traditional Chinese medicine focussed on correcting imbalances and strengthening the body's resistance rather than attacking a pathogenic invader.[57] For example, in the treatment of beri-beri,[58] Chinese medical men stressed sleeping upstairs to avoid damp, taking more beans and potatoes instead of rice, using purgative rather than tonic drugs and cultivating of a healthy mindset as a 'mind diseased [might] be prejudicial to the patient'.[59]

The difference in conception of disease and its causes between Western and Chinese medical theories is most clearly illustrated by contrasting attitudes towards excrementitious filth. From the municipal perspective, nightsoil harboured disease-causing germs, exuded noxious odours and must be systematically removed from human habitation and destroyed. From the Chinese perspective, nightsoil was a valuable source of manure to be accumulated in water-tight vessels and sold to farmers. Such a practice aroused strong condemnation from the municipal authorities and Western medical men who considered it 'unscientific from [a] health point of view, as it [helped] to increase, if not actually cause, the incidence of Diarrhoea, Dysentery, Typhoid and diseases of that group'.[60]

Given the strength of traditional Chinese medicine in colonial Singapore, its easy accessibility to the populace and the vital role it played in everyday life, the bulk of the immigrant Chinese plebeian class remained unsympathetic towards Western-imposed concepts of disease.

Non-compliance of the Chinese populace to sanitary reform

I will now attempt to demonstrate that standard assumptions of mass passivity of the Asiatic classes are misleading and that they were capable of knowledgeable and at times skilled strategies to prevent control over their daily practices and management of the environment from being totally wrested from them. In countering the expanding scope of institutional power over various aspects of urban life, the most common strategies usually took the form of apparent acquiescence to municipal demands as a façade for unobstrusive non-compliance. The municipal reports and summons notebooks abound in examples of such forms of 'passive resistance'.[61]

In March 1895, vexed by the inefficient scavenging of the city, the municipal commissioners ordered the distribution of notices among the inhabitants of Chinatown to prohibit the deposition of household refuse on the streets.[62] All occupiers were required to provide themselves with dust-boxes placed outside their houses to facilitate municipal scavenging and to prevent refuse from being left to fester in the streets. At the end of the year, it was reported that the order was 'a dead letter' and the acting municipal engineer protested that 'no [scavenging] scheme [could] be expected to keep the Town clean when all classes of the community persist[ed] in the practice of throwing into the streets at all hours of the day, office, house and kitchen refuse'.[63] As a rule, offences against the order could not be brought home to the offender as he had to be caught in the act of throwing out refuse before the municipal authorities could prosecute.[64] Hence, unless 'a tell-tale envelope betray[ed] its origin',[65] the ubiquity of the offence and the anonymity of the offender amidst the masses rendered municipal surveillance ineffective. Even when a sanitary offence could be traced to a particular house, it was often difficult to single out the actual culprits. For example, in a municipal summons case, conservancy overseer Michael Gabriel found nightsoil 'overflowing with maggots' coming down a drain at the back of a sundry shop at Telok Blangah Road. The case against the owner of the shop had to be withdrawn as the upper floors of the shophouse were inhabited by a large number of coolies 'for whom [the shop-owner] could not be held responsible'.[66] At the police court, municipal summons cases were considered minor offences or 'odd jobs' often delegated to be heard by the most junior magistrate and dismissed after a perfunctory hearing and the occasional fine.[67]

In most cases of filthy premises, illegal erection of cubicles, overcrowding and the non-removal of nightsoil, given the difficulty of distinguishing the guilty from the innocent, summons were frequently served on the principal tenants in the hope that they would find ways and means of exerting control over sub-tenants to abate nuisances. At best, however, this resulted in a perfunctory cleansing of premises, minor repairs and the temporary removal of partitions rather than more permanent sanitary improvements. Once municipal vigilance was averted, '[c]ubicles pulled down by order of the Sanitary Department [were] frequently put up again at the earliest opportunity, improvements effected to drains and latrines [were] not properly maintained and soon [in a state of] disrepair'.[68] Minor fines were unlikely to deter principal tenants from erecting cubicles and overcrowding premises as their gains from increased rents were more than adequate to offset occasional fines. Regulations against overcrowding were especially difficult to enforce as overcrowding occurred chiefly at night whilst municipal officers only had powers of entry under certain conditions between sunrise and sunset. Dr. W. R. C. Middleton, the municipal health officer, doubted the efficacy of introducing night surveillance and felt that it would only lead to more 'prosecutions, convictions, fines – and little if any improvement'.[69] In his view, prosecution as a means of preventing overcrowding was futile:

If an occupier were prosecuted often enough to instil into him an observance of the law it would only lead to those who were put out of the overcrowded houses seeking shelter elsewhere and adding to the numbers of inmates in houses that already contained their share, or sleeping on the five-foot-ways where, if they ran less risk of contracting Phthisis they would run more of contracting other diseases, e.g., Bronchitis and Pneumonia.[70]

Despite the proliferation of municipal laws to strengthen the system of surveillance, the Asiatic plebeian classes were capable of exploiting 'openings' in the system and thwarting the realization of sanitary objectives. To actively challenge the imposition of sanitary laws was often beyond their power and resources, but by alternating temporary observance with rampant disregard of the laws, and by executing non-compliance and non-cooperation in anonymity, they were sometimes able to escape the grasp of sanitary law.

The efficacy of sanitary reform depended not only on the stringency of sanitary law but also on the quality of its enforcers and those directly involved in working out its implications on a day-to-day basis. Given the meagreness of municipal pay, police constables, sanitary inspectors and peons were often open to the offer of bribes from those who were willing to pay a small sum to avoid trouble with the police courts.[71] Sanitary ideals at higher levels of municipal administration were hence seldom translated into equally visionary zeal at the level of enforcement. Instead, in practice, these ideals were often subject to pecuniary negotiation between enforcer and offender.

Evasion: a counter-strategy to municipal disease control

Another range of Asiatic counter-strategies – those of evasion and conceal-ment – were most clearly demonstrated in the attempts to evade municipal disease control. Dr Middleton warned that 'in dealing with cases of infec-tious disease, the sanitary staff [had] to be on the alert to all the dodges practised by the natives to conceal such cases'.[72] If a 'dangerous infectious disease' were traced to a particular dwelling, the resulting inconveniences included domiciliary visits by sanitary inspectors, isolation and disinfection of the dwelling and belongings, and the possible imposition of quarantine regulations, all of which meausres were perceived by the Asiatic population to be irksome interferences with their livelihood routines. Death from infectious diseases also entailed *post-mortem* examination which was highly disliked and seen as 'an intolerable interference with the religions and sentiments of different classes of the [Asiatic] [c]ommunity'.[73] The Asiatics were hence keenly conscious of the consequences of detection and avoided it at all cost. Strategies resorted to were highly imaginative as illustrated by the following catalogue:

[T]he existence of cases of infectious disease [were] carefully concealed, the patients surreptitiously removed in jinrikishas or gharries to the hospital or to sick-receiving houses. Or they [might] be taken in a moribund condition or after death and deposited on the street or any convenient piece of vacant ground from which they [were] removed to the hospital or cemetery by the Police. Every device is resorted to, to prevent the authorities from tracing the houses from which such cases were removed, such as changing jinrikishas two or three times between the house and the hospital, giving false addresses, or declaring that the patient had newly arrived in the town and had been picked up on the street or five-foot-way.[74]

Deaths from 'dangerous infectious diseases' were also frequently misreported as resulting from other innocuous causes. As the cause of death in a large number of cases (more than 60 per cent of deaths in the first decade of the twentieth century) was not medically certified but dependent on a perfunctory inspection of the corpse and enquiries from friends, it was inevitable that a significant proportion of misinformation evaded detection.[75]

The negotiation of power

In their daily conflicts with the municipal authorities over the sanitary condition of the environment, the Asiatic plebeian classes were hence neither powerless nor inert. Whilst they lacked formal power, they possessed a 'power' based on the strength of their own systems of health and medicine and in specific strategies of evading and inflecting municipal control. How effective were these forms of 'passive resistance' in protecting or advancing the interests of the Asiatics in the long run? Local counterstrategies were not capable of revolutionizing the colonial pecking order or introducing large-scale structural changes into society. However, these strategies might effect immediate, *de facto* gains for the plebeian classes, and occasionally, passive non-compliance on a large scale and over a substantial period might force the colonial authorities into revising their goals and strategies. Fourteen years after the municipal ordinance of 1896 came into effect, a commission set up to enquire into the efficacy of various aspects of municipal administration including sanitary regulation concluded that the system of sanitary surveillance had failed to improve health conditions in the city:

In a place like this Colony ... where there is so much hostility to sanitation, and so little belief in its utility on the part of the bulk of the population, it is not surprising to find that the effect of these sanitary measures is not reflected in the mortality returns, and that they have not given many visible beneficial results, but that on the other hand they have only created dissatisfaction and discontent.[76]

The Commission acknowledged the crucial role the Asiatic classes played in thwarting municipal sanitary measures:

All the good effects which may be expected from the measures taken by the Health Officer are, as a rule, counterbalanced by the action or inaction of the persons whom they are intended to benefit ... [I]t is of very little practical use to depend on the assistance of the portion of the public accustomed to insanitary surroundings, or to expect any help from them in the execution of sanitary reforms.[77]

Given the difficulty experienced in forcing compliance on an unwilling Asiatic public, the Commission concluded that 'larger scale' measures which were 'more automatic in their operation' were necessary if any improvement in sanitary conditions were to be realized.[78] It urged the Health Department to 'initiate . . . schemes outside its routine duties' even if these involved 'questions of finance and engineering' including large-scale anti-malarial work, the establishment of public dispensaries and putting into effect some of Professor Simpson's proposals for improving the health of the city.[79]

The commissioners' report marked a significant turning point in municipal goals and strategies. Whilst the system of surveillance was not abandoned, the municipal authority increasingly turned to other schemes and strategies to secure the health of the city. During the second decade of the twentieth century, anti-malarial work was inaugurated with a vote of twenty thousand dollars,[80] a modern system for the removal and treatment of sewage adopted,[81] a hospital for the isolation of infectious diseases opened[82] and the work of opening up back-lanes pursued with increased vigour. The 'power' of the Asiatic classes and their counter-strategies of evasion, concealment and non-compliance, were at least in indirect ways, partly responsible for forcing the hand of the municipal commissioners.

Power over the sanitary conditions of the urban environment was hence neither the prerogative of the municipal authority nor the people but had to be continually negotiated in the quotidian flow of activities. The relationship between the 'powerful' and the 'powerless' was one of dynamic adjustments and readjustments: the former's control of the central ground was seldom directly contested but concessions had to be made at the margins as the latter continually stretched the bounds of the possible and chiselled away at the corners of existing structures in what Scott describes as 'long-run campaigns of attrition'.[83]

NOTES

1 See G. Balandier, The colonial situation: a theoretical approach, in I. Wallerstein (ed.), *Social change: the colonial situation* (New York 1966) 34–61; R. J. Horvath, In search of a theory of urbanization: notes on the colonial city *East Lakes Geographer* 5 (1969) 69–82; D. Simon, Third World colonial cities in context: conceptual and theoretical approaches with particular reference to Africa *Progress in Human Geography* 8 (1984) 493–514; A. D. King, Colonial cities: global pivots of change, in R. Ross and G. J. Telkamp (eds.), *Colonial cities: essays on urbanism in a colonial context* (Dordrecht 1985), 7–32

2 Simon, Third World colonial cities, 499

3 *Ibid.*, 506–8

4 T. G. McGee, *The southeast Asian city: a social geography of the primate cities of southeast Asia* (London 1967) 72

5 A. Giddens, *The nation-state and violence: volume two of a contemporary critique of historical materialism* (Cambridge 1987) 7

6 *Ibid.*, 9

7 *Ibid.*, 11

8 Giddens defines this as the 'strategies of control within defined contextual boundaries' in A. Giddens, *The constitution of society: outline of the theory of structuration* (Cambridge 1986) 288

9 In the context of peasant resistance in Kedah, West Malaysia, J. C. Scott calls this form of passive protest the ordinary 'weapons of the weak'. They include tactics ranging from foot-dragging, dissimulation, false compliance, pilfering, feigned ignorance, slander, arson, sabotage to concealed, publicly unacknowledged strikes and combined action amongst the poor. See James C. Scott, *Weapons of the weak: everyday forms of peasant resistance* (New Haven, Conn. 1985)

10 F. B. Smith, *The people's health 1830–1910* (London 1979) 195–6

11 C. Gordon (ed.), *Michel Foucault: power/knowledge, selected interviews and other writings, 1972–1977* (Brighton 1980) 150–1

12 Quoted in G. Harrison, *Mosquitoes, malaria and man* (London 1978) 138

13 Municipal government in the Straits Settlements *The Colonial Office Journal* 4 (1911) 220

14 Between 1896 and 1910, tuberculosis/phthisis averaged 1,313.8 deaths per annum, beriberi, 1,137.3 and remittent fever, 934.2. For every death from a 'dangerous infectious disease', there were over 14 deaths from fevers, phthisis and beriberi (W. R. C. Middleton, The working of the births and deaths registration ordinance *Malaya Medical Journal* 9 (1911) 47–50)

15 Most of the diseases in this group were considered 'dangerous infectious diseases' subject to compulsory notification under the municipal ordinance of 1896

16 Administrative report of the Singapore Municipality (hereafter ARSM) 1895, Health Officer's Department (hereafter HOD) 44

17 Experimental proof of the association of beriberi with the continuous consumption of polished, decorticated white rice as a staple article of diet was established by H. Fraser and A. T. Stanton, both researchers at the Institute of Medical Research in Kuala Lumpur. See H. Fraser and A. T. Stanton, *An inquiry into the etiology of beriberi* (Kuala Lumpur 1909); The etiology of beriberi *The Lancet* 17 December 1910

18 Public Records Office: CO 273/103, 1 June 1880 Report of the Committee on the cause of the outbreak of the disease known as 'beri-beri' in the Criminal Prison, Singapore paras. 3–4; similar views were expressed in M. Simon, Some remarks on the nature and causes of so-called 'beri-beri' or peripheral neuritis, in the tropics, and on its place in the 'nomenclature of disease' *Journal of the Straits Medical Association* 3 (1891–2) 59; A. Bentley, *Beri-beri: its etiology, symptoms, treatment and pathology* (Edinburgh 1893) 23–5

19 Simon, Nature and causes of 'beri-beri', 57

20 ARSM 1896, Report on a special investigation into the system of death registration (hereafter SISDR) 12

21 Reported cases represented only a small fragment of the actual number of cases given the widespread practice of concealment and the underreporting of infectious disease

22 Districts A, B and C were traditionally Chinese-dominated areas, with the Chinese comparising 87.1 per cent, 89.6 per cent and 96.4 per cent of the district population respectively, according to the 1901 census. District D, the traditional domain of the Malay sultan, had by the turn of the century been infiltrated by increasing numbers of Chinese, overspilling from the densely packed traditional Chinese quarter. District E, the harbour area was considered the 'gateway to disease' as many infectious diseases were imported from ships calling at the port of Singapore

23 The statistics are compiled from ARSM, Health Officer's Report (hereafter HOR), 1985–1925

24 This was comparatively high, bearing in mind that the Chinese accounted for about 75 per cent of the total municipal population during this period

25 SISDR, 9–10

26 At the turn of the century, wells were found in the great majority of houses and compounds in Singapore, most of which contained contaminated water (SISDR, 6–7). A survey of existing wells within municipal limits conducted in 1902 returned 3,877 wells of which 85 per cent were considered 'obviously bad' and unfit for domestic use (ARSM, HOD, 154)

27 SISDR, 7

28 ARSM 1893, Report on and estimates for, the disposal of nightsoil at Singapore and for the improvement of the surface drainage, 61

29 *Ibid.*

30 *Ibid.*

31 *Ibid.*

32 W. J. Simpson, *Report on the sanitary conditions of Singapore* (London 1907) 11

33 Thomas Stamford Raffles, the founder of the British factory and settlement of Singapore, issued a set of instructions for the systematic laying out of the town of Singapore in 1822 (reproduced in C. B. Buckley, *An anecdotal history of old times in Singapore* (Singapore 1984) 81–6)

34 Simpson, *Report on sanitary conditions,* 11

35 *Ibid.,* 12

36 *Ibid.,* 28

37 *Ibid.,* 14

38 *Ibid.,* 16

39 *Ibid.,* 25–7

40 ARSM 1905, Appendix L Correspondence with government on the subject of high death rate in Singapore (hereafter CGHDR) 107

41 ARSM 1896, 17

42 K. Hintze, Sanitäre Verhältnisse und Einrichtungen in den Straits Settlements and Federated Malay States (Hinterindien), *Archiv für Schiffs- und Tropen Hygiene* 10 (1906) 525

43 E. Chadwick, quoted in P. Stallybrass and A. White, *The politics and poetics of transgression* (London 1986) 126

44 Gordon, *Michel Foucault,* 78–108

45 ARSM 1894, Appendix J Contribution to the cost of the police force (hereafter CCPF) 68

46 C. G. Garrad, *The Acts and Ordinances of the Legislative Council of the Straits*

Settlements from the 1st April 1867 to 7th March 1898 vol. II (London 1898) 1556 and 1559; CCPF, 66–8

47 CCPF, 68

48 CCPF, 68

49 ARSM 1896, 11

50 The Institution later moved to Wayang Street in the 1890s. Both locations were within the traditional core area of Chinese settlement (Su xiao xian, Tong ji yi yuan yan ge shi lue (A brief history of the Thong Chai Medical Institution), in *Tong ji yi yuan da sha luo cheng ji nian te kan* (Souvenir magazine of the opening ceremony of the newly completed Thong Chai Medical Institution Building) (Singapore 1979) 117)

51 Li bai gai, Yuan zhen ji zu (A brief description of diagnosis at the Thong Chai Medical Institution), in *Tong ji yi yuan da sha luo cheng ji nian te kan* 155

52 Li song, Xin jia bo zhong yi yao de fa zhan, 1349–1983 (The development of Chinese medicine in Singapore, 1349–983) *Singapore Journal of Traditional Chinese Medicine* 10 (1983) 49

53 *Directory of associations in Singapore 1982–1983* (Singapore 1983) 36; *Brief account of the Kwong Wai Shiu Hospital* (notes by the secretary of the Hospital, Singapore 1988)

54 List compiled by Mr Chen Chin Ah of the Thong Chai Institute of Chinese Research, Singapore, from *Xin jia bo cha yang hui guan bai nian ji nian kan* (Souvenir magazine of the centenary celebration of the Char Yong Association, Singapore) (Singapore 1958)

55 J. A. Jewell, Theoretical basis of Chinese traditional medicine in S. M. Hillier and J. A. Jewell, *Health care and traditional medicine in China 1800–1892* (London 1983) 232–4

56 Li bai gai, Yuan zhen ji su, 155–6

57 Jewell, Chinese traditional medicine, 234

58 The Western view of beriberi has been discussed earlier

59 R. MacLean Gibson, Beri-Beri in Hong Kong *The Journal of Tropical Medicine* 4 (1901) 98

60 Chinese gardeners and disease *The Malayan Medical Journal and Estate Sanitation* 1 (1926) 33

61 Phrase used by the municipal president, Alex Gentle (ARSM 1895, 15)

62 ARSM 1895, 15

63 ARSM 1895, Engineer's Department, 23; ARSM 1896, Engineer's Department, 28

64 ARSM 1895, 16

65 *Ibid.*

66 Municipal summons note book (hereafter MSNB) commencing 5 July 1912, case 5821 (6 August 1912)

67 The experiences of young cadets appointed as supernumerary magistrates to deal with municipal summons cases in the early twentieth century are recounted in E. W. F. Gilman, *Personal recollections, speeches and newscuttings, c. 1920 to 1931 while a civil servant in Malaya* (Rhodes House, Oxford (hereafter RH): Mss.Ind.Ocn.s.127) and F. K. Wilson, *Letters home, Jan. 1915 to Dec. 1916* (RH: Mss.Ind.Ocn.s.162)

68 Proceedings of the Legislative Council of the Straits Settlements 1910, Report of the Municipal Enquiry Commission (hereafter MEC), C200

69 CGHDR, 118

70 ARSM 1909, HOR, 38

71 See for examples, MSNB commencing 5 July 1912, cases 6578 (2 October 1912) and 9166 (13 December 1912). The widespread incidence of bribery, extortion and corruption within the police force is also well attested to by the evidence of senior police officers and anonymous petitions from the public accusing specific officers of corrupt practices (PRO: CO 273/259 13 December 1900 Swettenham to Chamberlain)

72 W. R. C. Middleton, The sanitation of Singapore *Journal of State Medicine* 8 (1900) 702

73 Middleton, Births and deaths registration ordinance, 50

74 SISDR, 10

75 An investigation into the accuracy of death returns in 1896, for example, revealed that no less than 87 cases of infectious disease (cholera and enteric fever) were within six weeks returned under other names (SISDR, 3)

76 MEC, C199

77 MEC, C200

78 MEC, C200

79 MEC, C200–2

80 ARSM 1911, HOR, 38

81 ARSM 1911, HOR, 42

82 ARSM 1912, HOR, 6

83 Scott, *Weapons of the weak*, 298

9

Ideology and the landscape of British Palestine, 1918–1929

GIDEON BIGER

Imperialism – the imposition of rule by one people on other peoples and areas – is an ancient phenomenon. World history is a continuous list of such conquests. Until about forty years ago, almost 40 per cent of the world's land area was under imperial rule and over half of today's independent states were effectively founded by external imperial forces.

Imperial activity in its various forms was one of the principal factors in the shaping of the geographic landscape, in the territory controlled by, and even sometimes in the home territory of, the controller. The connections between the imperial territory and the centre, the mutual interactions between the ruling population and the ruled, the economic, social and political contacts that were part and parcel of imperial rule were all influential in the creation of new human landscapes alongside, or in place of, those that existed prior to the process of outside domination. This resulted, in some cases, in the total changing of areas beyond recognition. The geography of such an area often differed greatly from that of an adjacent independent territory, or from one that was under the control of another imperial power.

Despite the influence of imperialism on landscape formation, geographic research has largely ignored this factor. Against the myriad of historical, economic, social and political studies dealing with the rise, maintenance, functioning and decline of the various empires, there has been a paucity of both empirical and theoretical geographic research. Most of the research written in English up until the early 1950s dealt with the description and analysis of geographical phenomena in a spatial context in geographic regions under colonial control,[1] but it did not deal with the connection between the authority, with its world-views and methods of rule, and the landscape which it created. Later, there were attempts at geographical classification based on different types of imperialism.[2] Most of the studies in this field however, deal with the process of creation and shaping of the empire and its mode of operation, rather than with its influence on the land and its role in landscape formation.[3]

In 1951, Harrison-Church called for geographic research on colonies.[4] He claimed that whereas the British empire was large, there was very little activity by British geographers in studying the geography of the colonies, especially when compared with their colleagues in France and the United States who were working on their topic on a systematic basis.

His call was a cry in the wilderness. The process of decolonization, which shaped the political geography of the world in the last thirty years and which was then at its peak, did not allow for new geographical research into now former colonies. The vanishing imperial societies themselves, because of guilt feelings concerning the negative aspects of colonial rule, and also the geographers in the new states, turned a blind eye – as far as possible – to any constructive aspects of the imperial power in the formation of the human landscape.

The reawakening of interest by geographers in the landscapes of empire came from two directions. On the one hand, there was the development of literature in political geography with the objective of investigating imperialism,[5] mainly from a Marxist perspective.[6] On the other hand, research in historical geography dealt with the analysis and study of colonial landscapes formed in the past.[7] The discussion of the topic in political geography declined rapidly and disappeared, while research in historical geography has recently found theoretical expression in Meinig's analytic approach.[8] Meinig calls on historical geographers to conduct research into the question of how regions change in the context of the domination of one people over other peoples, throughout history.

Such research can be conducted both in a specific manner (the study of a dominated area in all its aspects) or in a more thematic fashion (by studying a single element in many areas under domination). The question, 'how do areas under imperial rule differ from their situation prior to such rule?' can be answered by a series of cross-sections. In contrast, it is possible to raise another question: 'by which means did a region change as a result of its conquest by an imperial power?' In this case, the study of imperialism will consist of a set of processes, each with geographic implications. Research of this nature in Kenya[9] revealed the role of the British regime in the creation of that area's segregated landscape.

Thus historical geographical studies dealing with an area once under imperial rule must discuss the spatial influences that such rule has on the dominated territory and its inhabitants.[10] Even though the basic political and economic forces of each imperial rule might be the same, local features have occasionally resulted in differences in specific places. A good understanding of the connections between the political and spatial processes was achieved by Hugh Prince, when he stated that 'knowledge of the political and religious institutions mediating change are the first steps towards understanding such processes as colonisation, land clearing, urbanisation, trade, investment, education, sanitation, etc.'[11]

Imperialism and development: the case of Britain

Palestine, which was ruled by Britain from 1917 to 1948, was the last of the many countries acquired and administered by Britain. Over centuries, and particularly during the second half of the nineteenth century, Britain administered approximately one-quarter of the world's land areas, encompassing 36 million square kilometres, as a series of dominions, Crown colonies, dependencies and administrative territories. The extent of the involvement of the British government and its representatives in the dominated areas differed from region to region, according to the differences in the nature of the rule and the aims of the administration, the characteristics of the local population, and the characteristics of the territory itself. Towards the beginning of the twentieth century, the British goverment and its representatives adopted a general set of rules regarding the form and character of the operation of authorities in the territories under their control. Although these principles were never implemented as rules dictating operations, there was in effect a scheme of authority acceptable to all those involved in imperial administration. At the height of imperial rule in the late nineteenth century, Lord Cromer – a modern imperialist and British representative in Egypt – declared that 'whatever the country – whether India, Burma or Egypt – the recipe must always be the same: attention to the laws of sound finance, and in particular to the importance of low taxation, efficient fiscal administration, careful expenditure on remunerative public works, and a minimum interference in the internal and external traffic of goods'.[12]

This approach, which dealt primarily with those protectorates in which there was a certain level of local authority, was called 'the 'Cromerian System'. It was acceptable to the group of new colonial leaders that crystallized at the end of the nineteenth and the beginning of the twentieth centuries, around the personality and viewpoint of the energetic Colonial Secretary, Joseph Chamberlain. This group included Lord Curzon (Viceroy of India and, in effect, Governor of India), Lord Milner (Governor of South Africa and later of Egypt), Cromer and others. At the end of the First World War, those men reached positions of high authority in the British government; at the war's end, Curzon was British Foreign Secretary and Milner was Colonial Secretary.

The process of colonial development differed from place to place, both because of the specific local conditions and because of the difference in the approach adopted by the authorities and as a function of the character of the population being governed. At the same time, throughout the empire as a whole, there was the influence of general development ideas, ideas developed by Chamberlain in particular, with regard to underdeveloped areas, paralleling the 'Cromerian System' in the indirectly controlled territories. It was possible therefore to speak of 'The lines taken while Britain got hold of an Oriental area. Efficient public health services, opening of primary schools, the maintenance of law and order, improving roads, railroads, ports, telegraphs and telephones for better communications, draining the marshes and planting the trees.'[13]

The desire to develop the colonial territories found acceptance amongst the majority of the British public during this period. It was not possible, however, to convince the British government that it was necessary to use the revenue of the British taxpayer to implement these policies. There was strong opposition to the use of government funds for this purpose. This was partly the result of the long-standing British viewpoint with regard to colonial politics, namely that, since the colonies themselves also benefited from the colonial administration, they should be entirely self-supporting. Internal means should be found to support local development, while financial support from the British government should be used only for security costs and the maintenance of imperial soldiers. This approach, different to that of French colonial policy, led to a situation in which development carried out by the British government between 1920 and 1929 involved little financial outlay. Instead of government aid for the construction of railways and ports, construction was provided by means of a loan, while during the same period the British government spent only one million pounds on health and agricultural research throughout the empire as a whole.[14]

Thus it was natural that a rich country with natural wealth, local resources and efficient manpower was more likely to undergo development than a poor territory. Under this *laissez-faire* attitude, all government activities in the colonies came almost entirely from local revenues, through loans from Britain (to be repaid), or by financial investments from private individuals or companies in Britain or elsewhere. In principle, this was the way things were run in Britain itself, and Sir Herbert Samuel (as Palestine's first High Commissioner) 'did his utmost to create a similar system in Palestine, and beyond this no more could be expected'. Taking into consideration that the tax burden in Britain was far higher than that in the colonies, it follows that the level of services that could be provided by the authorities was necessarily more restricted. This chapter examines the operation of these processes in Palestine, a case-study reflecting the latter years of British imperial rule.

Britain in Palestine

Palestine was conquered by the British army during the First World War, resulting in British control of the region until May 1948. A civilian administration was established in July 1920. The new authorities had to choose a policy of administration which would express the aspirations of Britain herself and would enable implementation to take place along these lines. In part, the problems faced by the British were of a general nature, such as the financial and administrative involvement in the development process and in defining the role to be played by the local inhabitants. But Palestine was not just another British colony. Palestine had been granted to Britain by the League of Nations as a mandate territory. On receiving the mandate, Britain entered a unique colonial situation. Palestine now officially became a country of Jewish immigration from throughout the world. This immigration was not composed of British citizens (that is, it did not involve the establishment of colonies by the ruling power). This meant that the activities of the Palestine government in aiding the immigrants and in attempting to control their development could not be aided by the British government itself. The British authorities were faced with unique problems, not the least of which were defining the objectives within the country, the methods by which the principles of the mandate concerning the establishment of a Jewish homeland could be applied, the treatment of the non-Jewish population and the overall problems concerning the funding of development. These development ideas were mainly administrative, intertwined with economic, social and political motives.

Initially the British activities in the region were typical of the rapid 'take-off' period which characterizes all new administrations. This process depended on the help of the local Palestinian population as well as aid from the wider Jewish world. During the first ten years of its existence, the British administration tried to co-operate with the Zionist movement and the Arab population in an attempt to build a modern state in Palestine. This attitude was brought to a halt following the bloodletting that took place between Arab and Jew in the summer of 1929. It became clear to the British authorities that their policy of attempting to create a political a synthesis between the local Jewish and Arab populations had failed, and that it would not be able to realize its objectives of creating a unified single society in Palestine. British government policy then switched from one intent on creating a unified society to one of imperial representation in a colony, similar in nature to the British administrations in other colonies. This chapter deals only with the first decade of British rule in Palestine, namely that period during which the social and economic development of the country stood at the centre of its activities.

The implementation of British policies and the development of Palestine

Since this chapter deals with the activities of the British imperial power in Palestine and not with the overall changes undergone by the landscape during this period, the following section will concentrate on the specific implementation of British policies while ignoring the roles played by the resident Arab and Jewish populations. In doing so, the chapter highlights some aspects of British activities which influenced Palestine, namely the delimitation of boundaries for Palestine, the development of the communications infrastructure throughout the region, afforestation, provision of agricultural services and other governmental activities.

Boundaries

During the four hundred years of Ottoman rule, Palestine did not constitute a separate administrative unit. Before the British occupation of Palestine, the southern part of the territory was under direct rule from Istanbul (Constantinople) while the extreme south and the area to the east of the River Jordan were part of the Damascus province. Northern Palestine was part of the Beirut province. Palestine – the 'Holy Land', that portion of the earth 'known' to every scholar – had no valid areal definition at that time.[15] The First World War coincided with the rise of national movements in the Middle East, and both Jews and Arabs had their ideas as to the future designation of the area. The French government was influential in the creation of the new Palestine, while the main agent for change – namely the British government – had its own views as to the boundaries to be defined. These views were aired in Britain herself, in Paris (where the peace conference marking the end of the First World War took place), in Egypt and in Palestine. Each party's claim was backed by geographical, strategic, historical and political arguments.

In the southern area (Fig. 9.1) the agricultural potential of northern Sinai and the northern part of the Negev influenced the Zionist organization to lay claim to the area around El Arish. The British authorities in Egypt claimed this same area for similar reasons, while the British delegation in Paris backed the Zionist demands. Other Zionist territorial aspirations included a gateway to the Red Sea and control over the natural resources of the Dead Sea. A major topic discussed by the British delegation in Paris and their representatives in Egypt was the fate of the Bedouin tribes in the area. Any new boundary line would have to cut across the Bedouin migratory grounds and thus damage their entire way of life.[16]

Geographical claims for the eastern border were based on the fact that the area east of the River Jordan had long been the granary of Western Palestine. This constituted the main Zionist argument for controlling this

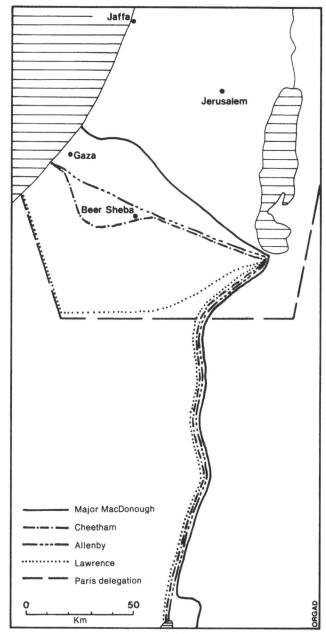

Jaffa

Jerusalem

Gaza

Beer Sheba

Major MacDonough

Cheetham

Allenby

Lawrence

Paris delegation

0   50

Km

ORGAD

Fig. 9.1   Proposals for the southern border of Palestine. *Source:* MR/761 in Map Library of PRO.

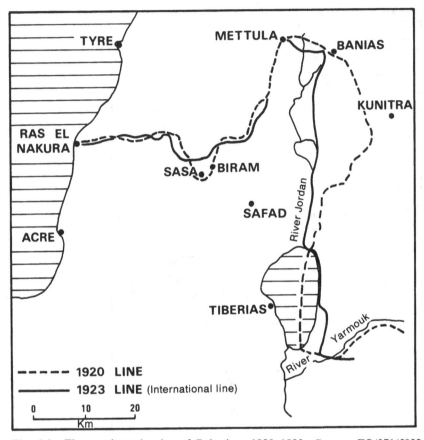

Fig. 9.2   The northern border of Palestine, 1920–1923. *Source:* FO/371/5032 document 1375/22/44 in PRO.

area and was backed by Lord Balfour (the Foreign Secretary) and other prominent British politicians. The Zionists also argued that the existence of a Christian population consisting of friendly tribes, as well as some Jewish settlement, all east of the river were important factors promoting their claim, as was the area's potential for the construction of a hydroelectric power station.[17]

The most difficult problems arose in the north (Fig. 9.2) partly because the line had to be drawn across an inhabited area (the eastern and southern boundries mostly crossed desert areas) and partly because this would also constitute the dividing line between the British and French areas of influence and control in the Middle East. The Zionists and the British wanted water for irrigation and hydroelectric power. Thus they attempted to include the Jordan basin and the southern part of the River Litani

(Kassamiya) within Palestine. The French authority, which controlled Syria and Lebanon, claimed roads that run between the Hauran (the Golan Heights) to the east and the sea to the west. Britain demanded control over the Yarmouk Valley in order to construct a proposed railway from Iraq and the Persian gulf to the Mediterranean Sea. Other geographical problems involved the agricultural lands of the villages in the frontier area and the future of the various ethnic minorities' (Jewish, Shi'ite, Sunni) settlements.[18]

In addition to these geographical claims, there were a number of strategic, historical and political arguments that were employed. Before it was known that Palestine would become part of the area under British control, the British army had wanted to construct a strategic line between Egypt and Palestine. Later on, the army requested a defensive line in the east, against possible attack from France or Turkey. These strategic claims were rejected by the British government because, as Winston Churchill – then Colonial Secretary – stated: 'In Palestine, the political and economic arguments are more important than the strategic ones.'

Another important argument was the historical one, originating, rather surprisingly, in Britain and not from within the Zionist camp. The British Prime Minister, Lloyd George, invoked the Biblical formula 'From Dan to Beer Sheba'. The French authorities accepted it as the limit, but it proved difficult to establish the precise location of ancient Dan. George Adam Smith's historical geography of the Holy Land was used as evidence for this point, as well as in support of the claim that the River Jordan had never been used as a boundary between the two parts of Palestine.

Political arguments were raised by the French, demanding the retention of the Sykee–Picot line, this having been agreed to during the war.[19] The Arabs claimed the whole of Palestine as an Arab state, referring to correspondence with the Sheriff Housain in 1915.[20] Later, Trans-Jordan claimed the southern part of Palestine, this having been ruled from Damascus until the British occupation. Because of these many conflicting claims, it took ten years to complete the process of boundary delimitation in Palestine. British Palestine (Fig. 9.3), as it eventually took form, did not correspond with any former territorial unit, mainly because of the need to achieve equitable solutions based on compromise.

In the south the conflict was between Zionists (Jews and others) and the British authorities in Egypt. A decision was finally taken to adopt the pre-war administrative line between Egypt and Turkey, a compromise that had been agreed to in 1906. Palestine received a gateway to the Red Sea and the entire area south of Beer Sheba, while Egypt received the Sinai Peninsula.

The eastern border reflected the physical geography of the region, following the River Jordan and the lowest level of the Arava depression, but it did not constitute a real boundary between two states. Under British rule

Fig. 9.3   British Palestine. *Source:* Line map 48840(61) in map collection in the British Library.

up to 1946, Palestine and Trans-Jordan were both part of the same mandate, having the same High Commissioner. There were no obstacles placed upon the free movement of people and goods, and the boundaries did not cause any disruptive influence in the future development of the whole of the Palestine territory – both sides of the Jordan. The territory to the east of the Jordan river was eventually excluded from those areas in which Britain undertook the obligation to help create a Jewish national homeland, despite it having originally been part of Palestine.

In the north, Britain competed with France in an attempt to acquire a more economically sound territory for Palestine. This goal was not achieved, although the British did succeed in providing Palestine with a territory which contained most of her needs for her future development. The northern line, unlike the eastern and southern boundaries, was a real international boundary (Fig. 9.3).

This process of boundary making resulted in the creation of Palestine as a territory with a political identity, the first step towards an eventual independent state. From then on, the geographical development of Palestine and the various processes affecting landscape change all took place within a well-defined territorial unit.

Communications

Before the war, Palestine was characterized by the poor condition of all of its lines of communication. There were two separate railway systems – a 100 centimetre gauge railway from Jaffa to Jerusalem, and a 105 centimetre gauge railway from Haifa to Dera'a, via Tzemach, on the Hijaz line. Motor cars were not in use and there were only poor roads built of macadam without any asphalt covering. There was no telephone system, no uniform postal service and only a poor telegraphic system. The ports mainly consisted of small piers in sheltered positions.

The war brought some changes in its wake, mostly concerning the railway network. The Turkish army connected the two rail systems by means of a 105-centimetre gauge line from Lod to Afula, while destroying the Lod to Jaffa line. The line was also extended further south towards Gaza and Beer Sheba. The roads were improved for military use and made suitable for cars in dry weather, but the heavy rains of the 1917/18 winter destroyed most of them. The main road was the hilly, longitudinal, axis from Beer Sheba in the south to Nazareth in the north, linking the largest inland settlements. There were also two lateral routes, running roughly parallel to the railway: the road from Jaffa to Jerusalem and Jericho, and the road from Jaffa to Tiberias via Nazareth.[21]

As the British army advanced northwards from Egypt into Palestine, a standard gauge railway was built from Kantara on the Suez Canal to Haifa in the north, linking up Rafiah, Gaza and Lod. Although this line was primarily a military one, it did not reach Haifa until two months after the end of the war. The Jerusalem line was widened to a standard gauge and connected to the port at Jaffa. Another new line was constructed between Rafiah and Beer Sheba in order to help the Bedouin tribes in this region. As a result of this activity, both for military purposes and for the benefit of the civilian population, the railway networks were united. Nevertheless, two different gauges continued to exist: a standard line from Egypt to Haifa and the old Hijaz line of 105 centimetres between Haifa and Damascus. Jerusalem, Haifa, Beer Sheba and Jaffa became points on the international railway network, connecting Egypt with Turkey and Europe, via Syria. The British civil administration placed great importance on the maintenance and improvement of this system. This was partly because during this period, Palestine was served by railways rather than roads and also because the railway was a government controlled network which offered investors the prospects of future profits. Concentration on the railway network was a major characteristic of the first ten years of British rule in Palestine. Nearly three million pounds raised in loans were invested in the improvement of the Rafiah–Haifa and the Jaffa–Jerusalem lines (Fig. 9.4). A new line was constructed to Petah Tiqva, partly financed by Jewish investment. Branch lines were laid to quarries and factories, main lines were improved, unprofitable lines were closed, stations were added, new locomotives were purchased and a central shunting yard and workshop were constructed in Haifa. By 1929, Palestine had acquired a railway system that was one of the best in the Middle East.

In contrast to the railway system which was mainly built for military purposes, the road network was constructed mostly for the benefit of the civilian population. The British Palestinian government had to fight the Colonial Office over its road construction plans. The Colonial Office, which had the power to veto any development scheme, as it had to authorize the expenditure, viewed the construction of new roads as a major threat to the government railway system. Government revenue from the railway would be reduced if money was spent on roads because of competition. The Colonial Office requested that this money should be spent to cover the railway's debts or alternatively that Palestine itself should contribute a larger proportion of the cost of maintaining the British garrison in the country. In Palestine, the railway department insisted on institutional protection and rejected rapid development of the road network. A number of roads were constructed in the north for reasons of strategy and accessibility during the early years of British control. The first asphalt surface – on the Jaffa–Ramla road – was laid only in 1925. Even then, the Colonial Office was informed that the work had been done for security reasons.

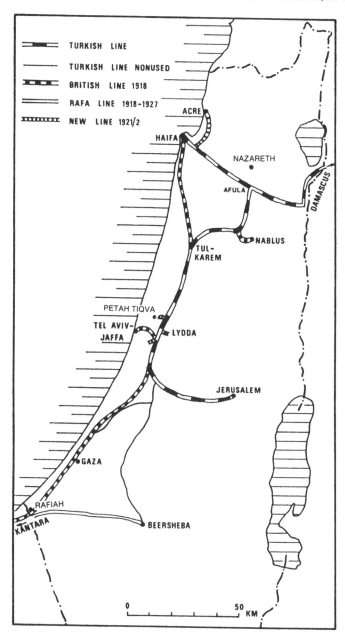

TURKISH LINE
TURKISH LINE NONUSED
BRITISH LINE 1918
RAFA LINE 1918-1927
NEW LINE 1921/2

ACRE
HAIFA
NAZARETH
AFULA
DAMASCUS
NABLUS
TUL-KAREM
PETAH TIQVA
TEL AVIV-JAFFA
LYDDA
JERUSALEM
GAZA
RAFIAH
KANTARA
BEERSHEBA

0   50 KM

Fig. 9.4   The development of the railway network in Palestine, 1917–1929. *Source:* Railway Department Reports, 1920–9.

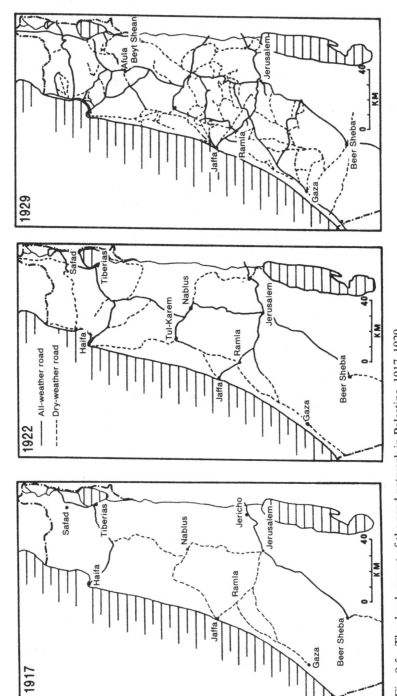

Fig. 9.5   The development of the road network in Palestine, 1917–1929.
Source: Yearly Reports of Public Works Department of Palestine.

Eventually, though, a network of good roads was constructed by the British administration in Palestine (Fig. 9.5). By 1929, nearly 900 kilometres of all-weather roads had been constructed, 150 kilometres of them laid with asphalt. Together with hundreds of small roads constructed by the local population with the help of government advisers, the road system now connected most of the towns and large villages of Palestine. The region was now connected by good roads to Lebanon, Syria and Trans-Jordan, although the connection to Egypt continued to be by rail alone. The major internal arteries were the same as before but were now 'all-weather' roads, suitable for use by the thousand private cars registered in Palestine in 1929.

Another significant transport development constructed by the civilian government was the deep-water harbour at Haifa. The decision to locate a new harbour at Haifa was made in 1923, although the necessary loans needed to finance its construction were only received in 1927, and only minor work was undertaken until 1929. However, the spatial impact of the decision was already being felt. The pattern of the location of industry and trade in Palestine began to change. Whereas before the war, most of the country's economic activity was located around Jaffa and Jerusalem, this now began to move towards the Haifa area. For Palestine, there were some disadvantages in the location of the harbour at Haifa, not least its distant location with respect to the main urban areas of Jaffa, Tel Aviv and Jerusalem, and its remoteness from the sources of the country's major exports – citrus fruits from the coastal plain and, after 1930, potash from the Dead Sea. The coastal plain, which was partially cultivated during this period, was served by the old port at Jaffa. The British decision to construct the new harbour was based on three important advantages of this area, although only one of these – namely, the good natural site – was of advantage to Palestine. For the British, the location of Haifa enabled good access to the inland regions of the Middle East under British control (Trans-Jordan and Iraq) and it also provided a maritime outlet for the future oil pipeline from northern Iraq to the Mediterranean Sea.[22]

The British administration also developed a telecommunications system (Fig. 9.6). In 1929, the telephone network linked 3,977 private telephones alone (4 per 1,000 inhabitants),[23] and it was also connected to Egypt. A good telegraphic system connected all the towns and large villages in Palestine with the rest of the world, and the government operated the best postal service in the Middle East (letters were delivered daily in Jerusalem),[24] post being transported by boat, train, cars and horses, as well as, from 1927, by air. Travelling times were reduced – instead of three days to reach Nazareth from Beer Sheba before the war, it took only six hours in 1925.

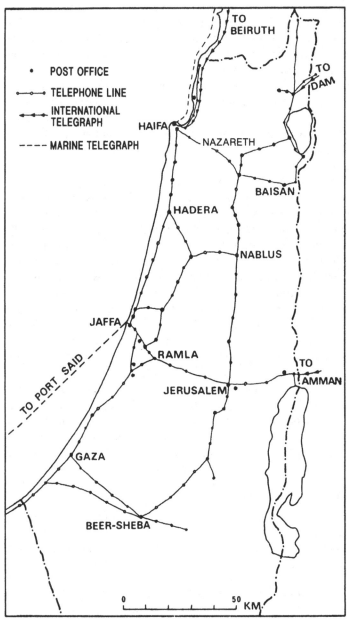

Fig. 9.6 Telecommunications in Palestine, 1929. *Source:* The Blue Book of Palestine 1929.

Afforestation

Palestine, which in the past had been full of forests, was left with few trees by the time of the British occupation, because of the long history of cutting them down for domestic needs and because of over-grazing in young forests. The war cut off Palestine from all the sources of coal and oil supplies and the Turkish army cut down all the trees within walking distance of all railway tracks. Immediately following the complete conquest of Palestine, the British military administration issued an order forbidding the cutting down of carob and oak trees, a ban on the uprooting of these trees' roots and the imposition of fines and imprisonment of offenders,[25] and the military administration also began activities to reafforest the country. Between 1918 to 1920, more than two million trees were planted and afforestation nurseries were established throughout the country.

The establishment of the civilian administration gave a boost to afforestation activities. The civilian administration viewed the beautification of the country and the rehabilitation of land as playing an important role in the country's development. A 'Forestry regulation' was issued, including the designation of various areas as forest regions under government control. By 1927, nearly 150 forest regions covering approximately 61,000 *dunam*, had been established. The afforestation activities were enlarged, three more nurseries being established and millions of trees planted every year. Trees were planted alongside the route leading up to Jerusalem and in a joint campaign with the railway administration, trees were planted at all the railway stations. The Jewish tree festival in the month of Shevat (January/February) became a governmental festival throughout the country. By 1929, some ten million trees had been planted by the British administration. It also constructed afforestation experiments in desert areas and in damaged forests.[26]

Even though damage continued to be caused to the forests by shepherds, wood hewers and charcoal makers, Palestine's mountains and sand dunes slowly began to receive a greener tone.

Drainage and drying of swamps

Government attempts to dry Palestine's swamps were undertaken in an attempt to eradicate the malaria plague as well as for the purpose of creating new fertile agricultural land in places where the swamps prevented cultivation from taking place.

The administration prepared plans for the drying of the Kishon marshes near the city of Haifa and initial planning for the drying of the Sanur marshes near Jenin.[27] The Beit Shean project, in which government lands were transferred to local residents, was prepared for cultivation by draining the region. The town of Beit Shean was completely purified from malaria. Swamps to the north of Acre were prepared for cultivation. The marshes of Jericho were dried and a large land area was introduced to the work cycle. Between 1925 and 1927, about £240,000 was spent on the eradication of the malaria in Haifa, draining and drying the marshes of the Kishon and turning them into productive agricultural land.

Agricultural activities

Palestine was not a colony in the sense that no British settlers came to live in the country. Government activities were therefore mainly concerned with assisting the local farmers. A Department of Agriculture and Fisheries was established. 'It appointed a number of technical officials, placed agricultural advisers in all the districts and ordered them to continually investigate the villages. At the outbreak of a disease and a plague of insects, they will immediately notify the occurrence and take measures to prevent their spreading.' It began a campaign to introduce a modernization and efficiency into the agricultural economy. Explanatory pamphlets in three languages were printed and distributed. Agricultural experts passed through the villages giving advice on everything concerned with the various crops; their growth and the possibilities of further improvements. The department organized an agricultural museum and agricultural exhibitions in rural centres. An experimental station was established at Acre. A veterinary service for the inoculation of cattle was also set up. These activities all helped to change the agricultural landscape of Palestine.

Other geographical activities

It is also possible to study in detail the many other activities which contributed to the development and changing geography of Palestine. However, a brief summary must suffice in order to demonstrate the comprehensive nature of development.

The indirect help given by the British administration to the rural and urban sectors was a further agent of geographical change. The security measures undertaken by the British forces led to the development of new Jewish as well as Arab settlements throughout the country. Existing settlements were enlarged. Measures aimed at assisting Jewish and Arab rural areas included technical help aimed at improving agricultural produce, revision of the tax system, low-interest loans for new development projects and the introduction of new agricultural implements and produce (honey and poultry). The health department was instrumental in cleaning the villages and in helping reduce the death rate, particularly that of small children; the 'anti-malaria' battle was won, eye disease was reduced and the

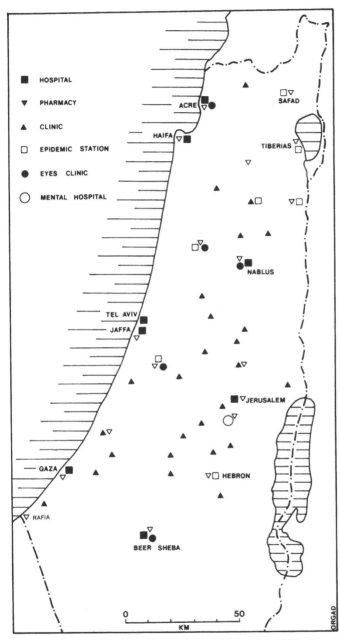

Fig. 9.7   Government health provision in Palestine, 1929. *Source:* Yearly Reports of the Health Department, 1927–9.

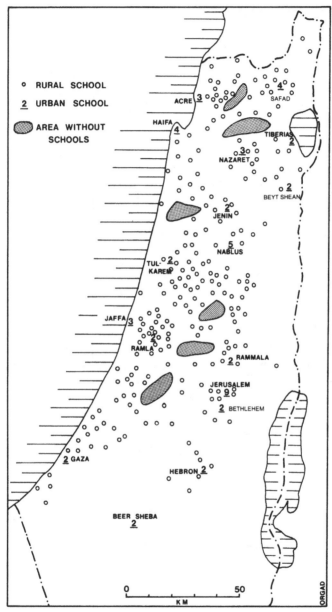

Fig. 9.8   Government schools in Palestine, 1929. *Source:* Yearly Reports of the Department of Education of the government of Palestine.

general health conditions of the local population were improved (Fig. 9.7). The government organized an education system for the rural areas by opening seventy-five village schools each year during the early part of its administration (Fig. 9.8).

Even though most of the government funds were channelled towards the rural areas, administrative efforts were mostly directed towards the urban areas, where the British influence was greatest. The first town-planning ordinance was published in 1921. This general ordinance, together with specific local ordinances for some towns in Palestine, was influential in moulding the character of cities such as Haifa, Gaza and especially Jerusalem.[28] These regulations and, what is even more important, the desire to work according to them, was a British characteristic which was followed in Palestine. Four master plans for Jerusalem[29] and other master plans for Haifa, Lydda, Ramla, Gaza and various other towns, were all drawn up within the first ten years of the British occupation. Even though the British did not build much, only some government buildings and army barracks, the local inhabitants were forced to build according to these laws and plans. Palestinian towns became more European in character and conception. The British administration also dealt with the water supply and drainage problems. Not one city in Palestine had a modern water supply before the war. British engineers helped to construct water systems in Jerusalem, Tiberias, Nablus, Haifa and other places. Modern drainage systems were designed and the general urban landscape underwent a period of change.

Conclusions

This chapter has described the many changes that the geography of Palestine underwent during the first decade of the British administration. In order to understand the speed with which this process took place during a period of disequilibrium, it is necessary to uncover the motives of the administrators themselves, working, as it were, 'backstage'. Analysis of British policy enables us to make a threefold classification:
1 The influence of the general administration,
2 The influence of the British colonial administration,
3 The special influence of the British administration in Palestine, derived from the special relationship between Britain and this country.

By the influence of the general administration, we refer to the actions taken by every administration in governing a region and the resultant geographical changes. Modern government usually deals with three main areas: external defence, internal security (law and order) and finance. The geographical impact of these activities constitutes the influence of a general administration upon a landscape.

The British military operations against the desert tribes from the east and the south freed Palestine from long-standing enemies. These operations enabled new settlements to be established throughout the country as well as strengthening existing settlements, without any outside interference. The war itself, which can be viewed as an act of general administration, brought roads and railways to Palestine for civilian use. Indirect security measures did not have a direct influence on the landscape (except for the construction of a few very prominent police stations), but it enabled the population to work and move freely.

The British fiscal system was a contrast to the previous Ottoman one, as it used all the tax revenue for the development of the country. Until 1927, Egyptian currency was in use. The Palestinian pound was then introduced, both currencies being linked to the British pound – then one of the strongest world currencies – giving stability to the Palestinian economy. Most of these schemes were implemented in the first years of British control, until the general administration achieved its initial objective of ensuring effective control over the administered territory.

The colonial administration also left its mark on the landscape. The aims of colonial rule are often in conflict with the aspirations of the local population. Modern colonial rule creates political territories, consisting of well-defined, sometimes marked, boundaries and a capital city. This was indeed the case with Palestine, which was delimited for the first time with boundaries and a capital city, providing the necessary territorial framework for a modern state. The development of a deep water harbour at Haifa and the construction of a railway system, particularly to Egypt, reflected colonial rather than local needs.

Thus the British colonial character was moulded in the landscape. Most of the British high officials and administrators came from other colonial areas and brought their expertise with them. Agricultural development, education and health projects were all organized along similar lines to those in other British-controlled areas, such as Sudan and Egypt, by administrators with experience in those regions. Thus Heron (health) in Egypt; Sower (agriculture) in East Africa; Bowen (education) in Mesopotamia all brought their experience with them. The cadastral survey was carried out according to the same system used in Egypt. The afforestation of Palestine was copied from the system used in Cyprus, while the railway network was planned and organized by Holmes, who came to Palestine after a long term of service in Africa. This was therefore a classic case of diffusion of ideas by administrators from one colonial territory to another.

The effort involved in attempting to improve the railway system, opposed to the development of the road network, and the building of army camps outside the cities, all had a colonial character. When summarised, these policies can be seen as aiming to give indirect help to the local population while avoiding direct action. This policy was particularly directed towards the local Arab population, constituting approximately 85 per cent of the local population during this period.

Palestine was a new experience for the British empire. Here was a country, colonized by non-British white people, and whom Britain felt an obligation to help settle. The Jewish sector of the Palestinian population was European in orientation, and had its own specific economic and national objectives. On the other hand, Palestine was – and still is – the Holy Land, with all the historical and religious connotations that this had for the British. The administration had to develop a new system of treatment towards Jewish Palestine, in the Holy Land and part of the Mandate.

A significant geographical outcome of their differential treatment of Palestine was the mass immigration of Jews into the area. The immigration was initially carried out with the help of the British administration and was a major factor in landscape change, despite the fact that such a policy was against the wishes of the local Arab population and even some shades of public opinion in Britain herself. The special treatment afforded to Jerusalem was also part of this policy. The British administrator, with the Bible in his hands, attached special importance to Jerusalem, the capital of ancient Palestine. The new capital was established here, despite its location in an interior, non-central, mountainous region and by no means suited to function as a modern administrative centre. The special treatment given to Jerusalem included the cleaning and repair of old buildings, new master plans for the development of the city, construction of a modern water system, plantation and other public works.

It is clear therefore, that these three ideologies – the general administration, the British colonial rule and the special treatment of Palestine – form the necessary background for understanding the role played by the British administration in changing the face of Palestine during the second decade of this century.

NOTES

1 L. D. Stamp, Land utilization and soil erosion in Nigeria *Geographical Review* 28 (1938) 32–45; *idem*, The southern margin of the Sahara, *Geographical Review* 30 297–300; G. A. Price, White settlers in the tropics *American Geographical Society Special Publication* no. 23 (1939)
2 H. De Blij, *Systematic political geography* (1967), R. D. Dikshit, *Political geography* (New Dehli 1982); R. Muir, *Modern political geography* (London 1975)
3 R. L. Merritt, Systems disintegration of empire *General System* 8 (1963)
4 R. J. Harrison-Church, *Modern colonization* (London 1951)

5 S. Folke, First thoughts on the geography of imperialism *Antipode* 5, 3 (1973) 21–32

6 J. M. Blaut, Imperialism: the Marxist theory and its evolution *Antipode* 7 (1975) 1–19

7 F. Tennings, *The invasion of America: Indian colonialism and the cant of conquest* (Chapel Hill 1975)

8 D. W. Meinig, Geographical analysis of imperial expansion, in A. R. H. Baker and M. Billinge (eds.) *Period and place* (Cambridge 1982) 71–8

9 J. Overton, The colonial state and spatial differentiation: Kenya 1895–1920 *Journal of Historical Geography* 13 (1987) 267–82

10 A. J. Christopher, *Colonial Africa* (Beckenham 1984)

11 H. Prince, Real imaginary and abstract worlds of the past *Progress in Geography* 3 (1971) 1–81

12 R. Oven, *The Influence of Lord Cromer's Indian experience on British policy in Egypt 1883–1907* (Oxford 1965) 113

13 H. Samuel, *Great Britain and Palestine*, the second Loussian Wolf lecture (1935) 12

14 B. Porter, *The lion share* (London 1975)

15 G. Biger, Where was Palestine? Pre-World War I perception of an unknown familiar area *Area* 13 (1981) 153–60

16 M. Brawer, Geographical factors in the delimitation of the Israel–Egypt boundary *Research in the Geography of Israel* 7 (1970) 125–37 (Hebrew)

17 A. Elsberg, Fixing the eastern boundary of Palestine *Zionism* 3 (1974) 230–50 (Hebrew)

18 G. Biger, Geographical and political factors in the delimitation of the northern boundary of mandatory Palestine, in A. Shmueli, *et al.* (eds.), *The lands of the Galilee* (Haifa 1984, Hebrew) 427–42

19 Y. Nevrakivi, *Britain, France and the Arab Middle East* (London 1969)

20 E. Kedourie, *In the Anglo–Arab labyrinth* (London 1977)

21 GS (I), EEF, *Military handbook on Palestine* (Cairo 1917–18)

22 S. Reichmann, The evolution of land transportation in Palestine 1920–1947 *Jerusalem Studies in Geography* 2 (1971) 55–90

23 File no. 2/218 in Israel State Archives

24 H. Samuel, *Report of the High Commissioner on Palestine Administration 1920–1925* (Jerusalem 1925)

25 *The official Gazetta*, 1918–19, various notices

26 U. Cohen and G. Biger, British activities for preserving Palestine's landscape and nature, in E. Schiler (ed.), *Zeev Vilnay book* (Jerusalem 1987) 285–302

27 Palestine High Commissioner to colonial office, 3 May 1921, file CO/733/3 in Public Record Office

28 G. Biger, The contribution of the British to the development of Jerusalem 1918–1925 *Researches in the Geography of Israel* 9 (1976) 175–200 (Hebrew)

29 H. Kendall, *Jerusalem – the city plan* (London 1948)

10

Ideology, identity, landscape and society in the lower colonies of British North America, 1840–1860

GRAEME WYNN

Often perceived from afar as a unit, the three lower colonies of British North America (New Brunswick, Nova Scotia and Prince Edward Island) were deeply divided in politics, economy, ethnicity and religion.[1] Separate jurisdictions, until the Confederation of Canada in 1867, each had its own Lieutenant-Governor, Council and Assembly, although their combined population, at mid-century, was little more than half a million. In commerce, all three colonies had strong links to Britain, the British West Indies and the northern States. But within this matrix, economies cross-cut political boundaries. Most of New Brunswick's people lived in the Saint John–Fundy catchment and looked to Saint John as their commercial capital (as indeed did many settlements on the Nova Scotian side of the Bay of Fundy); on the northern (gulf) shore of New Brunswick, and in Prince Edward Island, however, Halifax held greater commercial sway than Saint John. Across the colonies, fishing, shipbuilding, lumbering and farming underpinned local life, but the first three were, largely, separate spheres that gave markedly different casts to landscapes, economies and societies in those areas in which they dominated. Further, although perhaps eight of every nine inhabitants of this region in 1860 were native-born, most identified strongly with the ethnic, religious and other traditions of their forebears, who in the process of settlement had made the colonies a patchwork quilt of different 'allegiances' – Acadian, Loyalist, pre-Loyalist, Palatinate, Yankee, Scots, Irish, English – fragmented by adherence to one or another of a dozen different religions.[2]

Reflecting these patterns, diversity, complexity, fragmentation, heterogeneity have become the leitmotivs of scholarly reflection upon the region. For all its modest size, it is commonly portrayed as a place without a 'unifying configuration of physical features', a world of 'islands, peninsulas, and river valleys' marked by particularism, small-scale competition, and cross-purposes. Within provincial bounds 'local interests were thoroughly schooled to grudge ambitions to others'; considered as a whole, and compared to territories of similar or greater size elsewhere, this region appears 'unusually divided against itself'. Even the surrounding sea – which some have seen as a unifying influence, if only by implying that Maritimers (as residents of the area have come to be known in the twentieth century) are people who 'smell of salt to the Prairie' – 'provides a matrix rather than a focus'.[3]

Such variety offers a stern challenge to cohesive interpretation. How can we treat the region *qua* region and reflect its complexity without becoming mired in the endless detail of richly textured reality? The answer offered below is decidedly experimental.[4] It combines a series of descriptive vignettes drawn from many corners of, and facets of life within, the three colonies ('Dimensions of complexity'), with more general, broadly theoretical, insights derived from recent literature on the sources of social power, the processes of state formation and the development of modernity ('Towards interpretation') to throw light on the connections between ideology, society, the state and the landscape in this corner of the new world.

Dimensions of complexity

In a restive city

In the mid-1850s, Saint John was a bustling commercial and manufacturing centre, a place of compact neighbourhoods and more than 30,000 residents.[5] Reflecting the rapid growth of the city, and its position at the western end of an active trans-Atlantic trading system, approximately three-quarters of its 4,200 family heads had been born in the British Isles; native New Brunswickers accounted for most of the remainder. On the peninsula that formed the heart of the urban area almost three of every five households were headed by men (or women) of Irish birth. In the neighbouring parish of Portland (north and west of the peninsula), well over one-third of all residents had been born in the Emerald Isle. Most had lived in the city for a decade or more. They were, generally, migrants from the northern counties of Ireland, strongly Protestant in religion and generally sympathetic to the British monarchy. By 1850, many held white-collar, artisanal, and semi-skilled jobs. If relatively few lived in the city's better neighbourhoods, they were otherwise widely scattered through the town. In contrast, the significant minority of Irish Catholic households in Saint John included a disproportionate number of unskilled workers. Recent arrivals, refugees from the Irish famines of the 1840s, they were crowded into substandard housing in two small areas of the city: York Point in the northwestern corner of the peninsula and the wharf area of Portland, directly across the Mill Pond (and Portland Bridge) from York Point.

The decade that produced this distinctive, divided ethnic geography was a tumultuous one for residents of Saint John. The economy of the city was buffeted by recurrent crises of confidence, price instability and fluctuating demand in the colonial timber trade. Prophecies of economic collapse were common. As barriers to trade with the United States fell, with the abolition of old imperial trading policies, artisans and farmers who produced for local markets saw their livelihoods jeopardized. Between 1840 and 1842, the number of tailors in the city fell by 85 per cent, the number of shoemakers by almost the same proportion. In the middle of the decade provincial farmers, fishermen and manufacturers petitioned for protection. Tariffs of 10–20 per cent were levied on a range of manufactured goods. But they did little to stave off depression. And the difficulties were only exacerbated by the arrival, between 1845 and 1849, of over 30,000 new-comers, almost all of them Irish.

The city was divided by a series of increasingly violent confrontations among its citizens.[6] In 1841, 1842 and 1843, Irish Catholics clashed with members of newly established Orange Lodges in the streets. In March 1844, Squire Manks, a prominent local Orangeman, shot and killed a Catholic Irishman in York Point and the year ended with a week of disturbances involving Catholics and Orangemen. On St Patrick's Day 1845, celebrating Catholics were fired upon by Orangemen, and general rioting ensued. These sporadic confrontations occurred against a background of robberies and assaults spurred by poverty and privation that the city's magistrates and watchmen were powerless to control. Matters came to a head on 12 July 1847, when Orangemen followed a band playing sectarian songs into York Point. They were met with sticks and stones and driven back across the bridge to Portland. Adding firearms to their numerical reinforcements, they marched again into the Catholic ghetto. Shots rang out and the ensuing general mêlée only ceased with the calling out of the military at midnight. The weeks that followed were marked by vengeance assaults and murders. Although 1848 was relatively quiet, violence erupted again in 1849. Early on 12 July heavily armed Orangemen from Portland, Saint John and neighbouring Carleton marched through York Point to greet brethren from the St John Valley. Outnumbered by a jeering Catholic crowd, they dipped their banners beneath a pine bough arch erected on their route. But 600 strong after the arrival of the valley men, they returned to York Point. Five hundred Catholics lay in wait. In a hail of bullets and bricks, possibly a dozen or more people were killed and hundreds wounded before the Orangemen emerged to continue their march through the city. Troops stationed in Market Square earlier in the day, moved in to seal off the Catholic ghetto and prevent further skirmishes.

At the back of these confrontations lay a complex changing grid of sentiment, affiliation and power in the rapidly expanding Irish population of Saint John. Earlier, secular organizations that sought to advance the interests of all those of Irish descent in the city were challenged by new, more aggressive and increasingly partisan representatives of sectoral interests. Early in the 1840s, the Saint John Sons of Erin had celebrated St Patrick's Day before a 'harp surrounded by shamrocks and orange lilies entwined to form the motto United We Stand: Divided We Fall'; a decade later they supported Daniel O'Connell's vision of a Catholic Irish state.[7] At the same time, Saint John Catholicism became more self-conscious with the consecration of New Brunswick's first Catholic bishop, and the growth of ultramontanism in the province. On the Protestant side, the Orange Order expanded in tandem. As late as 1840, prominent citizens filled leadership roles in both the Sons of Erin and the emerging Orange movement. Two years later, the so-called Protestant Conservative Association claimed to have enrolled 600 men at a single meeting. By 1846 the city had ten active Orange lodges, some with more than 100 members. Newspapers also provided a ready source of partisan opinion, ranging from support for repeal of the Union of Great Britain and Ireland, through denunciations of Catholics who questioned clerical authority, to the vigorous and skilful promotion of the Orange cause.

Faced with lawlessness, violence and confrontation, the Common Council sought ways to control its city. At the beginning of the decade an eleven- or twelve-man 'watch' was charged with keeping the city's nighttime peace. Employed by the Common Council, but underpaid, understaffed and overwhelmingly Irish Protestant, this watch was described by the Grand Jury as 'lamentably inefficient either for the preservation of good order or the prevention of crime'.[8] The elected aldermen who served as magistrates for their wards, and their constables, were little more effective in policing the city and apprehending criminals. Lamenting the dangers of being abroad at night, the editor of a leading city newspaper suggested that citizens follow the lead of New Orleans residents, and carry weapons if they ventured out in the dark.[9] By the end of 1841, a 400-strong 'Mutual Protection Association' patrolled city streets. Through the following years the question of how best to police the city was hotly debated. Few doubted the need for improvements, but tough decisions about the distribution of power and control had to be made. To increase the size of the watch would mean drawing recruits from 'classes whose feelings, sympathies and prejudices all make them partisans on one side or the other, and so increase in place of putting down the agitation'.[10] There were proposals to appoint stipendiary police magistrates and police forces responsible to the province's Executive Council – but these foundered on the reluctance of Saint John's Common Council to relinquish powers. Other possibilities were mooted and rejected. Crime and violence continued. Then in 1847 the Lieutenant-Governor established a permanent police force under a stipendiary magistrate in Portland. Two years later, a stipendiary magistrate with administrative responsibility for the police force had been appointed in Saint John. Constables' salaries were raised, and their number increased. Uniformed, armed and accorded new status as part of professional bureaucracy, they brought order to the city as the economy improved and the influx of immigrants waned.

A local parable

When Patrick Medley wrote to the printer of the *St Andrews Standard* in March 1840, he described himself as a settler in southwestern New Brunswick whose aim was to describe life in his neighbourhood.[11] He told a neatly conceived tale. 'Dennis Snug' and 'Slouch' were its protagonists. The former was a diligent farmer who worked hard on his land, took care of his stock and lived in comfort. Slouch was also a farmer – of sorts. Not content to devote his energy to cultivating his backcountry acres, he spent his winters 'trivin away at all kinds of lumberin'. To feed his oxen engaged in the heavy work of getting out timber, Slouch took his hay to the woods, and starved his cattle at home. 'Between hauling provisions and river drivin, and the likes of that', there was precious little time to attend to the farmer's winter chores. When spring came, very little seed was put in the ground. Overworked in the woods, Slouch's oxen were too weak to plough, so Slouch had to

wait till Dennis Snug got his work all done before he could borrow his oxen to plough a place for to rase a few prataes on, and then there was little or no manuer, and to tell the truth how could there be! for the poor starvin critters of cattle got nothing to ate but a trifle morning and evenin and then turned out of doors in the cauld.

To make matters worse, the surveyor found fault with many of the logs Slouch had cut; together they amounted to very little. Provisions consumed in the woods also turned out to be more costly than anticipated. At the end of the day Slouch had to sell some of his land to pay his debts.

Advice for colonial farmers

Anxious for guidance, many an 'improving farmer' of the mid nineteenth century Maritime colonies turned to one or more of the several texts on agriculture that circulated in the region before Confederation. There was plenty of choice. Works by agricultural chemists renowned in Britain and the United States were widely available, and those who preferred home-grown instruction could turn to John Young's *Letters of Agricola* (Halifax 1822) or James Dawson's *Contributions toward the improvement of agriculture in Nova Scotia* (Halifax 1856). Less voluminous, but in its way no less important, there was also James Ross' *Remarks and suggestions on the agriculture of Nova Scotia.*[12]

Published in Halifax in 1855, this pamphlet claimed distinction. Its author was a practical farmer. His remarks and suggestions were 'not the fanciful productions of fireside speculation, but the sober conclusions at which ... [he] had arrived, during a moderately long life entirely spent in the laborious pursuits of Agriculture'. Critical of gentlemen farmers who made 'ruinously profuse' expenditures, as well as of 'imported knowledge' advanced without local experience, Ross sought to redeem Nova Scotia's agricultural reputation, and to provide useful guidance for ordinary settlers who depended upon their farms for their livings.

Not surprisingly, much of his booklet offered very specific advice. This was a manual to which farmers with a bent for improvement, settlers perplexed by the challenges before them, and colonists in need of basic information might turn for help. But Ross' booklet also sought to set the earnest farmer on a distinct and demanding course. It was firmly against display. Show was not evidence of prosperity. Economy in farming was a great virtue. 'He who acts judiciously', he wrote, 'will prefer commodious barns and stables for his crops and cattle to a fine house.' More than this, Ross was critical of those archetypal Nova Scotian jacks-of-all-trades who divided their attentions among several pursuits as whim or weather dictated. Recognizing that necessity obliged the settlers of young countries to turn their hands to 'a great variety of labour', and that this both encouraged the 'versatility which is naturally inherent in man' and called 'many of his best energies into action', he nonetheless insisted that such multifaceted activity was a 'misallocation of energy' in most parts of mid-nineteenth-century Nova Scotia. In his mind there was no question that the 'skillful farmer will ... allot his time' and complete the jobs assigned to a particular period on schedule.

On Ross' favoured farm, years and days were divided into three. As the seasons turned there came times for sowing, for cultivating and for harvesting. This everyone knew; but Ross would allocate specific numbers of days to these tasks: no more than three weeks should be given to making hay; two weeks were sufficient for the grain harvest. So, too, it was important to observe the divisions of the day with undeviating attention, never permitting 'the labours of one part to interfere with those of another'. The small jobs that demanded attention on every farm should be taken care of before 8.00 am; between 9.00 and 1.00 and 2.00 and 6.00, the plough or cart should be yoked, and an acre of land ploughed or a certain quantity of manure carried to the fields. The evenings were for care of the stock and 'those numerous small pieces of labour of which every careful farmer will find abundance in a well-regulated establishment'. Holding himself as a model, Ross concluded:

Unless on very extraordinary occasions I never permit any thing to interrupt the completion of work so arranged ... Knowing that there is a reasonable time for every part of my work, I never say, if I can get this or that done, but having made my arrangements, I endeavour to put them in execution in their proper season, and the man unaccustomed to systematic effort can scarcely form an accurate conception of the amount of labour which can thus be done, even by an individual.

On a colonial farm

Day by day through 1853, John Murray recorded the business of life on the 112-acre (45.3 hectares) farm near the town of Pictou that he called New Rhynie.[13] Home to a family of eight this was one of the better farms in its county. Valued at $700 in the census of 1851, it had approximately 60 acres (24.3 hectares) under the plough, its stock included 22 cattle, 25 sheep and a couple of horses, and its occupants lived in a large new house.

A native of Aberdeenshire in his sixtieth year, John Murray was a man of standing in his community. After departing Scotland in his early twenties, he had spent two years in Halifax and fourteen years in the West Indies before purchasing the Pictou property that he named after the parish of his birth. At approximately the same time he married Scottish-born Jane Irving of Mount Thom, who was twenty years his junior. Three years later the first of their five sons was born. John Murray quickly began to play a significant part in local affairs. He was a founding member of the Pictou Agricultural Society, and a member of its managing committee in 1839. Through the next several years he judged society ploughing matches, purchased the new seed wheats that it imported, and took several premiums for stock at its annual exhibitions. In the 1850s, Murray was elected an elder of Pictou's First Presbyterian Church and discharged the duties of that office 'with exemplary fidelity and conscientious devotedness' until his death in 1873. A neighbour of long-serving County Sheriff John Harris, and associate in church and agricultural society of many of Pictou's leading citizens, Murray was by wealth and connection among the loosely defined elite of the mid-nineteenth-century Pictou countryside.

With a productive farm and easy access to Pictou, John Murray made frequent trips to town, to sell his produce, to purchase the services of a blacksmith or to acquire goods for his household. He also travelled relatively frequently to local mills, and to neighbouring farms. On average, such off-farm journeys occurred more than twice a week (although their incidence was highly variable and clearly seasonal). Not every such trip brought a sale or purchase. On occasions Murray went to town to assess the state of the market (before deciding, for example, to slaughter one of his animals); on others he returned home with the goods he had hoped to sell. But his meticulous accounts recorded sales on almost a hundred days of the year; purchases of goods were just as frequent. In all, the value of farm produce sold in 1853 was almost $70. Total recorded expenditures, which included the 'Parsons Stypends', a stud fee for 'Jets mare', school fees and supplies for the older children, taxes, wages for servants and various sums for spinning, weaving and the use of a threshing machine, amounted to less than $46.

Omitted from the accounts, but just as central to the operation of New Rhynie farm, were the reciprocal exchanges that enmeshed John Murray and his family in a web of mutual interdependence and more or less formal obligations. Some of these were straightforward enough – simple payments in kind to neighbours in return for certain goods or assistance with a specific task. Thus on 6 June, Murray gave 2 bushels (72 litres) of oats to John Curry for his help in killing a pig. Some were acts of neighbourly charity, as for example when Murray took oats and wheat to the mill 'for Alexandr McKay that got his house burnt fully'. Communal responsibilities – the three days of statute labour on the roads required of John Murray in June, or the less formal 'breaking of roads' after major snowstorms – formed a third category of obligations. Others ensued from complex, continuing relationships, built perhaps on proximity and friendship, whose rich nuances it is now impossible to uncover. Favours were done, favours were returned; in a loose way there may have been some informal reckoning of the balance of obligation; but in the end it was the mutual identification of a general equilibrium of give and take (and perhaps in the case of youthful labour simply the chance of change and company) rather than the strict tally of shillings and pence that defined these associations. So in October Murray noted that his two teenage sons and his pony were assisting 'J. W. Harres Shurref at Threshing'. Far more intricate, because it was close and continuing, was Murray's relationship with the mason James Dawson. On several occasions Dawson borrowed Murray's pony; in February he helped to slaughter an ox; in April he added a chimney to the Murrays' house. A few days after the chimney was finished, Murray got four loads of manure from Dawson, and took him 1.5 bushels (54 litres) of potatoes. In June and September there was more manure from Dawson's stable for New Rhynie's fields. In rough return, Murray hauled several loads of sand for the mason, and in October delivered a load of cabbages to him. None of this figured in John Murray's accounts. But several years later his son would marry James Dawson's daughter.

For Jane (Irving) Murray, life on New Rhynie farm followed a rhythm very different from that of her husband. Because the New Rhynie journal is generally silent about the activities of the sabbath, we might assume that she and her children joined John in the family pew of the Prince Street Church on most Sundays of the year. We might also assume that she accompanied her husband to the funerals of two neighbours and on a few of his numerous trips into Pictou. But in the journal record of market visits, trips to the mill and the incessant round of work in the fields and barns, the only three references to Jane Murray note her absence from New Rhynie. By implication her place was in the home.

Although Jane Murray was relieved of the chores of spinning and weaving by hired help, her days were full. Innumerable chores demanded her daily attention, from cooking and washing to churning butter, tending the hens, ducks and turkeys, caring for the garden and putting up provisions. In the autumn there were the additional demands of the harvest, with its long days, large appetites and extra mouths to feed. Jane's mother came from Mount Thom for extended stays, and occasional visits from Mount Thom relatives broke the routine and brought news. Servants, hired for a month or two at a time, helped with the care of the family's four young children. Still, Jane's was a life of closely circumscribed horizons, an existence structured (as was John Murray's) by a demanding routine of work, but lived within much narrower spatial limits and devoid (by contrast) of opportunities for casual social intercourse.

Houses of treatment and correction

Travelling in the lower provinces in the summer of 1862, Andrew L. Spedon remarked upon the progress of Saint John, a 'once dirty, insignificant hamlet' now 'swallowed up by the magnificent city, whose wealthy and elegant edifices, designed by the lights of science and projected by the hand of industry ... [stand] as the unmistakable evidence of ... wealth and prosperity'.[14] Especially worth remark were the 600-feet (180 metres) long suspension bridge, spanning the river just below the falls, and beyond it, in the parish of Lancaster on the west bank of the St John, the Provincial Lunatic Asylum – 'a splendid building,' set amid 40 acres (16.2 hectares) of well-cultivated ground, and possessing 'a fine view of the city and surrounding country'.

Six years before, American humorist Frederic S. Cozzens had witnessed in Halifax a 'luckless pilgrimage', upon which 'the jolly old rain poured down'.[15]

There were the 'Virgins' of Masonic Lodge No. –, the Army Masons, in scarlet; the African Masons, in ivory and black; the Scotch-piper Mason, with his legs in enormous plaid trowsers ... the Clerical Mason in shovel hat, the municipal artillery; the Sons of Temperance, and the band. Away they marched, with drum and banner, key and compasses, BIBLE and sword, to Dartmouth, in a great feather, for the eyes of Halifax were upon them.

Their purpose was to lay 'the corner stone of a Lunatic Asylum', a feat accomplished by Lieutenant-Governor LeMarchant and framed by artillery salutes. The 85 acre (34.4 hectares) site shared with those of both the New Brunswick institution, and the asylum and house of industry erected outside Charlottetown in the late 1840s, a picturesque situation near the province's major city.[16] And the building was just as impressive as its Lancaster counterpart. Built of brick in Georgian style and called Mount Hope, it stood on rising ground with noble views of village, farm, wood and harbour. Its grounds included lawns, grain fields, a garden and a nursery of trees and shrubs. Altogether it offered 'very decided proof of provincial advance' and was a 'credit to the country'.

The similarities of site, situation and structure among the lunatic asylums of the three colonies were not accidental. Each was intended to provide a carefully designed and closely controlled environment in which the minds of the afflicted could be coaxed back to sanity. Advocates of new facilities for the accommodation of the insane in the colonies after 1835 borrowed heavily from the doctrines of American reformers that depended, in turn, upon the pioneering works and writings of William Tuke in England and Philippe Pinel in France. By these lights, hospitals for the insane 'should not be near a large city, nor within half a mile of any street which is, or will likely become, a populous part of the town'. They 'should be so elevated as to command a full view of the surrounding country', and be situated 'where the scenery is varied and delightful'. At best the asylum should command views of 'a navigable river bearing on its basin a variety of water craft, public roads thronged with the evidences of life and business, but not so near as to be exciting, [and] a populated and cultivated country'; its buildings, surrounded by ornamented grounds, 'should be in parallel lines and as nearly in a right line as they can be'.

According to current wisdom, work, play and worship were the cornerstones of life in the asylum, but of these work was the most vital. It occupied the mind, and stayed the patients' morbid, melancholy inclinations; it taught industry and fostered useful skills; it drained the excess energy that produced frenzied behaviour. Field and dairy, threshing floor and wood pile, workshop and sewing room were 'as indispensable as the strong rooms have been for the refractory in times past'. But work was most useful if closely integrated into a highly organized routine. 'Without system there ... [would not] be success', reported the Nova Scotia commissioners appointed to consider construction of an asylum in 1846.

Of Acadians and 'people of colour'

Two pictures open Frederic Cozzens' *Month with the blue noses*. These, asserts the author, are 'the first, the only real likenesses of the real Evangelines of Acadia'.[17] Travelling with the lines of Longfellow's poetry firmly in mind, and seeking to write an 'Evangeliad', Cozzens was disappointed by his first glimpse of Chezzetcook, the largest Acadian settlement in the vicinity of mid-nineteenth-century Halifax. Its cottages were 'not the Acadian houses of the poem, "with thatched roofs and dormer windows projecting"' but comfortable homely looking buildings of modern shapes, shingled and un-weather cocked. There were 'no cattle visible, no ploughs, nor horses' but the impoverished boat builders and coopers who worked on the shore were 'simple, honest, and good tempered enough'.

At daybreak, a few women from Chezzetcook might be seen, fleetingly, in Halifax – the two who provided the frontispieces for Cozzens' book were among them, though, he took great pains to explain, it was no easy task to capture them on daguerreotypes, for 'as soon as the sun is up [they] vanish like the dew'. 'A basket of fresh eggs, a brace or two of worsted socks, a bottle of fir-balsam', these things comprised 'their simple commerce'. To sell them, they walked the 22 miles (35 kilometres) from Chezzetcook, and then returned on foot. This journey was 'no trifle' agreed Cozzens, 'but Gabriel and Evangeline perform it cheerfully, and when the knitting-needle and the poultry shall have replenished their slender stock; off again they will start on their midnight pilgrimage, that they may reach the great city of Halifax before daybreak'.

In response to Cozzens' surprise that a 'mere handful' of Acadians should live so near the colonial capital, yet remain so isolated, that their 'village of a few hundred should retain its customs and language, intact, for generation after generation', his informant explained that this was so because Acadians 'stick to their own settlement; never see anything of the world except Halifax early in the morning; never marry out of their own set; never read ... and are ... *so slow*, so destitute of enterprise, so much behind the age'.

Visiting Halifax in 1850, agricultural chemist J. F. W. Johnston found T. C. Haliburton, the author of *Sam Slick*, presiding over a court in which 'a perfectly black man' sat 'in the box as a juror'.[18] For Johnston this was evidence enough that in British North America 'people of colour' enjoyed 'the same political privileges as are possessed by other classes of her Majesty's subjects'. Some of them were industrious owners of small farms. Yet most of those in Halifax were in humble employment; they were described 'as indolent, as hanging about the towns and as suffering much from the severity of the winter'. Had Johnston travelled along the coast, he might have added detail to these fleeting impressions. Not far from Halifax was a 'Negro settlement', where 'log house[s] perched on ... bare bone[s] of granite that stood out on a ragged hillside'. These 'scare crow edifices', observed a visitor in 1856, were

all forlorn all patched with mud, all perched on barren knolls or gigantic bars of granite, high up like rugged redoubts of poverty, armed at every window with a formidable artillery of old hats, rolls of rags, quilts, carpets and indescribable bundles, or barricaded with boards to keep out the air and sunshine.[19]

They were inhabited, explained his guide, by 'a miserable set of devils'. During most of the year, he continued,

they are in a state of abject want, and then they are very humble. But in strawberry season they make a little money and while it lasts are fat and saucy enough. We can't do anything with them; they won't work. There they are in their cabins, just as you see them, a poor woe-begone set of vagabonds; a burden upon the community; of no use to themselves, nor to anybody else.

Measuring progress

When the legislature of New Brunswick passed 'An Act to provide for taking a Census' (23 Vict. c49) in April 1860, they followed a relatively well-worn path. Decennial censuses had been taken in the United States since 1790, and in England and Wales since 1801, and increasingly detailed enumerations of New Brunswick had been conducted in 1824, 1834, 1840 and 1851. In contemporary minds, the census would 'afford much useful information, dispel many erroneous ideas, and form the basis of most important legislation'.[20] Yet the nature of this 'useful information', and the means of its collection, remained to be decided. At minimum, appropriate geographical divisions had to be established, enumerators chosen and appointed, and rules, regulations and schedules drawn up.

By the end of February 1861, the 'schedule of enquiries' had been finalized. Data were to be collected in six broad categories, but in very different quantities: five questions pertained to the fishery, a dozen to manufacturing, sixteen to population and forty-eight to farming. Taken as a whole, this was a far more ambitious effort to identify salient characteristics of the developing colony than its 1851 predecessor, but its emphasis was clearly economic; in essence it documented production.

In July, census enumerators were appointed for the first of the colony's 160 census districts. Paid at the rate of 10 shillings (50 pence) per day, these were desirable appointments; unsolicited applications for them began to reach the Governor in September 1860, and a further three dozen were received in the next ten months. But most appointments were by recommendation; hard on the heels of the election in June, Members of the Assembly were asked for nominations from their districts. Hardly surprisingly, relatives, friends and supporters were prominent among those whose names were forwarded. So Samuel Freeze and S. Nelson Freeze were appointed in Norton and Sussex parishes on the recommendation of their local member, the husband of Miriam Freeze. Another Assemblyman left no room for ambiguity in promoting his brother-in-law, among others, for a position: 'I wish to inform you the men that I want appointed to take the census of the county . . . these are the men that I want appointed and please appoint the same'.

According to the enumerators' instructions, the census was to be taken 'with the least possible delay' to 'represent the state of the country as it existed on the 15th August'. Yet the enumerator for Upper Queensbury, 15.5 miles (25 kilometres) upriver from Fredericton, was not appointed until 3 September; at least one set of blank schedules was sent to the wrong post office, where it lay for weeks; and several enumerators failed to realize that Schedule VI was on the back of Schedule V. Even when appointments were made and schedules provided on time, the enumeration was some-times slowed by the difficulties of travel, illness and the absence of people from their homes. One enumerator did not begin work until 20 November. Slightly more than half the returns reached Fredericton by the end of October, but eight remained outstanding into 1862, and the last was not received until 23 January. Generally, enumerators spent twenty to forty working days at their task, but the enumeration of Moncton required two and a half months. Such a protracted process clearly jeopardized the accuracy of the census: how many could recall, accurately, the state of their farms on 15 August, two or three months after that date? how many simply assumed the date of the census taker's visit to be the point of reference in reporting the number of births or deaths in the preceding year?

There were also problems in the formulation of census categories. Schedule III required a return of 'Pork, slaughtered, pounds', but enu-merators were not informed whether they should enter 'the quantity to be slaughtered the fall of 1861 or what . . . [they found to] have been slaughtered on the 15th of August . . . or what . . . [had] been slaughtered in the fall of 1860'. Those enumerators who sought clarification of an ambigu-ity in the agricultural schedule were instructed to enter their own estimates of 'the number of males and females that could do the work if steadily/ continuously employed'. When the returns were finally compiled, many essentially arbitrary decisions had to be made. Scores of the enumerators' farmer-lumbermen and fishermen-farmers had to be shoehorned into one of the compilers' occupational categories: professional; trade and com-merce; agricultural; mechanics and handicraft; mariners and fishermen; miners, labourers and miscellaneous. So inconsistent were the returns on 'race' that the category was omitted from the printed tabulation. And 'owing to misconception on the part of a large number of Enumerators', 'it became necessary in the abstract to include the Baptists and Free Christian Baptists in one body and the adherents of the Church of Scotland, Free Presbyterian Church and Presbyterian Church of New Brunswick in another body'. By chance (or more confusion), Samuel Freeze of Kings County appeared on two enumerators' returns, his own and A. B. Smith's.

The returns were completed a month apart, Freeze's at the end of December, Smith's in January. Both should have referred to the state of Freeze's household and farm on 15 August. By Freeze's own account he was an Episcopalian tavern keeper with 30 improved acres (12 hectares) on a 200-acre (81 hectares) farm valued at $2,000, with a servant Matilda Hoggins and two labourers Richard Bigelow and John Golding. According to Smith, Freeze was an innkeeper, a farmer and a Free Baptist; with 40 (16 hectares) of its 200 acres improved, his farm was worth $2,400; his servant was Emily Driscoll, he had only one labourer, Bigelow, and one of his sons, aged three by Freeze's count, was four. It was, reflected the census compilers, 'without doubt, extremely difficult to devise such forms of schedule as will tend to procure accurate accounts of the several matters which it may be considered advisable to embrace in the Census Returns'.

A royal tour

Although the weather was 'anything but agreeable', Haligonians crowded the waterfront and lifted their voices in 'thrilling and vociferous cheers which rang loud and long' as the Royal Squadron bearing His Royal Highness Albert Edward Prince of Wales anchored in front of the city in July 1860. Once ashore, the prince was mounted upon a 'fine high-mettled charger' to take his place in a long and magnificent procession that included fire companies; national societies; craft organizations; the Sons of Temperance; militiamen and representatives of Her Majesty's Forces. Through streets lined by soldiers and volunteers, thronged by thousands, and decorated with at least seventeen arches, as well as 'transparencies', flags, banners and evergreens, the parade followed the firemen who bore 'a trophy fifty feet high, surmounted by a colossal figure holding a hose-pipe'. Thirty-five hundred school children, 'dressed in white and blue . . . sang the National Anthem'. In the next three days, His Royal Highness reviewed troops, witnessed Indian Games (where he laughed 'heartily at the ludicrous scene' presented by a war dance), mingled with zest at a grand ball attended by 3,000, watched displays of fireworks, and (it was said) 'sat on his horse nobly and never flinched' when drenched by a sudden shower.[21]

Through early August, the prince's grand progress continued, to Saint John, to Fredericton, and then to Charlottetown, before he took ship for Quebec. Throughout, communities responded to the prince's visit with exuberant enthusiasm. There were levees, balls and formal presentations of addresses. 'National and Trade societies and Volunteers' – of which there were always 'great number', with members invariably comprising 'a fine-looking body of men' – paraded. Streets were decked with bunting and whole cities beautifully illuminated. Bonfires were built, bells rung and guns fired. 'People cheered and cannon roared.' Joy knew no bounds, even when planned firework displays 'were completely destroyed by the immense deluge of rain'.

Testaments to colonial loyalty were everywhere. At the grand ball in the Provincial Building in Charlottetown, one of the 'many beautiful devices' that graced the scene carried the message

> Thy grandsire's name distinguishes this isle;
> We love thy mother's sway and court her smile.

In Saint John, 5,000 'fancifully dressed' and flower-bedecked school-children added three verses to their rousing rendition of 'God Save the Queen', ending

> Hail, Prince of Brunswick's line,
> New Brunswick shall be thine;
> Firm has she been
> Still loyal, true, and brave
> Here England's flag shall wave
> And Briton's pray to save
> A nation's heir.

And on 7 August, when His Highness returned to the mouth of the St John River en route to Windsor, Pictou and Charlottetown, 'a party of stalwart though gentle firemen' unharnessed the horse from his carriage and pulled 'their "dear little prince" as they delighted to call him', across the suspension bridge to the wharf.

Even Bishop Medley (who excused his decision to preach to the prince and the congregation of Fredericton's Christchurch Cathedral on the text 'So, then everyone of us shall give an account of himself to God' by claiming 'a higher mission' than simply voicing the 'language of congratulation') could not avoid the pervasive spirit of loyal adulation. 'When we look round among the nations of the earth and consider the past and present conditions of countries favoured with a fruitful soil and a more genial climate than our own', intoned the bishop in the middle of a sermon full of eloquent warnings of the awful power of God's judgement

how inestimable is the price of our manly, rational, and constitutional freedom, how deeply should we cherish, how diligently should we guard and preserve, the integrity of our limited monarchy, the nice balance of our respective estates and realms, the just and merciful administration of our laws and the various expressions of freedom and safe-guards against license with which a gracious Providence has endowed us. Our monarchy, our language, our religion are rich in all the associations of the past ... Our sufferings and our joys are the common property of the empire. One year our bosoms throb with fear and sorrow at the massacre of Cawnpore, in another we hail the coming of a Prince ... sent forth by the love of the Mother of our country, to consolidate the affection of a distant empire and to bring nearer in loyalty, love, and friendship the claims which science and commercial action have already united.

Little wonder that the young prince left the lower provinces with 'an endearing regard and sympathy' for their inhabitants in whom he found 'the love of freedom ... combined with a deep-rooted attachment to the mother country, and the institutions in which we have all been nurtured'.

Towards interpretation

However vividly these vignettes capture elements of the colonial scene, they also make it clear that we cannot come to grips with the past simply by retrieving fragments of it. The facts do not speak very eloquently for themselves. Many and various as they are, these sketches provide a terribly incomplete rendering of the diversity of land and life, economy and society in the mid-nineteenth-century Maritime colonies. Taken on their own they form a rather incongruent assortment of snapshots. Their number could be multiplied, almost endlessly, but this would only complicate the kaleidoscope. Still, we cannot retreat from our engagement with the variety, nuances and texture of life as it was lived. It is just such intricate, multifaceted reality that we must seek to grasp if we would justify and sustain the vitality of our enquiries. The challenge, for those who would make sense of places, is surely to confront their complexity in ways that can heighten understanding instead of simply adding to our store of factual knowledge.[22]

In the broadest, and most ambitious, of terms this means approaching the places and people in whom we are interested with a curiosity honed on the whetstones of contextual understanding, pertinent comparison and theoretical insight in an effort to locate them on those 'grand maps of history' that promise to illuminate and explain how the world came to be as it is.[23] In more concrete, mundane and circumspect vein, it means approaching the foregoing vignettes as a series of ethnographic jottings, each and all of which capture something of the assumptions, aspirations and attitudes, the classifications and regulations, the constructions and constraints, that gave shape and meaning to life in the settings that they encapsulate. Perceived thus, woven back into the rich contextual fabric of provincial existence from which they are drawn as so many threads, and considered against recent discussions of power, property, cultural change and nation building they reveal much about ideology, society and state in the mid-nineteenth-century Maritimes.[24]

Capitalism and the moral agenda in the colonies

By this light we recognize first the pervasiveness of what Max Weber called the ethos of modern Western capitalism – calculating, rational, 'sober bourgeois capitalism' – in the rhetoric of the colonies.[25] In its emphases on time discipline and the division and specialization of labour, James Ross' pamphlet on agriculture embodies the confidence of the Victorian age in the benefits of system and the possibilities of improvement. In rural Nova Scotia, the doctrines of Adam Smith and the practices of English factory owners would redeem both the productivity and reputation of the colony's most important industry. With farmers committed to unwavering, per-severing and above all systematic effort, with their energies no longer dissipated by want of calculation and design, individual profits and colonial prosperity were certain. Doubters need only consider the plight of Nova Scotia's negro settlers, which contemporaries attributed to the failure of these 'vagabonds' to work as industriously as they ought. In this context work and system were logical-enough cures for insanity. That the judicious application of labour and capital to the myriad activities of the farm required careful assessment of inputs, outputs, expenditures and returns escaped mention in Ross' pamphlet, but was well understood by other leading agricultural improvers of the period. Through books, newspapers and county agricultural societies the importance of system and science, and with it the importance of detailed record-keeping of the sort practised by John Murray of New Rhynie, was driven home to those colonial farmers who would listen.

Patrick Medley's parable spoke in different tones to much the same project. At base its message was simple: those who 'cleared land, developed their farms and lived on them were the real producers'. Farming was a stable enduring occupation; the well-managed farm was a 'permanent enterprise oriented to constantly renewed profit'. Lumbering was a preca-rious, adventitious, speculation, a lottery attractive for its promise of 'spectacular ad hoc killings'. As 'O'Leary', a stereotypical Irish ex-lumber-man who appeared in a series of letters from Paul Jones to the *New Brunswick Courier* had it, the lumber industry was 'a game of haphazard'.[26] Those who engaged in it might prosper temporarily, but in the end they would find 'their farms mortgaged [and] their houses ... tumbling down wid hardly a light of glass in them but stuffed wid ould rags hats and straw'. By contrast, the settler who invested in 'the bank of earth' would dwell in a 'neat and smart' house, and his 'ould wife' would always have '"siller" in her pouch'.

By the middle decades of the nineteenth century these were well-established themes in the formal discourse of colonial life. Given full and self-conscious expression in Thomas McCulloch's *Letters of Mephibosheth Stepsure* early in the 1820s, they were reiterated time and again in vice-regal pronouncements, in the columns of the provincial press, in the apological observations of travellers and in the reports and injunctions of the provincial and local agricultural societies whose formation and continued activity rested upon the fiscal support of the three colonial governments.[27]

Closely associated with these rhetorical and administrative attempts to influence the construction of aspirations and to entrench the ethos of rational capitalism in the colonies was a moral agenda. Here farmers and lumberers again found themselves the metaphorical centrepieces of a wider tableaux. They were 'as unlike each other *professionally* as ... the black Ethiopian and the White European *personally*'. Lumberers lived a 'toilsome and semi-savage life'; they were men of 'spendthrift habits and villainous and vagabond principles', strangers 'to every rational enjoyment'; their work was demoralizing and debilitating, they 'spent their winters in the woods and their summers lounging about the towns'. 'Happy' farmers, by contrast, followed 'a pursuit of innocence and peace', derived their pedigree from the patriarchs and stooped to no man; their characteristically neat cottages surrounded by rich and cultivated land betokened their elect status. Farmers were snug by virtue of their sober and industrious characters; profane, sabbath-breaking, gambling lumberers were poor, indifferent, lazy, idle, loutish and slovenly fellows – slouches indeed.

That these were thoroughly derivative arguments, drawn from the Ancients and the agricultural literature of Renaissance Europe mattered not a whit. Nor was it of much moment to those who used them that they caricatured – rather than characterized – the colonial scene. Their purpose was to cajole, not to describe. They sought to define a moral universe, to encourage particular forms of behaviour by portraying them as normal, appropriate and acceptable and to discourage others by representing them as inept and repugnant. Nor were they the only vehicles for the promotion of a sober, industrious, bourgeois mentality. The Temperance movement, which flourished in the region during the 1840s and 1850s was directed to similar ends.[28] Teas, picnics, processions and excursions brought the Sons, Daughters and Cadets of Temperance together, developed a sense of camaraderie among them, and heightened the visibility of their cause. In 1852, 9,000 people petitioned the New Brunswick Assembly for a ban on the import of alcoholic beverages; passed into law, their resolution imposed official prohibition on the colony through 1853, but quickly proved unworkable. 'Conceived in tyranny' as the Attorney-General recognized, it soon led to 'fanaticism and violence', and was rescinded in 1854.[29] Still, the Temperance movement had revealed its political power and its moral authority in 'urg[ing] upon other men, as good, such lines of conduct as are good for them, whether good or evil to the other people'.[30]

Sectarian and cultural ideologies

Nowhere were the contours of the emerging social order more clearly revealed than in mid-nineteenth-century Saint John. Here the endemic disorder, the ethnic rioting and the responses of city officials to both, reflected and helped to frame prevailing convictions about the nature of colonial society. Orangemen in New Brunswick shared with their brothers on both sides of the Atlantic an unswerving loyalty to the Crown and a fervent belief in the superiority of Protestantism. Irish famine immigration was resisted because it was seen as one part of a massive campaign to establish the authority of the Vatican around the globe: 'A great ... conflict is at hand', warned the Saint John *Church witness* in September 1853, 'between Protestant Truth and Popery leagued with Infidelity.'[31] And 'Popery', it was alleged, had formidable troops on its side: 'no one can deny', claimed the *Loyalist and conservative advocate*, 'that the lower orders of the Roman Catholic Irish are a quarrelsome, headstrong, turbulent, fierce, vindictive people'. One might as well attempt to 'wash the Ethiope White' as to 'tame and civilize' the native of Connaught and Munster. Time and again, editorials in the 'Orange' press played on these themes. Almost every contemporary newspaper carried anecdotes and 'Irish jokes' that mocked the Celtic newcomers as barbaric and/or ignorant. When economic arguments were added to these ethnic (or racial) slurs – for the destitute immigrants formed a large pool of cheap labour in direct competition with unemployed native labourers during the 'hungry forties' – they formed a powerful goad to Protestant Nativism. Fully half of New Brunswick's mid-nineteenth-century Orangemen were born in the colony; their membership in the Ulster-based organization was a measure of its attraction as a defender of Protestantism and British hegemony.

When Orange fervour spilled over into vigilante action and the provocation of conflict by ceremonial invasions of Catholic areas of the city, the processes of colonial justice generally ensured more lenient treatment of Protestant than Catholic participants. When the Worshipful Master of the Wellington Orange Lodge shot a Catholic Irishman in 1844, he was placed in protective custody rather than arrested, and quickly exonerated of any crime – on the claim of self-defence – by city magistrates. After riots in 1842, an all-Protestant force of special constables arrested several Irish Catholics who were subsequently convicted of rioting. Three years later, both Orangemen and Irish Catholics were arrested after fierce fighting in the streets, but the all-Protestant Grand Jury refused to bring their co-religionists to trial. And in 1849, when the clash of Orange and Green forces was widely anticipated, and the route of the Orangemen's approach to York Point was cleared well in advance, garrison troops were deployed not to ward off conflict by barring entrance to the Catholic district, but to seal off the riot once the Orange parade had moved through the area, and thus to allow the procession to continue, unmolested, through the core of the city. Here the sentiments and convictions of Saint John's powerful majority were made clear: Irish Catholics were rejected for their cultural and religious differences; disparaged in the vigorous rhetoric of the day, they became legitimate targets of attack as the authorities turned myopic eyes on the provocative actions of vehement nativist Protestants.

Elsewhere, anti-Catholic ideology also ran strongly during the 1840s and 1850s. In Prince Edward Island, 'Romanism' became an issue in the elections of the late 1850s which turned on 'a serious and most unaccountable misunderstanding' over the place of the Bible in the educational system, and produced an all-Protestant administration in a colony whose population was almost half Roman Catholic.[32] In Nova Scotia, 'Popery' was seen as a threat to established values, denounced for undermining colonial attachments to Britain, and criticized as an obstacle to colonial improvement.[33] 'The Popery of the [D]ark [A]ges ... [was] the Popery of the present generation.' Catholic countries published fewer books and had fewer miles of railway than Protestant ones; thus 'Protestant areas nourished "progress" while Catholic lands promoted ignorance and lethargy.' As the Scottish-born, Free Church minister, Reverend A. King, told a Halifax audience, and readers of his pamphlet *The papacy: a conspiracy against civil and religious liberty* (1859), Catholic priests nipped 'in the bud the first appearance of ... that assertion of liberty to think and act for himself, which belongs to man as a moral and accountable being'. Little wonder, in this view, that the Acadians of Chezzetcook were '*so slow*, so destitute of enterprise, so much behind the age'.

The development of political culture and consciousness

Beyond all of this lay various, more or less explicit, 'official' initiatives and strategies that worked to shape and confirm social and political identities in the three Maritime colonies. When colonial governments embarked on the construction of asylums for the insane, their actions implied more than the simple provision of facilities for the unfortunate. The new institutions were trophies in the landscape, placed there with appropriate fanfare to symbolize the progressiveness of their creators. They reflected the growing embrace of the asylum for treatment and correction, and were but single – if important – pieces in a larger mosaic of corrective institutions. By 1860, Halifax had a handful of penitentiaries, homes for juvenile delinquents, rescue homes for prostitutes and a poor's asylum, as well as Mount Hope.[34] Saint John, likewise, built a cholera hospital, a county gaol-house of correction, an almshouse-workhouse-infirmary, and an emigrant orphan asylum in the dozen years or so before the opening of its institution for the insane. Furthermore, these were significant *colonial* initiatives. In 1845 the erection of a combined asylum for the three provinces was mooted, but came to naught in the face of perceived 'difficulties' and the conviction that 'separate establishments' for each of the provinces would be more desirable.

By the mid nineteenth century, there was a growing consciousness, among their elites, of Nova Scotia, New Brunswick and Prince Edward Island as separate communities. Peter Fisher published the first *History of New Brunswick* as early as 1825, and according to D. C. Harvey, at least, Nova Scotians *as such* had begun to emerge between 1812 and 1835, when the record of literary achievement shows them 'rubbing the sleep out of their eyes and facing their own problems, in various ways, but with discernment and energy'.[35] Certainly there is no gainsaying the importance of T. C. Haliburton's writings here, with their portrayal of the Nova Scotian as half English, the best product of his race.[36] And of Prince Edward Islanders, it was proclaimed in 1853: 'removed as they are from all intercourse with the world, these narrow-minded provincials really fancy themselves *par excellence* THE people of British North America'.[37]

To a degree such identifications were fostered by the legislatures, sessions, courts and grand juries which formed the essential infrastructure of these mid-nineteenth-century colonial 'states'.[38] Communities returned representatives to provincial assemblies, and by and large they were administered through courts of session composed of justices of the peace (appointed by the Governor and Council) who served, on paper, to extend the influence of central government into local affairs. Although the Common Council of Saint John established in 1785 held legislative, executive and judicial powers, and a few other towns and cities in the region gained autonomy through incorporation after 1840 – namely, Halifax (1841), Fredericton (1848), Moncton, Charlottetown and Sydney (1855) – local government by sessions was the norm in both Nova Scotia and New Brunswick until the late 1870s. Moreover, many local officials – such as the supervisors of great roads and the Fredericton firewards in New Brunswick – were appointed by the Lieutenant-Governor in Council.

That these officials and institutions were tentacles of the 'state' is indubitable, but their strength in that capacity should not be overestimated. Technically subject to central control, most justices enjoyed a great deal of freedom in their conduct of local business; many indeed fell sorely short of the demands of their offices. Furthermore, proposals for municipal incorporation in the 1840s were rejected as 'encroach[ments] upon the liberties of the people'. Colonial politics revolved, for the most part, around the local distribution of government largesse rather than the development of a coherent political ideology at the centre and its implementation on the periphery. For all the social importance of politics, and its undoubted role in shaping the channels of power in individual communities, in the 1860s as in the 1830s it was essentially about local patriotism. As Joseph Howe noted in the 1840s, the central administrative duty of the government lay in 'dispensing the patronage of the County'; the fact that assemblymen were consulted about the appointment of justices of the peace was 'a substantial concession from the Crown to the People'.[39] New Brunswick lawyer and Assemblyman George S. Hill voiced similar conclusions more colourfully in arguing that 'The Russians under Peter the Great thought the privilege of wearing long beards the essence of liberty – our people judge it to consist in the right of sending members to the Fredericton legislature to get their by-road and school money – all beyond is a *terra incognita*, which they have no curiosity to explore.'[40]

The colonial 'states' role in staking out a cognitive territory with which its citizens could identify, and the severe limits to its effective authority were clearly revealed by New Brunswick's attempt to count its people in 1861. By mounting a census of the colony, the legislature instantiated conceptions of New Brunswick as a distinct and, in some sense, unitary territory; by establishing – whether *de jure* or *de facto* – the categories of that enumeration, it defined, however implicitly, those things that it held to be important; and, inadvertently or not, it served to blur distinctions (between Baptists and Free Christian Baptists, for example) that others held dear. Yet the fumbling, bumbling manner in which the census was conducted was a stark testament to the limits of central power. Without a professional bureaucracy to conduct the work, New Brunswick's authorities fell back upon traditional channels of patronage in selecting their census takers; among the motley crew of enumerators this produced few, apparently, were cowed by instructions to complete their work with 'the least possible delay'. In its many confusions and inaccuracies the census of 1861 stands as a measure of the very real constraints limiting the totalizing power of the 'state' in the mid-nineteenth-century Maritime colonies.

This is not to suggest that progress toward the development of more powerful and effective instruments of government was absent during the early nineteenth century. In both Nova Scotia and New Brunswick, the widespread disregard of statutes intended to control access to both land and timber was gradually restricted by refinement of both the relevant regulations and the means of their enforcement. So 'squatters' and 'trespassers' – official descriptions of those who claimed natural or moral rights of access to the abundant resources of the colonies – were evicted and fined, and provincials were increasingly forced to acknowledge the property rights legitimized by the state and conform to the terms of lease and sale that it established for them.[41] So, too, the development of a uniformed, bureaucratic police force in mid-century Saint John marked a significant extension of the effective range of control by centralized authority. But always there remained, through the mid-century decades, large segments of colonial life that lay, to all intents and purposes beyond the effective everyday sway of the colonial 'state'.

The colonies and societal classification: an analysis

To throw these patterns into bolder relief, it is useful to consider the mid-nineteenth-century Maritime colonies against the framework of societal types outlined in the writings of Anthony Giddens.[42] Primarily concerned to provide a perspective on the novel world of the late twentieth century, Giddens identifies several salient contrasts between traditional (so-called class-divided) and modern states. Foremost among them are those related to the range, scope and intensity over and with which economic and political power can be exercised. In traditional states, Giddens avers, 'the administrative reach of the political centre is low'; in modern ones it is extensive. So traditional states have frontiers, modern ones borders. In the former, most people live lives shaped by the rhythms of the seasons, structured by personal contact, and bounded by the limits of domestic production; they occupy relatively closed pockets of local order that combine to make of any more extensive territory a cellular, segmented space within which ruling groups – concentrated in the cities – are generally unable to influence the day-to-day lives of their subjects although they utilize their control of 'authoritative resources' (generally military power) to pacify and extract surplus production from the people. In the latter, new technological and organizational means of overcoming the barriers of time and space allow for interaction and the co-ordination of activities without face-to-face contact; here control over 'allocative' (economic) resources is critical; there is a high level of internal order; government is polyarchic, in the sense that it is responsive to the preferences of the people; codes of conduct tend to be spatially extensive, patterns of production are highly integrated, and administrators characteristically have an enormous capacity to shape even the most intimate features of their subjects' daily activities. Fundamental to this transformation has been the growing power to bridge distances; by the development of writing, and later other forms of information storage (which opened out the possibilities of social inter-action beyond those provided by the evanescent spoken word and heightened the prospects of central surveillance); by the acceptance of money (which enlarged the radius of exchange beyond that possible in a barter economy and led eventually to the commodification of everyday life); and by improvements in the transportation/communication system that produced time–space convergence (or a reduction in the friction of distance – measured in terms of time or cost – between places) and thus facilitated system integration.

   At base, this conceptualization rests upon an essential distinction between 'the state' as an instrument of government or power and 'the state' as 'the overall social system subject to that government or power' or, in other words, between the state as an administrative apparatus and the wider 'civil society' of which it is a part. Recognizing as much, the contrasts between traditional and modern societies sketched above can be associated with a decisive shift in the ability of the state, as instrument, to penetrate the civil realm. In traditional societies, substantial spheres of life 'retain their independent character in spite of the rise of the state apparatus'. Thus the classic contrast between city and countryside, markedly distinct places despite their interdependence. With the rise of the modern (nation)state (as an administrative organ), however, the distinction blurs then disappears; few areas or spheres of life survive beyond the administrative reach of the state, and those that do so 'cannot be understood as institutions which remain unabsorbed by ... [it]'.

Against this backdrop – barely and inadequately sketched as it is here – the contours of Maritime distinctiveness begin to stand out. On the face of it, the mid-nineteenth-century colonies embodied elements of both 'traditional' and 'modern' archetypes. There were close and finite limits to the administrative state's ability to penetrate and organize the colonial countryside; the provinces were a patchwork of social, ethnic and religious fragments; relationships built on barter and reciprocity integrated people into local worlds; many colonists remained remote from central authority, unaware of many of its dictates, and prepared in many cases to transgress against others with scant fear of detection. Colonial administrators, on the other hand, sought to shape and circumscribe the lives of colonial residents by defining property rights, requiring road work and administering justice. They took account of the colonists and their production. And they brought some of them under even more continuous surveillance in the asylums, prisons and rescue homes that were built during these years. These colonies were, moreover, territories with clearly identified boundaries, capitalist societies without limitations on the alienability of property, and places in which steam trains and steam ships had begun to revolutionize transport by 1840.

Yet it would be a mistake to simply classify the colonies, on this evidence, as 'transitional'. Rather than standing at some intermediate point in this conjectural framework, they form a special variant of the models identified by Giddens. Outliers of empire, they were neither modern states nor traditional 'class-divided' societies. Shaped by the encounter of post-Enlightenment Europe with a remote and essentially undeveloped wilderness, they were peculiar hybrids. They exhibited many of the forms of the modern administrative state, but lacked final sovereignty and were poorly integrated, by virtue of their colonial status and the difficulties of communication and time–space co-ordination across their territories. Janus-like, they present different appearances to our gaze. Viewed from their political centres – Halifax, Fredericton and Charlottetown – these small, and recent, colonial societies reveal many characteristics of the modern state and its substantial administrative apparatus. Courts and Assembly chambers provided a forum for the defence and advancement of civil and political rights; restrictions on the (male) individual's freedom to join organizations, express opinions, hold public office and vote were few; newspapers offered several alternative sources of opinion, provided readers with a steady flow of 'decontextualized' information and broadened the sense of membership in a political community; political leaders competed for support; and elections determined the composition of governments. Statutes applied colony-wide; surveyors of land and timber were appointed far and near to administer regulations framed in the capitals; institutions were created to adjust 'deviants' – criminal or insane – to the norms of 'acceptable behaviour'; public accounts were subject to scrutiny; and colonial census statistics began to provide the basis of that 'reflexive self-regulation' so essential to the administrative power of the state. Official documents, reports and correspondence reveal much about these aspects of colonial administration. Their real effects upon the rank and file of colonists are considerably harder to assess, however.

It is clear, nonetheless, that to view these societies from their peripheries – from the fields, camps, dories and kitchens of their predominantly rural populations – is to see them in a very different guise. Through most of the region, the sanctions of religion and the local community were more important than formal policing in maintaining 'order';[43] among people dispersed upon their farms, working in the woods in small, informal groups, or fishing in crews of one or two, there could be little surveillance over production. Although the colonies were defined and bounded spaces, they were hardly 'conceptual communities'; their people spoke different languages (Gaelic, reported mid-nineteenth-century visitors to Cape Breton, was more common there than in Scotland, French was the language of Acadian areas, Lünenburgers clung to their German, and so on), and worshipped in different churches; in these relatively new-settled places there was little shared history (or, yet, symbolic historicity of the sort that would be provided by celebrations of the loyalist centenary in the 1880s).[44] Even colonial government – the central structure of the administrative state – failed to override the profoundly fragmented and local character of colonial life; late in the 1850s, Lieutenant-Governor Sir Edmund Head of New Brunswick lamented the absence of 'public' (communal rather than individual, colonial rather than parochial) interest among the Assemblymen of his province.[45] However impressive in the colonial capitals, many tentacles of the administrative state reached but weakly into the provincial hinterlands. Distance – the formidable time and cost of movement in this new world – was a significant barrier to integration of the corners of colonial life with the centre.[46] Elaborated and refined through the first half of the century, the administrative apparatuses of these colonial states (which owed much to the model of the developing English nation state) cast longer, more solid shadows in 1860 than in 1800 or 1830, but their penumbras were wide, and they clearly failed to blanket several facets of colonial life.

All of this leaves us, finally, with an important and infinitely elusive question: how was the world made sense of by inhabitants of these three colonies? If there is an easy yet broadly accurate answer it is, simply, differently. Consider, for example, the black juror in a Halifax courtroom noted by J. F. W. Johnston. To Johnston and, no doubt, to many leading citizens of the province, his presence suggested the equality of political privileges in British North America. But the reality of circumstances in those nearby 'redoubts of poverty' – described by other travellers – which were home to many of the colony's black residents, surely conveyed a different message to those who lived in them. In the end we can do no more than allude to the range of this diversity, and recognize its complex manifestations. Yet in doing so we demonstrate, again, the magnificent, messy complexity of this 'naughty world' that can be illuminated but neither entirely boxed in nor completely explained by our theories of it.[47]

Consider first John and Jane Murray. In a very real sense, John Murray lived in several worlds. He was a colonial Briton, a resident of Nova Scotia subject to its statutes (which were themselves subject to disallowance in London) and entitled to participate in the election of a representative from his county to the House of Assembly in Halifax. But quite how strongly he felt his provincial identity, and whether it was as important to him as his identification with Pictou, Hardwood Hill, the small cluster of people who lived between Haliburton Stream and the harbour, or the larger congregation of First Church we will never know. Certain it is, though, that as a prosperous improving farmer, John Murray was a participant in the agricultural enlightenment, rational, observant and calculating in the operation of his farm. A devout Presbyterian, he gave a significant part of his time to the affairs of his church. A Scot in Pictou, he found his religious identity reinforced by his ethnic allegiance. Sometime overseer of district roads, prizewinner, judge, and committee member in the local agricultural society, he was both recipient and (indirectly) dispenser of small sums of provincial government largesse. Connected firmly to the market, he was also enmeshed in a web of local exchange and mutual interdependence. For Murray these, surely, were not separable spheres. Life involved a complex set of obligations, involvements, decisions and actions. It was centred on New Rhynie and essentially local, despite the wider economic, political and intellectual horizons to which many of Murray's contacts ultimately led.

And how much more decisively was this true of Jane Murray. If the state impinged little on the everyday existence of her husband, its presence was almost entirely absent from hers. She was not, of course, entirely removed from its orbit. In denying her a vote because she was a woman, the statutes of Nova Scotia limited her participation in political life and circumscribed her ability to influence the exercise of political power in her society. In ascertaining how much butter and homespun she made in the course of a year, provincial authority 'invaded' her home – but it did so in the person of a neighbour, and then only once a decade. So far as we can judge, and on balance, in short, 'state activities, forms, routines and rituals' neither greatly affected the conduct of life nor played much part in the constitution of Jane Murray's identity (or that of countless other mid-century Maritime colonists).

Recognizing as much, we are drawn to conclude that religion was a far more powerful influence upon everyday life in these mid-century colonies than was the influence of the provincial 'state'. Consider in support of this contention Bishop Medley (who was inclined to give credit to 'gracious Providence' for the constitutional freedom enjoyed by New Brunswickers). Or contemplate those who marched behind Orange banners through the streets of Saint John in the 1840s. Remember those Catholic Irish who resisted Protestant incursions into their districts of the city. Recall those Presbyterians who objected to the failure of provincial census takers to accurately record the nuances of their sectarian subdivisions. And bring to mind those Acadians of Chezzetcook whose houses clustered around their chapel. Generally closely associated with the tragic myths on which each of the ethnic/immigrant groups of the region built cohesion out of their pasts, religion was in some important sense the substance of the region's several mid-nineteenth-century cultures, the means of social integration through which collective memories were organized and constructed to form a far more effective basis for definitions of 'us' and 'other' than was provided by the still relatively feeble fabric of the colonial state.[48] Colonial politics in the mid-nineteenth-century Maritimes were what W. L. Morton described twenty years ago as 'limited politics': for most people they were secondary to the more compelling preoccupations of religion, ethnicity, business and survival.[49]

None of this is to deny the existence of what might broadly be called a 'political culture' in the Maritimes. But as Greg Marquis has argued recently, this was a culture with its roots firmly planted across the Atlantic.[50] Editorial and other contributions to the provincial press, and speeches made in councils and assemblies leave no doubt that 'the plain language of [Maritime colonial] politics ... came from English history'. In relatively recently settled colonies, whose people were deeply divided by religion, ethnicity and experience; whose economies were equally fragmented into distinct sectors (with different needs and interests); and which lacked the communications infrastructure to integrate scattered peoples into a single community, reference to an English past (and all that heritage stood for in terms of industrial ascendancy, imperial achievement and political tradition) provided a ready symbolic touchstone for the majority of the region's peoples. They were – as they never tired of reiterating in protest at the reduction of British preferential tariffs or tenancy in Prince Edward Island – British *subjects*, their rights, implicitly, those secured by the Glorious Revolution of 1688, rather than the 'rights of man' due 'citizens'.[51] Guard and cherish and preserve the integrity of 'our limited monarchy', the 'nice balance' of 'our respective estates and realms' urged Bishop Medley. 'Our sufferings and our joys are the common property of the empire.'

Time and again, colonials looked to developments in seventeenth-century England to give meaning to their circumstances. For Orangemen the references were specific and the images vivid. Ritual, songs, and parades celebrated the Protestant Succession, revered William III and commemorated the Battle of the Boyne. In lodge after lodge members of the order learned a particular version of the past that made much of loyalty to the Crown and the need for eternal vigilance against papist intrigue. For others, the lessons of history were framed in less explicit and extreme terms, but they were nonetheless important. If the tenants of Prince Edward Island were downtrodden, they could find hope in 'acquaintance with English history' which revealed 'the historic tendency of the race to throw off oppression'. When in 1868, Canadian initiatives were seen to threaten Nova Scotian liberties they were compared to the rule of the Stuarts. And seventeenth-century 'Country Party' rhetoric, suspicious of centralized government, opposed to taxation and attached to local custom, echoed through the pre-Confederation Maritimes.[52]

Of all of this there was no better symbol than the monarchy, and there is no better indication of its significance in the Maritimes than the events surrounding the visit of the Prince of Wales to the region in 1860. Although Halifax publisher William Annand made available a collection of Joseph Howe's letters and speeches in 1858 – and in the next decade or so the 'discontinuity' of (impending) Confederation spurred publication of documents pertaining to the early-eighteenth-century history of Nova Scotia, a new, political, history of New Brunswick and other reflections of a developing awareness of provincial distinctiveness – through the mid-century decades, 'loyal, true, and brave' Maritimers courted Victoria's smile.[53] To the degree that they did so, the 'colonial state' remained in some important sense a limited state.

And so, in the end, there is perhaps no better way of making sense of this kaleidoscopic picture than with a simile. Recognizing with historical sociologist Michael Mann that social life is built upon the overlapping skeletons of state, culture, and economy but that these are very rarely congruent, we might conceptualize the frameworks that gave pattern to life in the mid-nineteenth-century Maritimes as the parts of a wheel.[54] In this view, London appears as the ideological and emotional hub of the 'political culture' of the Maritimes; the colonial state becomes the rim, linking people together and exercising certain authority over them, but generally in a weak and provisional way, peripherally; and religion, ethnicity, and locality, coupled with those quintessentially Victorian doctrines of system, sobriety, thrift and toil (especially on the land) assume the position of spokes, giving shape and form and structure to the everyday lives, and identities of most colonists.

ACKNOWLEDGEMENTS

Written, initially, at the invitation of historians Allan Greer and Ian Radforth for a 1989 workshop intended to offer new perspectives on mid-nineteenth-century BNA, this chapter represents a first effort to bring to bear on the Maritimes some of the literature and ideas that have been the focus of discussion among human geographers at the University of British Columbia for some time now. I am especially indebted to my colleagues for the lively, challenging, but always pleasant and thought-provoking exchanges we have shared. I am also grateful to Greg Marquis for providing me with a copy of his unpublished 'In defence of liberty'. This chapter is substantially shorter than the original, which Drs Greer and Radforth hope to publish in Canada in a volume on state formation in BNA. I much appreciate their allowing me this opportunity to put a version of this work before a geographical audience. Equally, I am grateful to Drs Baker and Biger for agreeing to accept a rather longer chapter than it was possible to present at the Seventh ICHG.

NOTES

1 For the development of the region see G. Wynn, *Timber colony: a historical geography of early nineteenth century New Brunswick* (Toronto 1981); W. S. MacNutt, *The Atlantic provinces: the emergence of colonial society, 1712–1857* (Toronto 1965) 213–70; and J. Gwyn, 'A little province like this': the economy of Nova Scotia under stress, 1812–1853 *Canadian Papers in Rural History* 6 (1988) 192–225

2 A. H. Clark, Old World origins and religious adherence in Nova Scotia *The Geographical Review* 50 (1960) 54–72; A. H. Clark, *Three centuries and the island* (Toronto 1959); G. Wynn, Ethnic migrations and Atlantic Canada:

226   Graeme Wynn

geographical perspectives *Canadian Ethnic Studies/Etudes Ethniques au Canada* 18 (1986) 1–15

3 G. Wynn, The Maritimes: the geography of fragmentation and underdevelopment, in L. D. McCann (ed.), *Heartland and hinterland: a geography of Canada* (Scarborough, Ont. 1987) 174–246; J. G. Reid, *Six crucial decades: times of change in the history of the Maritimes* (Halifax 1987). Quotations from A. G. Bailey, Creative moments in the culture of the Maritime provinces, in *Culture and nationality: essays by A. G. Bailey* (Toronto 1972) 49; R. C. Harris and J. Warkentin, *Canada before confederation* (Toronto 1974) 169; MacNutt, *Atlantic provinces*, 268; C. Bruce, Words are never enough, in R. Cockburn and R. Gibbs (eds.), *Ninety seasons: modern poems from the Maritimes* (Toronto 1974) 62. See also G. A. Rawlyk (ed.), *Historical essays on the Atlantic provinces* (Toronto 1967) 1; D. C. Harvey, The heritage of the Maritimes *Dalhousie Review* 14 (1934) 29; J. M. Beck, The Maritimes: a region or three provinces? *Transactions, Royal Society of Canada*, Series IV 15 (1977) 301–13

4 K. S. Inglis, Ceremonies in a capital landscape: scenes in the making of Canberra *Daedalus* (Winter 1985) 85–126 was in my mind as I began this chapter, but a later, second, reading of it left me disappointed that it fell short in the task of interpretation

5 T. W. Acheson, *Saint John: the making of a colonial urban community, 1815–1860* (Toronto 1985) offers a full and careful treatment of most of the material summarized here; as elsewhere in this chapter, emphases and arguments are not necessarily those of the works from which information is drawn

6 *Ibid.*, 92–114. S. W. See, The Orange Order and social violence in mid-nineteenth century Saint John *Acadiensis* 13 (1983) 68–92

7 Acheson, *Saint John*, 100

8 *Ibid.*, 219

9 *New Brunswick Courier* (Saint John) 9 January 1841

10 Acheson, *Saint John*, 224

11 *St Andrews Standard* 14 and 21 March 1840

12 J. Ross, *Remarks and suggestions on the agriculture of Nova Scotia* (Halifax 1855); see also G. Wynn, Exciting a spirit of emulation among the 'Plodholes': agricultural reform in pre-confederation Nova Scotia *Acadiensis* 20 (1990)

13 Public Archives of Nova Scotia, MG.100, vol. 194, no. 16. New Rhynie Farm Diary and Accounts, 1853

14 A. L. Spedon, *Rambles among the bluenoses; or, reminiscences of a tour through New Brunswick and Nova Scotia* (Montreal 1863) 59–65

15 F. S. Cozzens, *Acadia: or a month with the bluenoses* (New York 1859) 33

16 H. M. Hurd (ed.), *The institutional care of the insane in the United States and Canada* (Baltimore 1916–17) vol. I, 427–97 and vol. IV, 37–119, 203–18. D. Francis, The development of the lunatic asylum in the Maritime Provinces *Acadiensis* 6 (1977) 23–38

17 Cozzens, *Acadia*, iv

18 J. F. W. Johnston, *Notes on North America, agricultural, economical and social* (Edinburgh 1851) vol. I, 7

19 Cozzens, *Acadia*, 40–1

20 A. A. Brookes, 'Doing the best I can': the taking of the 1861 New Brunswick census *Histoire Sociale/Social History* 9 (1976) 70–91

21 *The tour of HRH the Prince of Wales through British America and the United States, by a British Canadian* (Montreal 1860) 25–51

22 The arguments of this paragraph have, of course, been well made several times; see M. Bloch, *The historian's craft* (New York 1953); B. Bailyn, The challenge of modern historiography *American Historical Review* 87 (1982) 1–24; and G. K. Gilbert, The inculcation of scientific method by example *American Journal of Science* 3rd ser., 31 (1886) 286–7

23 D. Gregory, 'Grand maps of history': structuration theory and social change, in J. Clark, C. Modgil and S. Modgil (eds.), *Anthony Giddens: consensus and controversy* (London 1990) 217–33

24 See especially M. Mann, *The sources of social power. Vol. I. A history of power from the beginning to AD 1760* (Cambridge 1986); P. Corrigan and D. Sayer, *The great arch: English state formation as cultural revolution* (Oxford 1985); A. Giddens, *The constitution of society: outline of the theory of structuration* (Cambridge 1984); A. Giddens, *A contemporary critique of historical materialism. Vol. I. Power, property and the state* (London 1981); A. Giddens, *A contemporary critique ... Vol. II. The nation state and violence* (Berkeley 1987). This chapter is intentionally double-edged. Its first purpose is to look at the Maritimes through the lenses provided by the above-mentioned literature. Its second is to respond to the challenge of 'evolving an appropriate style, a mode of discourse' capable of wedding evidence and theory, of transcending the dualism between 'analytic' and 'narrative' history, and of dealing coherently with space, time, agency, and structure (on this challenge see P. Abrams, History, sociology, historical sociology *Past and Present* 87 (1980) 3–16). It may well be that I hit neither target squarely here, but it does seem important that the effort to ground Giddens' ideas (in particular) be made, because in much of his writing 'particular contexts are used more as passive illustrations than as active explanations' (N. Thrift, Bear and mouse or bear and tree? Anthony Giddens' reconstitution of social theory *Sociology* 19 (1985) 621)

25 M. Weber, *Economy and society* (Berkeley 1978) vol. I, 100–5; Corrigan and Sayer, *The great arch.* Afterthoughts 182–5; Giddens, *The nation state and violence* 123–33. Through this first section of the conclusion, I attempt to realize, in a way, Giddens' claim that theoretical concepts (such as those of structuration theory) are 'sensitizing devices', and that being theoretically informed does not mean 'always operating with a welter of abstract concepts'. Thus such 'theorizing' as there is here is deliberately unobtrusive; it draws most explicitly from Corrigan and Sayer, *The great arch.*

26 Paul Jones to the editors *New Brunswick Courier* 5, 12, and 26 January, 2, 9, and 16 February, 2 March 1850

27 G. Wynn, 'Deplorably dark and demoralized lumberers'?: rhetoric and reality in early nineteenth century New Brunswick *Journal of Forest History* 24 (1980) 168–87

28 J. K. Chapman, The mid-nineteenth century temperance movement in New Brunswick and Maine *Canadian Historical Review* 35 (1954) 43–60; E. J. Dick,

From temperance to prohibition in 19th century Nova Scotia *Dalhousie Review* 61 (1981) 530–52

29 W. S. MacNutt, *New Brunswick, a history: 1784–1867* (Toronto 1963) 351; J. Fingard, The relief of the unemployed poor in Saint John, Halifax, and St John's, 1815–1860 *Acadiensis* 5 (1975) 32–53

30 J. S. Mill cited in Corrigan and Sayer, *The great arch*, 129

31 See, *op. cit.* 79–80

32 I. R. Robertson, The Bible question in Prince Edward Island from 1856 to 1860 *Acadiensis* 5 (1976) 3–25; see also I. R. Robertson, Party politics and religious controversialism in Prince Edward Island from 1860 to 1863 *Acadiensis* 7 (1978) 29–59

33 A. J. B. Johnston, Popery and progress: anti-Catholicism in mid-nineteenth century Nova Scotia *Dalhousie Review* 64 (1984) 146–53

34 J. Fingard, Jailbirds in Victorian Halifax in P. B. Waite, S. Oxner, and T. G. Barnes (eds.), *Law in a colonial society: the Nova Scotia experience* (Toronto 1984) 89–102

35 D. C. Harvey, The intellectual awakening of Nova Scotia *Dalhousie Review* 13 (1933) 1–22

36 T. C. Haliburton, *The clockmaker; or the sayings and doings of Samuel Slick of Slickville* 1st–3rd ser. (London 1839–40); see also J. M. Beck, *Joseph Howe* (Kingston 1982). Vol. 1. *Conservative reformer, 1804–48.* Vol. 2. *The Briton becomes Canadian, 1848–73*

37 Robertson, Bible question 3

38 S. Oxner, The evolution of the lower court of Nova Scotia in Waite *et al.* (eds.), *Law in a colonial society*, 59–80. J. M. Beck, *The evolution of municipal government in Nova Scotia, 1749–1973* (Halifax 1973). H. Whalen, *The development of local government in New Brunswick* (Fredericton 1973) 11–39

39 J. Howe, *Lord Falkland's government* (Halifax 1842) 2–7

40 Cited in G. Marquis, In defence of liberty: 17th-century England and 19th-century Maritime political culture (unpubl. paper delivered at the Atlantic Canada Studies Conference, Edinburgh 1988) 9

41 For examples see G. Wynn, Administration in adversity: the deputy surveyors and control of the New Brunswick Crown forest before 1844 *Acadiensis* 7 (1977) 49–65; S. J. Hornsby, An historical geography of Cape Breton Island in the nineteenth century (unpubl. Ph.D. thesis, University of British Columbia 1986) 82–91

42 Sorely limited though it is, this discussion attempts to edge toward an exploration of the 'fundamental *reciprocity* between theoretical constructs and empirical materials' which some have found wanting in Giddens' writing (see, for example, N. Gregson, Structuration theory: some thoughts on the possibilities for empirical research *Environment and Planning* D.5 (1987) 73–91; for the reciprocity quote see D. Gregory, Thoughts on theory *Environment and Planning* D.3 (1985) 387). The most relevant of Giddens' writings to the discussion that follows is *The nation state and violence*, but see also *Power, property and the state* and *The constitution of society*. The extensive literature commenting on Giddens' project is also useful in explicating its evolving pattern; in a mountain of material see for example A. Callinicos, Anthony Giddens: a contemporary

critique *Theory and Society* 14 (1985) 133–66; H. F. Dickie-Clark, Anthony Giddens' theory of structuration *Canadian Journal of Political Social Theory* 8 (1984) 92–110; D. Gregory, Space, time and politics in social theory: an interview with Anthony Giddens *Environment and Planning* D.2 (1984) 123–32; *Theory, Culture and Society* 1 (1982) 63–113, a 'symposium' on Giddens; and D. Held and J. B. Thompson (eds.), *Social theory of modern societies: Anthony Giddens and his critics* (Cambridge 1989)

43 For another variant see J. Fingard, Masters, and friends, crimps and abstainers: agents of control in 19th century sailortown *Acadiensis* 8 (1978) 22–46

44 M. Barkley, The Loyalist tradition in New Brunswick: the growth and evolution of an historical myth *Acadiensis* 4 (1975) 3–45

45 A. R. Stewart, Sir Edmund Head's memorandum of 1857 on Maritime union: a lost Confederation document *Canadian Historical Review* 26 (1945) 406–19

46 For some discussion of the geography of movement in the region during these years see J. S. Martell, Intercolonial communications, 1840–1867, Canadian Historical Association, *Report* (1938) 41–61; and G. Wynn, Moving goods and people in mid-nineteenth century New Brunswick *Canadian Papers in Rural History* 6 (1988) 226–39

47 Cf. Mann, *Sources of social power*, 4, 'societies are much messier than our theories of them' and B. Kennedy, A naughty world *Transactions of the Institute of British Geographers* new ser. 4 (1979) 550–5

48 S. F. Wise, God's peculiar peoples, in W. L. Morton (ed.), *The shield of Achilles: aspects of Canada in the Victorian age* (Toronto 1968) 36–61

49 W. L. Morton, Victorian Canada, in Morton, *Shield of Achilles*, 311

50 Marquis, In defence of liberty

51 J. M. Beck, 'A fool for a client': the trial of Joseph Howe *Acadiensis* 3 (1974) 27–44

52 Marquis, In defence of liberty, has more on these matters; any serious sampling of the contemporary press throws up examples

53 W. Annand, *The speeches and public letters of the honorable Joseph Howe* (Boston 1858) 2 vols.; T. B. Akins, *Selections from the public documents of Nova Scotia* (Halifax 1869)

54 Mann, *Sources of social power*, Chapter 1

11

Interpreting a nations' identity: artists as creators of national consciousness

BRIAN S. OSBORNE

This chapter is not concerned with landscape but is about ideology – the ideology of nationalism. To be sure, the 'nature of place' has contributed to a 'sense of place' and to that degree landscape is part of national identity. But this landscape is a more contextualized one than the externally observed landscape. Nor is it the academic construct produced by assembling distinctive component parts. The context of the lived-in landscape is the history and culture of those who fabricated it and live in it. It is their habitat. It is the landscape with people in it. Such landscapes are very germane to many peoples' senses of national identity.

Any attempt at understanding the development of national identities can learn much from Tonnies' classic formulation of the transition of societies from *Gemeinschaft* (community) to *Gesellschaft* (society).[1] Indeed, an important dimension of the relationship between the continua of community/society and nation state is the varying degree to which they are grounded in a sense of place.

Gemeinschaft is essentially local and immediate. As such, the lived-in landscape becomes a fundamental concept central to a people's sense of community, heritage and nationhood. It is the setting for the day-to-day economic, social and ideological activities that serve to unite a distinctive people. All these culturally significant actions serve to imbue that place with evocative associations and transform it into a symbolically charged repository of past practices and events. Some sites are elevated to the level of shrines or sacred places. Such a landscape supports the culture as its 'hearth' of origin, its home for the present, and its refuge for the future. The idea of this common residence in a particular place is often the matrix binding the nation together.

Gesellschaft requires that people identify with an abstraction – the territory of an artificially constituted unit, the state. Long-standing localisms are replaced by new centralizing structures of government, communications and bureaucracy that facilitate the functional interaction of the state.[2] Attempting to cultivate the sense of 'oneness' that was so strong in the pre-state *Gemeinschaft*, states consciously nurture shared symbolic constructs that reinforce the sense of membership to the group and the development of a sense of distinctiveness. Indeed, Zelinsky has argued that 'modern states could neither exist nor operate effectively without an adequate body of symbol and myth, whatever other excuses they may have for their creation'.[3] Through these symbolic contributions, states are able to establish identities and thus create collective memories or 'imagined communities'.[4]

This assertion of the primacy of the centrist identity over the local identity is always difficult, and is especially so with states that lack long-standing ties with place. Immigrant nations such as Canada find themselves composed of mixed populations whose origins are diverse, whose remembrances are of former identities, and whose associations with the 'New land' are recent and limited. The challenge of new nationalisms has been to nurture the process of developing new identities in the new locales and, indeed, to accelerate the cultivation of a distinctive national consciousness. How this is done is central to my concern here.

Art and national consciousness

Louis Wirth has argued that such 'mass-societies' with some degree of collective consensus are creatures 'of the modern age and are the product of the division of labour, of mass communication and a more or less democratically conceived consensus'.[5] Leaving the matters of division of labour and democracy aside, the adoption of shared beliefs and values is facilitated by the dissemination of symbolic content by mass-communications to large, dispersed and heterogeneous populations. Indeed, this relationship between media and cognition is crucial:

The ways we reflect on things, act on things and interact with one another are rooted in our ability to compose images, produce messages, and use complex symbol systems. A change in that ability transforms the nature of human affairs ... New media of communication provide new ways of selecting, composing and sharing perspectives. New institutions of communication create new publics across boundaries of time, space and status. New patterns of information animate societies.[6]

Gerbner is referring here to the electronic age of 'mass production of symbols and messages' as a 'new industrial revolution in the field of culture'. But long before the world of electronic visual imagery, another 'animation of society' took place: the world of mass-produced visual imagery in book illustrations, newspapers, magazines and posters.

The printed word requires literacy and even in the modern developed world, over 20 per cent of the populace are functionally excluded from its messages. Sound communication is accessible to all possessed of hearing. Visual communication is an even more immediate mode of communication and the nineteenth and twentieth centuries have witnessed the direction of visual symbols to the promotion of particular ideologies – nationalism being one of them.

Some have argued that artists have had a powerful role in this process as interpreters of the essence of the experience of nationhood. Often, artistic imaginations are rooted in their lived-in worlds and their creative responses may informally contribute new and insightful dimensions to their culture-group's identification with their locus.[7] Other artists, however, have gone further. They have attempted to nurture a sense of nationalism by intervening formally in the creation of a discrete cultural identity and national consciousness. They have marshalled their imagery in a campaign to create a national imagination.

For example, the second half of the nineteenth century witnessed the application of realism in art to the expression of nationalistic sentiments in Europe. Attention was directed to the land of the people, to the peasantry and its tasks, and to the documentation of its way of life in all its mundane forms. The local and familiar became national and significant. Thus, Jean-François Millet dramatized the 'noble peasant' of rural France in such well-known subjects as *The sower* (1850–1), *The angelus* (1854–9), *The gleaners* (1857), *Man with hoe* (1859–62). If these vignettes of peasant life are remarkable for their preoccupation with the vernacular and mundane, they are powerful statements of new ideological currents in nations that concern themselves with the social conditions of the people and their poverty and labour (Fig. 11.1). They are recorded with a heroic grandeur reserved formerly for more painterly subjects. They were symbolic of a new ideological climate, and easily appropriated by the masses to advance it.[8]

A more explicitly ideologically charged mission may be recognized in nineteenth-century Russia. A major reaction against the establishment-art of church icons and patriotic mythology is associated with the 'Wanderers' (*peredvizhniki*) whose travelling exhibitions served as commentaries and critiques of contemporary Russian society. For one of them, Kramskoi, 'only a social feeling gives an artist strength and also the assurance that the work of the artist is needed and valued by the public'.[9] Such works as Kasatkin's *The worker-warrior* (1905) or Repin's *Volga bargemen* (1873) demonstrate their application of social realism as a critique of conditions in Tsarist Russia, together with the propogation of themes that were vernacular, nativist and accessible to popular appreciation (Fig. 11.2).

Less polemic, but no less nationalistic, priorities may also be recognized in other contexts. Indeed, nowhere was the need for powerful central images more pressing than in the 'New Worlds' of colonial expansionism

Fig. 11.1 Jean-François Millet, *Buckwheat harvest summer* (1868–74).

Fig. 11.2 I. Repin, *Bargemen on the Volga* (1873).

where identities were often still tied to colonial mother countries, and often diffused by diverse immigrant origins. The United States may be taken as the classic case. A colonial origin, an immigrant population and a continental extent posed major problems for the establishment of a unitary nation state. Zelinsky has demonstrated how imaginative the nation builders

[234 Brian S. Osborne]

234   Brian S. Osborne

Fig. 11.3   Norman Rockwell, *The four freedoms* (1943).

have been in creating a national mythology.[10] But it was not all flags, anthems and monuments. If Frederick Jackson Turner, Oscar Handlin and others provided a historiography that focussed on frontier and immi-gration experiences, Fredric Remington, Charles Russell, Norman Rockwell, and thousands of artists working for the New Deal's WPA popularized them. They communicated their view of the subtleties of the essential 'American experience' in an accessible mode for the American people (Fig. 11.3). They helped create a popular mythology of nationhood.[11]

    For Australia, Heathcote has identified the persistent themes of conflict with the native population, and nature, and of the progress effected by

Fig. 11.4   Julian Ashton, *The prospector* (1889).

pioneers in their transformation of the wilderness into productive land-scapes.[12] Apart from documenting these processes and contemporary atti-tudes, many of these works also contributed to the creation of a 'bush ethos' at the core of Australian national character (Fig. 11.4). Early Australian nationalism was thus served by the 'eidolon' of the pioneer in the outback rendered in mythic terms to personify the values of egalita-rianism, labour and optimism. It was a simplification and idealization of a stereotype that suited the emerging nationalism in Australian art. This nationalistic school of painting helped Australians define themselves in terms of their distinctive isolated location, distinctive social history and

Fig. 11.5   Tom Roberts, *Bailed up* (1895).

distinctive landscape. Thus, one of these artists, Tom Roberts, declared his intention to seek out and render on canvass 'the earnestness, vigour, pathos and heroism' of Australian life (Fig. 11.5); he dedicated himself to a mission of 'self-discovery and self-determination' to create for Australians a 'shared vision of the landscape'.[13] For another Australian artist, E. Mitchell, his rendering of the land, 'my picture of reality', allowed access to the quintessential questions at the heart of the young nation's search for identity:

Somewhere within it, is there an idea of our country that is not only mine – the 'land geist' that belongs to the north and the west, to the centre and the cities, the beaches and the coast of all this island continent? Is there a reality or only a multitude of appearances? The 'land geist' – if it exists, and I believe it does – may have the form which will shape our molten nation, may be a spirit whose breath will inspire. But spirit and form are things we can refuse – things we may not even see or feel if our eyes are blind and our intuition unawakened.[14]

More ideologically explicit appropriations of art were associated with those attempting to promote 'revolutionary' images of new states. Art became an important instrument of nationalistic propaganda in the Orwellian world of demagoguery created by Stalin, Hitler, Mussolini and Franco.

Thus, the aftermath of 1917 in the USSR required that social realism gave way to a 'revolutionary realism' that required art to put its shoulder to the wheel of change. Indeed, it may better be considered as the exercise of 'revolutionary romanticism' as the challenges of the 1920s required over-simplified idealizations and glorifications of the heroism and optimism of revolution. Artistic exercises in impressionism, 'Cézannism', abstract art or any other of the 'art for art's sake' schools were declared redundant and inappropriate because they were said to be bourgeois, non-Russian, individualistic, foreign, decadent, mystical and erotic. Such works were criticized for not addressing the apparent 'contradictions' in society, for retreating into the world of foolish fantasies and for avoiding the 'truth'. Clearly, 'truth' was to be defined ideologically and always with an eye to the best interests of the state. Not surprisingly, therefore, the 17th Party Congress of 1932 banned all art forms other than those approved by the state and in 1934 the preferred model of 'socialist realism' was defined by Zhdanov at the First All-Union Congress of Writers as:

knowing life so as to be able to depict it truthfully in works of art, not to depict it in a dead scholastic way, not simply as 'objective reality', but to depict reality in its revolutionary development. In addition to this, the truthfulness and historical concreteness of the artistic portrayal should be combined with the ideological remoulding and education of the toiling people in the spirit of socialism.[15]

Such a preoccupation with art in the service of the state produced a 'varnished' form in that the subjects of this 'official art' were removed from the grounded realms of reality to the fantastical realms of ideology (Fig. 11.6). the titles tell the tale: Gerassimov's *Stalin at the XVIIIth Party Congress* (1939); Kukrynisksy's *The end: the last days of Hitler in the bunker* (1947); and Latkionov's *Letter from the Front* (1947). To be sure, there was an accommodation of landscape themes as emotive statements of homeland and prideful documentation of industrial landscapes and economic achievement – and all in line with the required social realism.[16]

At the other ideological extreme, the National Socialism of Germany's Third Reich also rejected the impressionism, *Die Neue Sachlichkeit* (The New Objectivity), cubism and Dada that had been associated with the Weimar Republic. Here again they were considered to be inappropriate vehicles for proselytising the values of the state. Their abstractions, incoherence and cosmopolitanism were seen as interfering with the presentation of a coherent message of German national identity. An influential Austro-German painter of the day, one Adolf Hitler, forcefully expressed the power of art in motivating popular identification with the 'nation' and its mission:

art is more effective than any other means that might be employed for the purpose of bringing home to the consciousness of the people the truth of the fact that their individual and political sufferings are only transitory, whereas the creative powers and therewith the greatness of the nation are everything. Art is the great mainstay of the people, because it raises them above the petty cares of the moment and shows them after all, [that] their individual woes are not of such great importance.[17]

Fig. 11.6   A. I. Latkionov, *A letter from the Front* (1947).

For Hitler's Nationalist Socialist Germany, therefore, as in the Marxist–
Leninist Soviet Union, art served the state by reinforcing its mission and
providing psychic energy for the collective national consciousness. Its
importance to the 'Germanic Renaissance' was in illustrating a romantic
folk history, creating images of favoured national stereotypes, and generally contributing to a national solidarity (Fig. 11.7). Apart from perhaps
Mussolini's fascism, the demagogues of the twentieth century favoured an
image of their new nation states that was usually founded on artistic
realism, generally favoured bucolic and populist images, and always a
romanticized view of the past.

Fig. 11.7   W. Tscheich, *High Nazi morale amidst the rubble* (1944).

Creating a nativist image for Canada

Canada shared many of the problems facing nation states seeking centralizing rationales during the late nineteenth and early twentieth centuries and
artists were called upon here, too, to establish a distinct Canadian imagination. The problems were familiar: a very strong imperial connection, a
diverse immigrant population, an enormous continental extent and a substantial French minority acquired by conquest. Imperialism might have
served to provide an umbrella of identity but it did nothing for national
unity. In the decades following Confederation, however, Canadian arts and
letters began to exhibit a preoccupation with nationalist themes. Conven-
tionally, three major contributions may be noted: *Picturesque Canada*, The
Toronto Art Students' League and its offshoots and the 'Group of Seven'.
   Ironically, two Americans, Howard and Ruben Belden, were the first to

corner the Canadian market in the genre of picture books of countryside, county atlases, local histories and biographies of famous men. Their most influential publication, however, was *Picturesque Canada*. This pioneering venture in popularizing the visual images of rural and urban Canada consisted of 500 wood engravings accompanied by regional essays. It has been claimed that its publication and widespread circulation 'did more to kindle an interest and pride in Canadian scenery and Canadian pictorial art than any single agency up to that time'.[18]

Another critical agency in the development of a nationalist consciousness of place was the Toronto Art Students' League founded in 1886. Such artists as David F. Thomson, Robert Holmes, Duncan McKellar, Owen Staples, C. W. Jefferys, Thomas Greene and J. E. H. MacDonald – together with Gertrude Spurr, Henrietta Hancock and Mary Wrinch – constitute a group of cultural *provocateurs* that merit as much prominence as their more famed successors, the Group of Seven. Through their annual calendars, they popularized the domestic, vernacular and native Canada that was symbolic of a growing nationalistic spirit.[19] For at least one authority: 'Their work was the very core and marrow of Canada, and grew out of an informed consciousness of Canada, a great pride in its past, and an utter devotion to the beauty and bewitchment of its changing seasons. These are the real makers of Canada'.[20] Together with such spin-off organizations as The Mahlstick Club, The Little Billie Club, The Graphic Arts Club, the Canadian Art Club, and the 'Grip' group, the League was a major intellectual force in honing a sensitivity and pride for things Canadian.

This campaign was carried forward by the better-known Group of Seven. Its members' explicit patriotic and nationalistic theme was clearly expressed on the occasion of their first exhibition: it was axiomatic that 'an Art must grow and flower in the land before the country will be a real home for its people'; they favoured 'artists native to the land, whose work is more distinctive, original and vital, and of greater value to the country'; and they welcomed 'any form of Art expression that sincerely interprets the spirit of a nation's growth'.[21] Whatever their various other accomplishments, many Canadians looked for a national distinctiveness in the Group's renderings of Canada. Their work spanned the extent of Canada: Casson's villages and rural landscapes of Ontario; Jackson's Laurentian hills and Gaspé coasts (Fig. 11.8); MacDonald's Rockies and Carmichael's mining camps. Though not of the Group, Emily Carr's focus on the forests, oceans and peoples of the Pacific West must also be considered as part of this process of exploring and bounding the national psyche. For these artists and their contemporaries, the land held the key to establishing a distinctive Canadian identity.

This theme was repeated later by some of the delegates to the Conference

Fig. 11.8   A. Y. Jackson, *Nellie Lake* (1933).

of Canadian Artists held at Kingston in 1941. Clearly, A. Y. Jackson had
not lost his sense of mission:

> If we leave any trace of ourselves in history it will not be through the records printed
> in our blue books, but: did we create great music, or drama; did we create a great
> painting or sculpture, or literature, or a triumph in research such as Banting's
> discovery of insulin? These are the real values, and that is why the vision and
> courage of the artist is one of the nation's greatest assets. Where there is no vision
> the people perish.[22]

Further reinforcement was to come with the orgy of reproduction of the
Group's work on the centennial of Confederation in 1967. Subsequently,
many were to criticize the inappropriateness of this Central Canadian
vision of Canada at the expense of other regional and social realities.
Nevertheless, for many, the essential components of the Canadian 'land
geist' had been established by these artists and their predecessors.

Popularizing images of Canada

Inflated and self-conscious though the claims of patriotic artists may be,
there can be no doubt of the potency of art in energizing nationalist

imaginations. Nor can there be any doubt that such potency was not lost on those concerned with creating, redirecting, or even controlling, such nationalist imaginations.

To be effective proselytisers of ideologies, however, works of art should be considered as modes of communication interceding between the imagination of the artist and the imaginations of those who read, hear, see or touch them. They are communicators of meanings that are a complex mix of the artists' intentions, the audiences' interpretations and the stylistic conventions and preferences of the day. And if this process is to be effective, such images have to be communicated to the populace and be accessible to it logistically and intellectually. The question of access is crucial.

There is, therefore, a tension between 'high culture' produced by a cultural elite operating within some established set of aesthetic standards and a 'mass-culture' that produces works for 'mass consumption' and operates according to the criteria of popularity, acceptability and large-scale access to its works. It has already been demonstrated how the wide circulation of *Picturesque Canada*, the populist enterprises of the Toronto Students' Art League and the commercialization of the work of the Group of Seven did much for disseminating a consensus view of the essence of Canada. But there were also other agencies – and other images of national identity too.

Beyond the walls of art salons and art clubs – but not entirely isolated from them – was another world of commercial art. Several developments in printing and publishing in the nineteenth and early twentieth centuries allowed the mass production of visual imagery and ushered in what has been called the 'democratic art' of illustration.[23] New processes in printing and visual reproduction implied more than a mere increase in mass-circulated media and greater access to more information. There were also important sensory and cognitive implications as Canadians 'emerged from a world which had very few accurate visual self-images and entered one in which for the first time most people, at little or no cost, could see life illustrated in books, newspapers, magazines, photographs, and films'.[24] Through this new range of visual media, the 'commercial artist' had new means of accessing the Canadian public:

New and improved reproduction processes, the urgent needs of publicity, [the] modern god of commerce, the magazine pages, newspaper advertising, billboards, window-displays, booklets, folders, broadsides and all the munitions of industrial warfare, all call for the services of the illustrator, the designer and the painter. Certain it must be that commercial art, as it is invidiously called to distinguish it from easel art, reaching millions as it does through the medium of the printed page and the painted hoarding, possesses a dominating force of incalculable influence and worth to the agencies that employ it.[25]

Fig. 11.9 Election poster by the Toronto Lithographic Company, 1891: the first full-colour election poster campaign in Canada.

Clearly, Colgate was speaking here of commercial worth. Others were aware, however, of the potency of those images 'reaching millions' in other contexts.

The world of politics was quick to grasp the idea. The new technology of mass reproduction of imagery was applied to producing eye-catching and mass-distributed posters, the 1891 federal election seeing the first use of colour lithography in a political poster. The famous rendering of Sir John A. MacDonald, *The old flag, the old policy, the old leader* (Fig. 11.9), rallied Canadians to the Conservatives' cause of the 'National Policy' in opposition to the Liberals' platform of continentalism. Claimed as 'one of the best known images in Canadian history',[26] it and its seventeen companions in the campaign contributed colourful visual imagery to a nationalistic cause – and, incidentally, to Sir John's Conservatives' sixth victory since 1867. Soon, illustrators were called upon to provide images advertising not only political ideologies but also more prosaic products ranging from railway companies to dry goods.

The last quarter of the nineteenth century was also witness to an effusion of visual imagery in the new genre of illustrated magazines whose mission was clearly stated by Alexander Somerville of the *Canadian illustrated news*:

Description fails, however graphic, terse, minute, to convey the mind a correct or lasting impression. For this reason there always has been, and is, a strong desire to substitute sight for sound, the real for the imaginary. To meet this want, to portray objects as they are, or were, to present to the eye a lifelike representation, is the aim of an Illustrated Paper.[27]

Magazines such as *The Canadian illustrated news* (1869–83), *L'Opinion publique* (1870–83) and *The Dominion illustrated* (1888–93) provided their thousands of readers with pictures of Canadian people, places and events. Illustrated journals with a declared commitment to cultural nationalism continued into the twentieth century. The *Canadian forum*, for example, had a well-developed stable of artists and engravers to embellish their pages: A. Y. Jackson, F. H. Varley, Lawren Harris, Frank Carmichael, A. J. Casson, David Milne, Thoreau MacDonald and, of course, C. W. Jefferys. The other perennials of Canadian journalism – *Macleans* and *Saturday night* – also made use of the artistic talents of established Canadian artists, including the Group of Seven and C. W. J. With their combined circulation of hundreds of thousands of issues, therefore, the work of these Canadian image-makers entered the homes and minds of the society at large.[28]

The late nineteenth century had also been a period of considerable strengthening of Canadian literature. Such worthies as Charles G. D. Roberts, Charles Sangster, Duncan Campbell Scott, Charles Heavysege, Bliss Carmen and William Wilfred Campbell and Archibald Lampman 'gave imaginative shape to the land and voiced the aspirations of its people in North American accents'.[29] This continued into the first decades of the twentieth century and, increasingly, these works were being illustrated by Canadian commercial artists. E. T. Seton wrote and illustrated several volumes on Canadian wildlife and the outdoors; W. H. Drummond's French dialect poems were illustrated by Frederick S. Cockburn; successive volumes of Louis Hemon's *Maria Chapdelaine* were illustrated by Suzor-Cote, Thoreau MacDonald and Clarence Gagnon (Fig. 11.10). Indeed, this cross-over between the worlds of commercial art and easel art

Fig. 11.10 Clarence Gagon, illustration for Louis Hemon's *Maria Chapdelaine* (1933).

is demonstrated by McClelland and Stewart's commissioning J. E. H. Mac-Donald, Lawren Harris, A. Y. Jackson, Franz Johnston, C. M. Manly, while Ryerson Press contracted F. H. Varley. As for C. W. Jefferys, several Canadian, British and American publishers availed themselves of his talents although his particular association with Lorne Pierce and Ryerson probably resulted in his most prolific endeavours.

These developments in the generation, duplication and mass distribution of visual images added new power to those attempting to reconceptualize the world in which Canadians lived. The textual commentary on Canada's distinctive place and past were henceforth accompanied by mass-produced illustrations intended to excite particular and well-articulated responses in the Canadian populace. They were creating images of Canada and diffusing them widely.

Didactic nationalism

For many of the leading nationalist painters, therefore, work as commercial artists for newspapers, magazines and publishing houses provided the economic support to pursue their easel art. Not unexpectedly, the motifs in such artists' formal art often carried over into their commercial work. Indeed, given the explicit nationalist missions of some of the early pictorial magazines and Canadian writers, there was a considerable complementarity of interest. Other artists, however, moved beyond mere economic and intellectual convenience and co-opted the mass-circulation potential of the commercial media to proselytise their views of Canada that had exercised their imaginations as creative artists. Of these, two are worthy of particular attention: Thoreau MacDonald and C. W. Jefferys.

Thoreau MacDonald was the artist's illustrator. A decade of illustrating for the *Canadian forum* in the 1920s allowed the refining of his metier: indigenous flora and fauna, seasonal rounds, rural chores. The publication of *A year on the farm* (1934) served as textual and artistic repository of wise saws, rural wisdom, and the folklore of the countryside recorded for future generations. His illustrations for his father's volume of poetry, *West by east*, were again of such domestic images as apple picking, harvest fields, farmyards, homesteads, cemeteries and grist mills. Clearly, he was establishing a bucolic view of a mythical land of preferred values. Nevertheless, MacDonald's view was widely circulated and met with considerable approval in some quarters as it rendered:

the very core of Canada, and grew out of an informed consciousness of Canada, a great pride in its past, and an utter devotion to its changing seasons. These are the real makers of Canada. From all this it will be seen that Thoreau MacDonald is not only proud to be a Canadian but glad to be a regionalist. One must have roots, here, in the soil, and one's art must have metes and bounds, for how can one embrace the cosmos?[30]

The second example of didactic illustrator, Charles William Jefferys (1869–1951) also contributed to the furthering of Canada's national self-image through his contributions as artist, illustrator, historian and popularizer of Canada's land and its people. A man of considerable accomplishment as a formal artist of Canada, it has been claimed that Jefferys was 'the first to preach the doctrine of the pine and spruce as themes fit for the painter. If there is anything in the theory that Canada has developed an art expression peculiar to this country, we must consider Jefferys as one of the very first to initiate this movement.'[31]

Jefferys was not, however, concerned only with the interpretation of the land. His self-imposed task was to interpret, render in art form and disseminate to his fellow Canadians, images of the nation's historical experience (Fig. 11.11). As noted by Sutherland, 'the artistic nationalism, fine design sense and stiff but scrupulous draftsmanship of Charles William Jefferys could be found everywhere in the late nineteenth century and early twentieth century magazines.'[32] And not only in magazines. His experience as newspaper illustrator included the New York *Herald*, the Toronto

Fig. 11.11   C. W. Jefferys, *Clearing land about 1830* (1945).

Globe, and Toronto *Star*, experience that he was to put to good use later:

It dawned on me that here was the very best training for the job I wanted to do. I realized that yesterday was as alive today, and that the accurate and intensive observation of how people acted now and here was the best way to understand how they acted in the past. It was a simple lesson; but for me it was almost a revelation, and thenceforth, in imagination, if not in execution, my dead men were no longer dead.[33]

Jefferys was soon to breathe life into people of the past.[34] His first exercises in fictive illustration were for Marjorie Pickthall's several children's adventures and the somewhat whimsical drawings for *Uncle Jim's Canadian nursery rhymes*. Later he would provide imaginative settings for such Canadian literary classics as Kirby's *Golden dog*, Richardson's *Wacousta*, and Haliburton's *Sam Slick*.

As Jefferys developed his skills as historical illustrator, he became conscious of the need to rein in his artistic imagination to accommodate historical facts, without losing that opportunity to dramatize, fantasize and entertain. He expressed his objectives thus:

These drawings, therefore, must be considered as imaginative conceptions. I have based them on authentic data, and have not neglected minor facts; but I hope that they may also convey something of the spirit and the larger significance of the

248   Brian S. Osborne

Fig. 11.12   C. W. Jefferys, *The march of the North West Mounted Police across the Prairies in 1874* (c. 1930).

events they depict. I think if they visualize in some degree the life of our past, and arouse an interest in the common heritage of our country's history, their main purpose will be fulfilled.[35]

With these principles before him, Jefferys applied himself to illustrating G. M. Wrong's 56-volume *Chronicles of America* (1921), 32-volume *chronicles of Canada* (1922), and the 15-volume *Pageant of America* (1929). Recognizing the need to produce his own rendering of Canada's history, Jefferys produced his own *Dramatic episodes in Canada's story* (1930), *Canada's past in pictures* (1934), and the 3-volume study *The picture gallery of Canadian history* (1942, 1945, 1950). Together, these volumes contain several hundred imaginative drawings of what Jefferys considered to be salient events in Canada's history, as well as of factual reproductions of people, things and places of the past (Fig. 11.12).

Not surprisingly, concerned as he was with romanticizing and popularizing Canada's past, he welcomed the opportunity to participate in the design and production of history textbooks for schools. Concern had long been expressed over Canadian school textbooks that were often turgid, usually foreign in content and production and always unillustrated. In 1903, the Ontario Department of Education's Committee on History, chaired, incidentally by one George Wrong, recommended that:

> every student should be requested to purchase efficient working libraries of historical books (especially Canadian), sets of historical readers, and sets of historical illustrations – those means whereby alone a definite and intimate knowledge of the past and present of our people may be obtained and a sound and intelligent national spirit may be nurtured in each child.[36]

In 1909, the Eaton Company commenced production of *Public school readers* although the nurturing of the 'national spirit' was limited by the fact that these proved to be reprints of British and American materials. But they were illustrated and Jefferys provided visual interpretations of such themes as Hector and Ajax, Life in the desert, or Androcolus and the lion. While short on Canadian content, this exercise did serve to reinforce Jefferys' commitment to didactic illustration. The first Canadian history textbook, George Wrong's *Ontario public school history of Canada*, did not become available until 1921. And it was not until 1926 with its reissue as *Canada – a short history*, that Jefferys was able to demonstrate the power of historical illustration in education.[37]

Because of the very volume and omnipresence of his work, Jefferys has been said to maintain a 'tight grip on the collective memory of his countrymen'.[38] For his eulogist, Lorne Pierce:

> [Jefferys] for forty-five years has explored with utter devotion the history and tradition of Canada, its great characters and outstanding events. To tens of thousands of boys and girls the name Jefferys is synonymous with Canadian history. His work is truly monumental, and calls to mind the shield Hephaistos made for Achilles – the epic of his people wrought in gold.[39]

It is a bit much, but C. W. Jefferys would have loved it – especially the allusion to mythology. He was very much in the business of mythology.

If such assessments are at all reliable, it behoves us to consider what his images of Canada were. Simply put, if Jefferys, in concert with the fellow members of the Toronto Art Students' League and the Group of Seven had assisted in the creation of a 'national geography', Jefferys and others were engaged in creating images of a popular 'national history'.

And there were others involved. Adam Sherriff Scot, for example, also generated paintings with a romantic historical theme. His work for the Hudson's Bay Company, *James Douglas greets Governor Simpson, Fort Victoria, 1828*, is a good example of the genre. Or again, his eleven murals ringing the Gallery Bar in *Le Chateau Montebello* include such cultural events as planting the 'mai' and a sugaring party, together with the usual array of such Canadian heroes as Champlain, Cartier and Mgr de Laval, mixed in with the more populist themes as canal construction, rafting and rebels in the 1837 uprising. In the same vein, consider the piece by Charles W. Simpson (1878–1942), *Cartier taking possession of New France, Gaspé, 1534.*[40]

Taken together, these works were preoccupied with applying their artistic and historical imaginations to the creation of a glorious past. Jefferys and cohorts not only produced appropriate images of place but also populated it with an appropriate pantheon of domestic heroes such as Breboeuf and Brock, Montcalm and Wolfe, Champlain and Thompson. Some have seen in this the 'big man in history' approach. Then again, others would argue that his history is primarily of Central Canada, and even though Quebec and to some extent the West are part of his *œuvre*, a central Canadian imperialism and Whiggish faith in national progress are much to the fore. But it must also be said of Jefferys that if he did highlight his view of Canadian history with an array of the elite from French, British, Quebec and Central Canadian historical mythology, he did not ignore more populist themes. He also demonstrated an interest in the ordinary person's story, the commonplace and the vernacular. In his work, he attempted to demonstrate how history was made by men and women of all ranks and classes and he attempted to tell stories that would be of interest to the same groups also. To this extent at least, Jefferys would have agreed with Lenin that 'Art belongs to the people, it must have its deepest roots in the broad mass of the workers. It must be understood and loved by them.'[41]

Conclusion

The last decade of the twentieth century has opened with events that underscore Wilbur Zelinsky's claim that 'Nationalism is the reigning passion of our times. Here is what breeds the energy that drives the nation-state'.[42] Internationalism has been urged as ideologically desirable, economically necessary and technologically possible and yet nationalisms abound. Their survival reflects the potency of the concept and the efficiency with which it has been cultivated by nation states.

Nations have been moved by rhetoric, the written word, music and art and 'pressure-pot' nationalism has often demonstrated its capability for orchestrating powerful symbols to establish ideologies. In the context of the arts, therefore, it becomes important to determine when overt jingoism and didacticism in the cultivation of a 'peoples' art' becomes 'official art' and even propaganda.

The tension in the relationship between the arts and nationalism is perhaps well demonstrated by the case of Canada. On the one hand, we have the likes of Lorne Pierce, eulogizing the contributions of two artists-cum-illustrators – Thoreau MacDonald and C. W. Jefferys – for their contributions to Canadian national identity:

Two such men save a nation that still has far too many who patronize its arts and letters, who grow ribald over the colonial mind, or even deny that we are a people at all. You can't build a nation that way ... There are those ... who will paint and carve and make songs, and theirs will be the Canada in the days to come that men will live and die for. They will build a continuing tradition in order that we may have a continuing civilization here.[43]

But then, there have also been others who have reacted against blatant Canadian nationalism in art and letters. An editorial in *Tamarack*, a prominent journal of Canadian literature, complained that:

The dream of a distinctively Canadian culture still possesses us. It is an idle dream. In the first place, the fact of our sovereign independence doesn't make us a distinct nation, it only makes us a political unit. In the second place, we do not become more distinctly ourselves by pursuing singularity, by countering the Daughters of the Revolution with the daughters of the Empire, or Davy Crockett with Pierre Radisson. Such gestures, like the beaver-and-maple school of writing, are evidence of a provincialism, not nationalism.[44]

Or again, no doubt with a touch of scholarly mischief and whimsy, Roberston Davies asks rhetorically in one of his early farces;

Why do countries have to have literatures? Why does a country like Canada, so late upon the international scene feel that it must rapidly acquire the trappings of older countries – music of its own, pictures of its own, books of its own – and why does it fuss and stew, and storm the heavens with its outcries when it does not have them?[45]

Indeed, perhaps too much of it has been kitsch, too obvious an appeal to images, tastes, issues and ideas of a limited historical currency or less than widespread application. One writer has argued that 'historians should not be in the identity business at all. It is certainly true that to believe that history alone is enough to establish identity is to expect too much of it.'[46] Commenting on contemporary German identity, he goes on to say:

The point is that the observer of a national culture must presuppose some initial or underlying characteristics. Any meaningful concept of a national identity must posit a subsisting component, which requires description in terms of nonhistorical variables. We need to know history, therefore, to understand identity; but history will not suffice. If it did, countries would move in worn grooves, and trajectories would be predictable.[47]

Nowhere is this better demonstrated than Canada in the 1900s. As the nation comes face to face with a 'post' Confederation set of scenarios, some of the old myths of identity come to be questioned. Perhaps Gagnon's images of the *Québecois habitant* have done more to strengthen French

Canadian regional consciousness than Thoreau MacDonald's for Ontario. Have Jefferys' images of a Franco-Anglo-Celtic nation building become irrelevant to a Canada of Ukrainians, Icelanders, Mennonites and Finns – let alone the more recent Italians, Portuguese, West Indians, Indians, Pakistanis, Chinese and Vietnamese? What are the new signifiers of Canadian identity for a population that is possessed of an urban, post-industrial and cosmopolitan world-view?

If Canada is to develop a new concept of a nation state – with or without Quebec – there will be a need for new national images. Some might argue that while the traditional images of the 'national history' are now irrelevant to the modern condition, those of the essential verities of the 'national geography' still stand.

NOTES

1 For a recent discussion of this theme see J. Anderson, Nationalist ideology and territory, in R. J. Johnston, D. Knight and Eleonore Kofman (eds.), *Nationalism, self-determination and political geography* (Beckenham 1988) 18–39
2 For the classic formulation of this see Karl Deutsch, *Nationalism and social communication: an inquiry into the foundations of nationality* (Cambridge 1953)
3 Wilbur Zelinsky, *Nation into state: the shifting symbolic foundations of American nationalism* (Chapel Hill 1989) 13
4 Benedict Anderson, *Imagined communities: reflections on the origin and spread of nationalism* (London 1983)
5 Quoted in George Gerbner, Mass media and human communication theory, in Dennis McQuail (ed.), *Sociology of mass communications* (Harmondsworth 1976), 41
6 *Ibid.*, 38
7 Brian S. Osborne, Fact, symbol, and message: three approaches to literary landscapes, *The Canadian Geographer* 32 (1988) 267–9; also, Brian S. Osborne, The iconography of nationhood in Canadian art, in Dennis Cosgrove and Stephen Daniels (eds.), *The iconography of landscape* (Cambridge 1988) 162–78
8 For more on Millet, see Robert Rosenblum, *Nineteenth century art* (New York 1984); also Alexander Murphy, *Seeds of impressionism* (Boston 1984)
9 A. C. Wright, Literature, art, and propaganda in the USSR, *Mosaic* 2 (1969) 42
10 Zelinsky, *Nation into state*
11 See Brian W. Dippie, *Remington and Russell* (Austin 1982); also John Ewers, *Artists of the Old West* (Garden City 1965) and Alan Gossow, *A sense of place: the artist and the American land* (San Francisco 1971)
12 R. Leslie Heathcote, Pioneer landscape paintings in Australia and the United States, *Great Plains Quarterly* 3 (1983) 131–45
13 V. Spate, *Tom Roberts* (Melbourne 1978) 86
14 E. Mitchell, Spirit of the land, *Meanjin Papers* 2 (1945)
15 Quoted in Wright, Literature, art, and propaganda, 40

16 As late as 1960, N. S. Khrushchev declared, 'It is through the best works of literature and art that people learn to understand life correctly and to change it, that they master progressive ideas, form their characters and convictions as naturally and as unnoticeably as a child learns to speak. We need books, films, spectacles, musical works, paintings and sculptures which will educate people in the spirit of communist ideals.' Quoted in *ibid.*, 38. Not until the recent advent of Gorbachev's *glasnotz* has Communist art been freed from the fetters of nationalist and ideological imperatives

17 Adolf Hitler, 'Art and politics' speech at Seventh Nationalist Socialist Congress, Nuremburg, 11 September, 1935, quoted in Keith D. Wills, Introduction to William Yenne, *German war art, 1939–1945* (New York 1983) 15; see also Donald Egbert, *Social radicalism and the arts, Western Europe* (New York 1970)

18 William Colgate, *Canadian art: its origin and development* (Toronto 1943) 40

19 Their themes speak for themselves: *Wanderings over Canadian roads* (1896); *Canadian waterways* (1897); *Everyday life in Canada* (1898); *The impress of the century on the land and its people* (1900); *Canadian village life* (1901); *Sports and pastimes* (1902); *Canadian cities* (1903); *Characteristic landscape features of Canada* (1904). For more details see *ibid.*, 49–62

20 Lorne Pierce, *Thoreau MacDonald* (Toronto 1942) 7–8

21 Group of Seven, Catalogue, Exhibition of Paintings, May 7th–May 27th 1920, Art Museum of Toronto, 1920

22 Proceedings, Conference of Canadian Artists, Queen's University, Kingston, 26, 27, 28 June; National Gallery Ottawa, 29 June 1941

23 J. Burant, The visual world in the Victorian age, *Archivaria* 19 (1984–5) 110–21

24 *Ibid.*, 121

25 Colgate, *Canadian art*, 224–5

26 Brian Murphy, Poster power in elections, *The Archivist* 16 (1989) 6–7

27 Quoted in Fraser Sutherland, *The monthly epic: a history of Canadian magazines* (Toronto 1989) 58

28 For more on this theme see *ibid.*

29 George Parker, *The beginnings of the book trade in Canada* (Toronto 1985) 260

30 Pierce, *Thoreau MacDonald* 7–8

31 Colgate, *Canadian art*, 45; for more on Jefferys, see also Charles Stacey, *Charles William Jefferys, 1869–1957* (Ottawa 1977); Osborne, Iconography and nation-hood, and his Recording a nation's heritage: illustrators as fabricators of Canadian identity, in his *No vacant Eden: literary and artistic images of Canadian landscapes* (Toronto 1990); Dennis Duffy, Art-history: Charles William Jefferys as Canada's curator, *Journal of Canadian Studies* 11 (1976) 3–18

32 Sutherland, *Monthly epic*, 23–4

33 Queen's University Archives, Lorne Pierce Collection, C. W. Jefferys, Some personal notes

34 Osborne, Iconography and nationhood

35 C. W. Jefferys, *Dramatic episodes in Canada's story* (Toronto 1930) i

36 Quoted in Paul Walker, C. W. Jefferys and images of Canadian identity in school textbooks (unpubl. M.A. thesis, Queen's University 1990)

37 Other ventures soon followed: Stewart Wallace's *First book of Canadian history* (1928); Lorne Pierce's ten-volume *Ryerson Canadian history readers* (1930); P. J.

254   Brian S. Osborne

Robinson's *Toronto during the French regime* (1933); E. C. Guillet's *Early life in Canada* (1933); J. E. Wetherell's *Three centuries of Canadian history*, several historical works by W. Perkins Bull (1937); E. C. Guillet's *Pathfinders of North America* (1939); Stephen Leacock's *Canada: the foundation of its future* (1941); and M. H. Long's *A picture history of the Canadian people* (1942)

38 Duffy, Art-history, 3
39 Pierce, *Thoreau MacDonald*, 4
40 For more details on other Canadian illustrators see the following: St George Burgoyne, Some Canadian illustrators, *Canadian Bookman* 1 (1919) 21–5; 27–30; Angela Davis, The hothouse of Canadian art: a 'Golden Age' at Brigden's *The Beaver* February/March (1988) 37–47; Paul Duval, The story of illustration in Canada *Provincial's paper* 26 (1961) n.p.; Sybil Pantazzi, Book illustration and design by Canadian artists, 1890–1940 *National gallery of Canada bulletin* 4 (1966) 6–9
41 Quoted in Wright, Literature, art, and propaganda 41
42 Zelinsky, *Nation into state*, 1
43 Pierce, *Thoreau MacDonald*, 4
44 Sutherland, *Monthly epic*, 204–5
45 Robertson Davies, *Leaven of malice* (Harmondsworth 1985) 164
46 Gordon Craig, Facing up to the Nazis, *New York Review of Books* February (1989) 15
47 Charles S. Maier, *The unmasterable past: history, holocaust, and national identity* (Cambridge Mass. 1989), quoted in *ibid.*

12

Collective consciousness and the local landscape: national ideology and the commune council of Mesland (Loir-et-Cher) as landscape architect during the nineteenth century

ALAN R. H. BAKER

Nation and commune

On 14 July 1881 at Mesland, a rural commune 20 kilometres to the west of Blois, in the Loire valley, the *curé* took down the national flag which was flying from the church's bell-tower and threw it on the ground in the commune's newly laid-out public square. Flying the flag had been arranged by the local municipal council, which – no doubt correctly – interpreted the *curé*'s demonstrative action as a deliberate insult both to the national emblem and to its own local authority. On 7 August the council voted, by ten to one, to prohibit the *curé* and his clergy from processing on the streets of the commune without the prior approval of the mayor.[1] That specific conflict epitomizes the general battle waged between church and state in France during the nineteenth century and suggests that by the early 1880s at Mesland the former was seriously losing ground to the latter.

In 1863 one fifth of the population of France was unable to speak French; and in 1864 a school inspector in Lozère reported that at one village school he had visited not a single pupil could answer correctly questions like 'Are you English or Russian?' or 'In what country is the département of Lozère located?'[2] A sense of national identity emerged in France only gradually during the nineteenth century: the French nation had to be created.[3] While improvements to roads and the development of a railway network promoted the integration of the French space-economy,[4] it was the schools which 'taught hitherto indifferent millions the language of the dominant culture, and its values as well, among them patriotism. And military service drove those lessons home.'[5] French national consciousness was a political construction, one of the significant achievements of a centralized government making effective use of a local – and increasingly democratic – institution, the commune. The creation of a national ideology, of a new collective consciousness, was a complex, protracted and discontinuous process (not least because it came to involve in France a battle between two *remote* institutions – the established church and the parvenu state – for the minds of *local* communities). The creation of that French national consciousness carried with it, at least by implication, the construction of a French national landscape. Moreover, the landscape itself could be used iconographically to reinforce a sense of national identity. This chapter examines the impact of an emerging French national ideology upon the landscape of one rural commune in the Val de Loire and explores the extent to which the abstract notion of the French *state* came to be concretized, iconographically, within the landscape of that *commune*.

Within the nested spatial hierarchy of administration established in France in the 1790s, the commune was the smallest unit: it was the level of administration at which the citizen most frequently encountered the state and at which the state most directly confronted individuals. Communes were grouped into *cantons*, *cantons* into *arrondissements*, and *arrondissements* into *départements*. Central administration in Paris had – via the prefect of the *département*, he in turn through the sub-prefect of each *arrondissement* and he in turn through the mayor of each *commune* – an effective channel of communication to and from each of the local administrations in the 37,000 or so communes of France. Each commune council was, therefore, a potentially significant node in the administrative network of France: through it could be transmitted – and received, no doubt with varying degrees of clarity and willingness – the message of national ideology emanating from Paris and through it could be exercised considerable central control over a local democracy.

The role and responsibilities of commune councils were frequently modified during the nineteenth century and their composition and functioning became increasingly democratic.[6] Nonetheless, some basic characteristics persisted throughout this period, among them the activities of councils in changing local landscapes and in receiving and interpreting the message of a national ideology. Broadly speaking, commune councils functioned at three levels. First, they were able to take executive decisions about some strictly local and short-term matters (such as the administration of property owned by the commune, and the implementation of leases and the collection of rents on such properties). Secondly, they debated other local issues which had medium- or long-term implications (such as the commune's annual budget, the purchasing, selling or exchanging of property by the commune, its own building projects and plans for the widening of streets and laying-out of public squares): but any actions which a council wished to take in such matters had to be approved by the central administration, represented by the prefect. Thirdly, a commune council was able to express its views on issues initiated elsewhere – by central decree or by decision of some central administrative committee – but which might affect its own commune (such as development plans for roads and railways, the accounts of charitable institutions within the commune and the state of repair of religious properties in the commune). The commune council was essentially a debating chamber because it was the mayor's responsibility to be the commune's executive officer and simultaneously to act as the local agent for central authority (his responsibilities included, for example, being the local policeman, manager of the commune's properties, collector of the commune's revenues, keeper of its account and administrator of its markets, leases, sales and legacies). During the first half of the nineteenth century, mayors (and their *adjoints* or deputies) were appointed by the prefect from among council members who were themselves elected by an electoral corps comprised of the principal taxpayers in the commune. Within rural communes, each council was made up of a dozen or so members (the precise number being a function of the size of a commune's population). In 1848 local democracy was significantly extended with the introduction of universal male suffrage and with mayors being elected by the councils (instead of being appointed by the prefect) in communes with less than 6,000 inhabitants (which, in effect, meant most communes and certainly all rural communes). Throughout the nineteenth century, local commune councils were thus both taking their own initiatives and responding to those coming from the central administration. Minutes of the meetings of commune councils and files of correspondence between their mayors and the prefects of their *départements* provide, both explicitly and implicitly, an almost continuous record of those initiatives and responses. In so doing, they permit investigation of the extent to which an emergent national ideology came to be expressed in local landscape.

Living in Mesland, 1830–1914

In 1911, when the last population census was held in France before the First World War, the commune of Mesland – on the margin of the Val de Loire and la Gâtine tourangele – was home for 803 people, one-third of whom lived in the *bourg* and the others in the 47 hamlets and isolated farms scattered throughout the commune's territory (Fig. 12.1). Among the 803 there were only five people – two men and three women – who had been born in Mesland at least eighty years earlier, in or before 1831.[7] Their lives spanned a period which saw fundamental political, social and economic change in France as a whole and could be expected to have structured the experiences of the entire community of Mesland and particularly of those

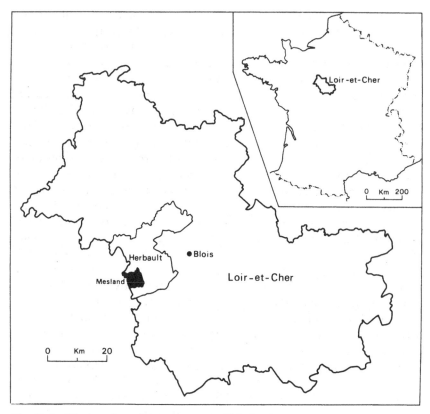

Fig. 12.1   The location of the commune of Mesland.

five octogenarians – and most especially of the two men, Théodore Chaput and François Collesse.

Théodore Chaput was born at Mesland on 24 June 1829: his father, Nicholas, then aged thirty-five, owned and operated a water-mill on the River Cisse just to the south of the *bourg* of Mesland. Thédore's mother, Anne, twenty-nine, his brother, Nicholas, four, and his sister, Anne, three, made up the family of five at the time of his birth but two more brothers and another sister were added by 1835, to bring the completed family to eight. Théodore was recorded as living at Le Moulin with his family in the 1841 and 1846 censuses but not in 1851: he was then twenty-two years old and might well have been away from his *pays* performing military service for his *patrie*. In 1851 his elder brother, Nicholas, twenty-six, was recorded as being the miller and their father as the *propriétaire*. By 1856 Théodore, twenty-six, was back at Le Moulin, working as a baker; his brother Nicholas was running the mill, one of his younger brothers was a barrel-maker and the other a farm-labourer. By 1861 Théodore, then aged thirty-one and a *cultivateur*, had married Eugenée Marsotteau, twenty-one: she had been a *domestique* at Le Moulin and now with Théodore formed a separate household, living in one of the houses at Le Moulin. By 1866 they had produced a son, then aged four, and a daughter, then only a few months old. They lived in one of the houses at Le Moulin; others were occupied respectively by Théodore's father, seventy-two, living on his income as a *rentier*; by Nicholas, the miller, and his wife; and by an unrelated day-labourer and his wife. In 1872 there were altogether eleven people living in the cluster of houses at Le Moulin. By 1876 Théodore, forty-six, had become a *propriétaire-cultivateur* and employed a living-in male *domestique*. During the next twenty or so years he farmed his property (which was most probably small and largely planted as vine-yards), at first with a farmhand and then with his son, Henri. By 1901 Théodore, seventy-two, had retired and was still living with his wife at Le Moulin – as indeed they were still in 1911, when he was eighty-two and his wife seventy-one. Thus, apart from a year or so in his early twenties, Théodore Chaput played as a child, and worked and retired as an adult in the commune of Mesland, living as part of the family and household community at Le Moulin.

François Collesse was born at Mesland on 18 February 1831: his father, then aged thirty-two, was a woodcutter (*fagoteur*). His mother, Marie, twenty-one, and his elder brother, René, three, made up the family of four at the time of François' birth but another two brothers and three sisters were to follow, so that by 1846 François, fifteen, was one of a family of nine, with children ranging in age from eighteen years to one year, all living in one of the two houses at Gros Bois, in the heavily wooded western part of the commune. François' parents were still living at the same house in 1851 and 1856, his father as a *fagoteur* and his mother as a sewing-maid (*lingère*) but François himself was not then living in the commune: during his early twenties, perhaps connected with military service, François was acquiring experiences beyond Mesland. By 1861, aged thirty, he was back, working as a *fagoteur* and living with his wife and newly born son at La Morandière, a farm to the southeast of the village. There they remained for about ten years, with François working as a *journalier* (day-labourer) and producing a second son. By 1872 François had moved, with his family, to live again at Gros Bois where he worked variously as a *journalier* and as a shepherd (*berger*) for the next ten or so years. He acquired some land and by 1886 was a *vigneron* (wine-producer); living with him and his wife was his *vigneron* son, Henri, and his daughter-in-law. In 1896, aged sixty-five, François, now a *propriétaire-vigneron*, was living at Gros Bois with his wife and two of their grandsons (aged two years and one year respectively). By 1901 François was no longer working:

as a *rentier*, he continued to live with his wife at Gros Bois and they were both still there in 1911. Thus, apart from a few years, during his early twenties, François Collesse – like Théodore Chaput – played, worked and retired in the commune of Mesland, living first at Gros Bois, then at La Morandière and then again at Gros Bois. Coming from a more modest family background than did Théodore Chaput, François Collesse progressed from *fagoteur* to *journalier* and ultimately to *propriétaire-vigneron*. Both men, apart from a short period during their twenties, had life-worlds which were structured essentially within the commune of Mesland. But between 1830 and 1914 that structure itself underwent changes which they must have noticed and which also involved them personally, though differentially.

The population of Mesland when Théodore Chaput and François Collesse were young children was a little over 600. It increased only gradually to about 650 when they were in their twenties, during the 1850s; reached 700 by the early 1870s, and then grew a little more quickly, reaching 750 by the mid-1880s and 800 by the early 1890s; then ensued a decline to about 750 during the 1890s and a return to 800 by 1906 (Fig. 12.2). So Théodore and François saw the population and their commune increase by almost one-third during their lifetimes. The *bourg* of Mesland absorbed more than half of that increase: in 1846 the 177 people living in the *bourg* represented 28 per cent of the commune's total population but by 1911 the 293 people in the *bourg* constituted 36 per cent of the total. The *bourg*'s population increased by 66 per cent between 1846 and 1911, while that of the hamlets and isolated farms increased by only 30 per cent and that of the commune as a whole by 28 per cent. Théodore and François witnessed the significant growth of Mesland's *bourg*, a manifest centralization of the commune's population.

Economic changes within the commune were less dramatic, probably less perceptible. When Théodore and François were in their twenties, in the 1850s, 75 per cent of the population of Mesland were dependent for their livelihood on agriculture; by their early forties, in 1872, it was 71 per cent and by their early sixties, in 1891, it was 72 per cent. Most farms in Mesland were owner-occupied, and there was a barely discernible shift towards tenant farming (from about 70 per cent owner-occupation in 1856, to 66 per cent in 1891). The second most important employment was woodcutting (on which 8 per cent of the population were dependent in 1856 and 6 per cent in 1891). A not insignificant proportion of the population (6.5 per cent in 1852, 3.8 per cent in 1872 and 10.3 per cent in 1891) lived on incomes from property and capital. More noticeable to Théodore and François would have been the changing number of people at Mesland dependent on artisanal occupations: in the 1850s 7.3 per cent of the population were dependent on the crafts of barrel-makers, stonemasons, long-sawmen,

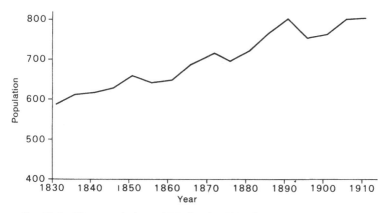

Fig. 12.2   The population of Mesland, 1831–1911.

dressmakers, seamstresses, shoe- and clog-makers, millers, ploughwrights and farriers but by 1891 the figure had fallen to 2.5 per cent. Paralleling this decline in rural crafts was a marked growth in Mesland's service sector: in 1852, there was a wool merchant, a rag-and-bone merchant, a wood merchant, one café and two bakeries (a range of activities on which 2.6 per cent of the population depended for a living); by 1891 the commune's hotel, café and diverse retail outlets (for food, clothing and furniture) provided a means of existence for 5.1 per cent of the population.

But potentially more striking to Théodore and François might have been the emergence in the commune during their lifetimes of a growing army of public employees. At the time of their birth, the only 'official' in Mesland was the priest. By the time Théodore and François reached their eighties, the commune had acquired a range of public employees: a *garde-champêtre* (local 'policeman' and the council's odd-job man); two schoolteachers; three road-menders; a licensed tobacco dealer, a postmistress and two postmen. In 1852 1.4 per cent of the population of Mesland depended for their livelihood upon such forms of 'public' employment; by 1891 the figure was 3.3 per cent. To the solitary figure of the priest, the uniformed representative of the church, had been added a set of officials, many of them also uniformed and all of them representatives of the state, whether they were employed by the commune council itself or by some other public agency. As more and more of these officials travelled the roads and tracks of Mesland, they signalled to its population quite explicitly the extension of the state within the commune and implicitly the declining role of the church (which, in the 1890s, was apparently unable to provide Mesland with either *curé* or a *désservant*). For Théodore Chaput and François Collesse those

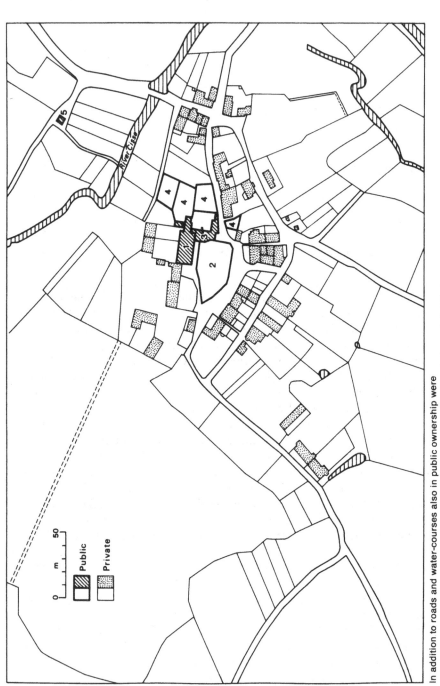

In addition to roads and water-courses also in public ownership were

1 Church 2 Cemetery 3 Prebytery 4 Gardens and grounds of presbytery 5 Public fountain

Fig. 12.3 Ownership of the landscape of Mesland, c. 1830

Fig. 12.4   The church and *bourg* of Mesland, viewed from the banks of the River Cisse.

ambulatory signals must have been reinforced to the extent that during their lifetimes the state replaced the church as the creator and manager of public space and landscape within Mesland.

The public landscape of Mesland in the early nineteenth century

Cadastral plans and registers of Mesland compiled between 1819 and 1825 confirm that the limited public landscape of Mesland in the early nine-teenth century had been created dominantly by the church (Fig. 12.3).[8] The only public buildings in the commune were the church and the presbytery. The imposing twelfth-century church, located in the centre of the *bourg*, was situated atop the western slope of the Cisse Valley, its tower dominating not only the village but the valley and beyond (Fig. 12.4). Adjacent to the church was the presbytery and its gardens, and the large cemetery. This complex of church buildings and spaces dominated the centre of the *bourg*: it was a semiotically powerful assemblage. The only other public components of Mesland's landscape c. 1830 were the water fountain, opposite the *bourg* on the other side of the valley, and the commune's roads, tracks and streams.

The strong hold which the church held upon the collective consciousness of the inhabitants of Mesland in the early nineteenth century is illustrated by the response of many of them to a storm which threatened the commune during the evening of 2 June 1840: they rang the church bells for a long time in an unsuccessful attempt to safeguard their crops from being damaged by wind, rain and especially hail. Their action was reported to the prefect by another Mesland farmer whose crops had suffered considerable damage from hail and flooding not only that year but also in 1839: he was convinced the ringing of church bells actually made the storms worse. He accused the bell-ringers of wanting 'to retain their superstition and old customs', even though the *curé* of Mesland was himself opposed to the practice of bell-ringing to ward off storms. The central administration of the *département* was strongly opposed to the practice and on 15 June 1840 the prefect wrote to the mayor (not, it should be noted, to the *curé*) of Mesland to remind him that ringing church bells during storms had been repeatedly prohibited, the last occasion being the prefect's circular to mayors dated 28 June 1839.

This incident reveals the continuing – but also the confused – influence of the church upon collective consciousness in Mesland until at least the middle of the nineteenth century. Rejection of the traditional authority of the church during the Revolution of 1789 was manifested at Mesland, as in many French communes, by deliberately damaging its most powerful landscape symbol, the church building itself. At Mesland, damage done to the roof of the church during the Revolutionary period and failure – in effect, unwillingness – to undertake repairs led to the gradual collapse of the roof within the walls, exposing the rest of its fabric to the elements in the early 1800s. From at least 1807 Mesland lost its separate religious existence and was no longer served by a priest. Then, in 1813, some of the commune's residents had the church repaired at their own expense and the *désservant* of the adjoining commune of Monteaux was paid by them to come to Mesland to conduct services on Sundays and *fête* days. In 1816 the commune council's petition to the civil and ecclesiastical authorities for Mesland once again to be provided with its own *curé* was accepted. The ideology of Catholicism had survived the Revolutionary challenge.

The *mentalité* of the people of Mesland had not been totally transformed but nor was it unaffected by the Revolutionary upheavals. The gradual evolution of collective consciousness is evidenced, for example, by the way in which the terms *paroisse* and *commune* are both used interchangeably in council minutes and in administrative correspondence until the 1820s, after which references to the *paroisse* of Mesland are rare: the ideal of a civil *commune* seems to have replaced that of the ecclesiastical *paroisse* by the 1830s. No doubt this changed conceptualization was reinforced by the laws of 1831 and 1837 relating to municipal administration, but its foundations had been laid earlier. An inventory in 1815 of the commune's papers, furniture and effects shows how life in the locality was being moulded into that of the developing nation. Among the papers were copies of the *Bulletins des Lois* of all of the national governments from 1789 and 1815 and of the *Journal de la Préfecture* from year X to 1815, both being instruments by which central authorities impacted upon local administration; they also included copies of the *Code Civil* and of tables for converting the weights and measures of Loir-et-Cher, both means by which a national system of law and mensuration were intended to replace local practices; and also copies of instructions on how to conduct the military call-up and lists of conscripts from year X to 1814, indicating how able-bodied young men of the *pays* were called to serve their *patrie*. Also among the papers was a statement of the costs incurred by the commune in acquiring a flag to fly from the bell-tower of the church, with the dual symbolic significance of demonstrating *both* the domination of the church by the state *and* the integration of the commune into the state.

Flying the new flag was a dramatic and an immediate statement.[9] Other changes linked to the growing power and role of the civil commune proceeded more slowly and less perceptibly. In the aftermath of the Revolution the presbytery had been confiscated and sold to one of Mesland's *notables*, M. le Comte de Montlivault. In 1820, as a benefactor to the commune and as its mayor, the count sold the presbytery to the commune for less than its market value so that accommodation could be provided by the commune for its *désservant*. Thus the commune council having been given responsibility for the church and the cemetery, now acquired an additional, formerly ecclesiastical, property. But the council's role as property owner during the early nineteenth century was limited. In 1832 and again in 1836 the council discussed the idea of moving the cemetery to a new site and investigated suitable sites to purchase but, on both occasions, took no definite action. As for an office, the council did not have its own *mairie* during the early nineteenth century; instead, it rented a house in the *bourg*. It made its own presence literally visible in the commune only in a small way from February 1828, when it decided to erect a board for the public display of official notices.

So, when Théodore Chaput and François Collesse were born at Mesland, in 1829 and 1831, the transformation of a collective consciousness from one based on religion and locality to one based on reason and nationality had made only a modest and intermittent impact upon the commune's landscape. More substantial changes were to be made during their lifetimes.

Schoolhouses and council offices

Maison d'école

The *loi Guizot* of 1833 required each commune in France to establish a boys' primary school and to provide a minimum annual stipend of 200F to its teacher as well as free education to *indigents*, whose families could not afford to pay the *contribution scolaire*. Nationally, its impact was significant. In 1833 some 30 per cent of France's communes were without a boys' school; by 1840 the proportion had fallen to 11 per cent. Within Loir-et-Cher, 37 per cent of communes were without a primary school in 1833; by 1838 the figure had fallen to 15 per cent; and by 1848 to 7 per cent.[10]

At Mesland, the commune council responded rather slowly to the provision of adequate accommodation for a schoolteacher and his pupils. At first, it simply rented premises in the *bourg* but even the mayor recognized their insalubrity, declaring them in 1845 to be a health hazard for both teacher and pupils. In the spring of 1843 the council had decided to purchase a house in the *bourg* and to convert it into a classroom and living accommodation for the schoolteacher. But it would be five years before conversion was completed. On 25 June 1843, at the public enquiry into the council's proposal, conducted by M. François-Martin Boitou, *juge de paix* of the Canton d'Herbault, ten people expressed their views: only five of them were able to sign their own names, but eight of them were in favour of the purchase of the house because of its location in the *bourg* and because it was reported to be in suitable condition. One objector said a better property should be found, the other perspicaciously observed that the house – which consisted of two inter-connecting rooms (only one of which had a fireplace) and with a long-loft above – was too small for the purpose intended. On 2 July 1843 the council agreed unanimously to acquire the house for 1,800F; necessary repairs and improvements estimated to cost 900F were expected to bring the total cost of 2,700F. The council agreed to meet 1,800F of that cost by raising a special tax of 450F for each of the four years 1845–8; it hoped that the government would provide the remaining 900F, the commune being unable to do so because it had in recent years spent 6,000F on acquiring and repairing the presbytery and because in 1839 and 1840 the commune had suffered considerable hail damage, so that a higher level of taxation was not considered to be reasonable. On 8 August 1843 the prefect instructed the architect of the *arrondissement* of Blois to prepare plans and estimates for converting the house and on 22 February 1844 he sent the proposed plans and estimates to the mayor of Mesland, at the same time authorising the council to impose an additional tax to meet the costs involved. On 17 March 1844 the council considered the revised estimates, which now totalled 5,134F35 (purchase of the property 1,800F; legal costs 360F; repairs and improvements planned and required by the architect 3,034F35). The council agreed to raise 2,500F of this sum from local taxation (500F annually 1845–9) and hoped that the balance could be met by grants from central government. The prefect, however, on 1 April 1844, pointed out that the maximum grant for establishing a schoolhouse was usually not more than 25 per cent of the total cost: in Mesland's case this would be 1,285F59, so that a further 1,350F76 would have to be raised by the commune. On 12 May 1844 the council voted 3,425F (two-thirds of the total cost) to be met from a special tax of 695F annually 1845–9. In effect, this would have increased the commune's normal budget (of 1,632F in 1844, of which half was expended on road maintenance) by more than 40 per cent. On the advice of his Comité supérieur pour l'Instruction Primaire in the *arrondissement* Blois, on 1 August 1844 the prefect petitioned the minister for a grant to Mesland of 1,709F. Although on 21 August the minister acknowledged receipt of this request, and although the prefect wrote to the minister on 18 March 1845 asking him to come to a decision as soon as possible, it was not until 19 January 1846 that the minister informed the prefect that a grant of 1,000F had been allocated to the school project at Mesland. In practice, this grant was worth only 760F because to the total cost of the project had to be added in the spring of 1846 interest of 240F on the purchase of the house, agreed as long ago as January 1843. On 22 March 1846 Mesland's council accepted that the total cost of the school project, 5,374F35, would have to be met, after receipt of the 1,000F from the Ministry, by a local tax of 875F annually from 1847 to 1851.

During 1846 the schoolteacher, M. Dubois, moved into Mesland's newly purchased schoolhouse and teaching began there in October, before the repairs and improvements had been completed. Bad weather during the winter of 1846–7 was blamed by the architect for delays in converting the property as planned. On 29 March 1847, M. Dubois complained to the Inspector of Schools that the mayor had promised the work would be completed by the end of October 1846 but that, far from being completed, work on the house was only just about to begin again. 'In the meantime', complained M. Dubois, 'I have only one room, if you can call it that, no cellar, no kitchen, nowhere to store even *un quart de vin*, no *lieux d'aisance* for my children, nor for myself.' He asked the inspector to press the mayor to hasten the work's completion, otherwise he would be bound to ask for a teaching post elsewhere. In April the prefect wrote separately to the mayor and to the architect, urging them to have the work completed without further delay. In fact, a further delay was in store because the council of Mesland had – without consulting the prefect – asked the architect to submit plans and estimates for the construction of rooms on the first floor, above the classroom, so that the schoolteacher would have better accommodation (including more than one fireplace) but also that there could be constructed, also on the first floor, a council office (*mairie*). Not until the summer of 1847 were these additional building works approved by the prefect and not until July 1848 were they completed, at a total cost of 5,531F03. From conception to completion, this conversion of a house into Mesland's first *maison d'école* had taken five years.

Ouvroir

In seems, however, that the patience of M. Dubois had been exhausted. By September 1849 the schoolteacher is the enterprising M. Amiot, who founded an *ouvroir* for young girls and ran it in conjunction with the school. He was concerned that schools in the countryside, because of their limited resources, brought girls and boys together and deprived the former of the advantages of a separate education suited to their special needs. On 1 September M. Amiot, with the support of both the mayor and the *curé*, petitioned the Minister of Instruction for a grant for the *ouvroir* and the following day, claiming the approval of the Inspector of Schools, he forwarded the petition to the president of the academy of Blois pleading with him to ask the minister to provide financial support for his *ouvroir*, to which he was optimistically sacrificing his own time and money. On 21 September the prefect wrote to the mayor declaring the administration's general support for the *ouvroir* which served to introduce girls at an early age to skills which would be useful throughout their lives, and to the habit of work and discipline. But the prefect wanted to know whether the council of Mesland had provided the *ouvroir* with appropriate accommodation or with financial support. That letter seems to have gone astray because on 2 February 1850 the prefect again wrote to the mayor asking for the information previously requested. So on 23 February M. Amiot wrote to the president of the academy of Blois, repeating that his limited income as a schoolteacher meant that he could not afford to subsidize for long the costs involved in running an *ouvroir*. But in this case, the *enthusiasm* of the schoolteacher for an *ouvroir* was not, it seems, matched by the commune council. The enthusiasm and philanthropy of a local individual counted for less than the compulsion and ideology of the state which required the setting-up of a school at Mesland.

Ecole de garçons et mairie

The house which had been converted into schoolroom, teacher's lodging and *mairie* by 1848 was soon to be inadequate for the educational needs of Mesland – as one witness at the public enquiry in June 1843 had predicted. A generation later an enlarged population – and one with an enhanced awareness of the potential benefits of primary schooling and so with a growing proportion of parents wishing to have their children attend school – deemed the classroom to be unsuitable. On 16 December 1873 the Inspecteur de l'Académie de Loir-et-Cher reported to the prefect approving Mesland council's decision of earlier that year to construct a new *maison d'école*: the existing schoolroom was *réellement insuffisante* and a new school was urgently needed. He noted that the commune had a population of 717, sufficient to oblige it by law to have a separate public school for girls. Similarly, the Inspecteur Primaire reported to the Inspecteur d'Académie on 16 August 1875 that Mesland only had a mixed school although it had a population of 717 and that the commune council had decided to create a girls' school; it was planning to do this by building a new boy's school, thereby releasing the present mixed school for use as a girls' school.

Plans and estimates for the new boys' school were discussed by the commune council on 19 July 1874; as the proposal involved buying a plot of land as well as building the school, the estimated cost of almost 29,000F was considered by the council to be excessive, so it called for new estimates based upon a modified proposal for a single-storey instead of a two-storey building. On 10 August 1874 the council approved modified plans for a *maison d'école et mairie*, at an inclusive cost of 27,520F, of which 23,871F was for the building. The council assumed that this included about 2,000F for the *mairie*, so that the cost of the school would be 25,520F; it further assumed that the state would contribute one-third of the cost of the school (or about 8,500F), leaving the commune to find about 19,000F, which it intended to do by raising a loan and paying it off over twenty-two years out of local taxes.

From August 1874 the building plans and estimates were negotiated between the local council and various central authorities until they were finally approved by the minister on 24 February 1877. In early 1875 the Conseil des bâtiments civils had suggested reducing the costs considerably, by not having walls and iron railings around the school's grounds and obtaining the necessary building materials closer to hand than was planned (in particular, if the stone were to be obtained from Onzain, only 4 kilometres away, rather than from Pontlevoy, some 20 kilometres away). The commune council insisted that a wall was needed between the school grounds and the road, because the proposed building plot was 1.5 metres higher than the road and a wall was necessary to retain the earth. The architect insisted that for *le moellon* (soft stone) he could only rely on the quarry at Gye and for *la pierre détaillée* on that at Pontlevoy but was otherwise willing to use local materials.

On 6 August 1875 the Conseil des bâtiments civils approved the changes made to the plans, noting that the estimated building costs had been reduced from 23,871F to 17,320F. On 30 September 1875 the prefect formally approved the project for a new school and *mairie*. But approval from the Ministry of Instruction was not forthcoming: on 8 January 1876 the minister expressed his concern about the excessive cost of the project for a commune of only 717 people and just for one school; he suggested that the plans should not include a *préau couvert* (a sheltered area in the school's yard) and that the building of the boundary walls should be postponed. The minister also expressed concern about the proposed school's proximity to the commune's cemetery. On 25 January 1876 the mayor – who was Nicholas Chaput, the miller and elder brother of Théodore Chaput – wrote to the prefect that it was virtually impossible to reduce the cost because the necessary raw materials had to be transported over considerable distances to Mesland (10 kilometres for sand, 12 kilometres for *le moellon* and 24 kilometres for *pierre dure*); he also argued that a shelter was very necessary as a place for the pupils to store their baskets and other belongings, but conceded that the retaining wall could be forgone. On 13 April 1876 the Inspecteur de l'Académie supported the mayor's case: the shelter in the yard would be small, would not add significantly to the cost and was absolutely necessary because in a school that does not have one, the clogs, clothes, parcels, baskets and rubbish of the pupils clutter up the classroom 'au grand préjudice de l'hygiène, de l'ordre et de la discipline'. The inspector also argued that the retaining wall was needed not only to hold back the earthen bank but also to stop children from jumping and running onto the public road. As for the cemetery, it was 150 metres away, which was an adequate distance, especially as it was to the north of the school (towards the fields, not between the school and the *bourg*). Despite these combined arguments of the mayor and the inspector, the minister on 15 July 1876 insisted to the prefect that the cost of the project had to be reduced. On 13 August the council agreed to forgo the enclosing wall but to keep the shelter in the project. At last, on 22 January 1877 the minister allocated a grant of 5,300F to the project and on 24 February formally approved the building plans and costs. The total cost of 21,600F, the Mesland council agreed on 15 April, would be met by grants of 5,300F from the minister and of 500F from the *département*, by using the 3,531F indemnity awarded to the commune for damages inflicted during the Franco-Prussian war of 1870–1, and by raising a loan of 12,260F (on 11 May 1877 the Ministry of the Interior authorized the commune to borrow 12,280F to be repaid over twenty-one years from local taxation).

When, on 4 June 1877, at the prefect's request, the mayor of the adjacent commune of Monteaux held a public enquiry in Mesland into the proposed purchase of the plot of land on a site for the new school, no observations of any kind were made: no doubt after four years of discussion within the commune, there was nothing more to be said! But some problems of a different kind did arise. On 9 June the prefect authorized the building work to be put out to tender, but by early November no bids had been received by the commune for the masonry work, for the roofing or for the building of the boundary wall. The commune council found it necessary to offer better terms for these items, and decided to meet from its own resources two-thirds of the additional cost of 1,610F and to ask the *département* to make it a grant for the remainder. In March 1878 the builder was having difficulty in obtaining *le moellon* needed because the quarries mentioned in the estimates were exhausted, so the commune agreed that the vault of the cellar could be in brick instead of stone.

Then, in June 1878, the commune council – with work well under way – decided that it did want the building to have a first floor. A supplementary estimate drawn up by an architect on 25 July 1878 (but not finally approved by the prefect until 31 January 1881) added 7,000F to the cost of the project. The architect pointed out that the initial plan was for a T-shaped building, with the classroom fronted by the schoolteacher's accommodation (comprising two small rooms looking onto the road, a study and a kitchen) and by a room which was to be the *mairie*. The plan included a loft over the whole building but with no staircase leading to it. The architect said the council had rightly judged that the teacher's accommodation would be inadequate and almost uninhabitable and had decided to add a first floor to the building, so that the schoolteacher's accommodation would be, on the ground floor, an entrance hall with staircase leading to, on the first floor, a dining room and a kitchen; on the first floor would be three bedrooms and a *grand cabinet avec placard*. The staircase would also give access to the loft which might in due course also be used to meet the needs of the schoolteacher.

Bad weather during the winter of 1878–9 delayed the school's construction. On 3 April 1879 the Inspecteur Primaire reported to the Inspecteur d'Académie that the existing schoolroom in use at Mesland was very small and overcrowded, 'les infants y sont littéralement entassés les uns sur les autres.' In his view, it was essential for the new school to be completed before the opening of the next school year, and he asked the inspector to plead with the prefect to press for completion of the work by warning Mesland's council that the existing schoolroom could not be used for the *rentrée* of the 1879–80 school year. The prefect wrote in these terms to the mayor on 9 April 1879. In July the council calculated that the total cost of the project had risen to 30,466F and that it needed to raise a loan of 19,350F; on 29 August the Ministry of Public Instruction authorized a loan of 19,300F, to be repaid over thirty-one years from local taxation.

During completion of the new school Théodore Chaput was serving on the municipal council of Mesland, not as one of the eleven elected members but as one of the seven *plus imposés* (most heavily taxed). In the summer of

Fig. 12.5   The boys' school and *mairie* of Mesland, completed in 1882. The council's offices are located on the ground floor and the schoolteacher's accommodation on the first floor.

1880 the council debated whether the water-pump planned for the school should be sited along the boundary with the road, so that it could also be used by the public, or located on the boundary between the schoolyard and the teacher's garden: by twelve votes to six it decided in favour of the latter.

Whether or not the new school was ready for use for the *rentrée* of 1879 is unclear, but it was certainly not until 30 December 1880 that the council approved the actual costs incurred in constructing it. After paying all the bills, the council discovered in the spring of 1881 that the funds set aside for the building had a balance of 3,500F. The council discussed how best to use this sum: on 12 November 1881 the mayor explained that one year after being completed the school did not have a covered shelter, a *boucher* (a place for sawing logs), a *buanderie* (laundry), enclosing walls or a water-pump: the cost of providing all of these was estimated to be 8,000F and the council agreed to use its balance in the school-building fund for those purposes and to seek a grant of 4,100F from the Ministry. The prefect was able to authorize such a grant on 4 February 1882 and the necessary construction works were contracted for completion by the end of September 1882, for *la rentrée* of that year. Thus the *maison d'école*, with its *mairie*, took nine years from its conception in the autumn of 1873 to its completion in the autumn of 1882 (Figs. 12.5 and 12.6). The construction of this boy's

Fig. 12.6   The classroom and yard of the boys' school of Mesland. The classroom forms the upright of the T-shaped building, to the rear of the council offices.

school released the old schoolroom for its exclusive use for girls – but it did so in the year when the *loi Ferry* introduced compulsory education for all children, boys and girls, between the ages of six and thirteen.[11]

Ecole de filles

During 1880, with the new boys' school nearing completion, the council obtained plans and estimates from an architect for repairing and improving the old mixed schoolhouse, to make it suitable for use as a girls' school and to provide appropriate accommodation for a schoolmistress. By January 1881 the Ministry of Public Instruction had agreed to make a grant of 3,000F towards the total estimated cost of 4,600F but when the mayor reported this to Mesland's council on 12 May 1881 he also pointed out that the schoolroom, which it was intended should be repaired, would be too small now that public primary schooling was to become free (as it was, by the law of 16 June 1881). The number of pupils then attending the school, forty-two, was certain to increase and to be too many for the classroom; furthermore, the two small schoolyards, each only 50 square metres, were inadequate, so that many children played in the street, which was dangerous. The council agreed to the mayor's suggestion that a new girls' school should be built, financed by the sale of the existing schoolhouse and by grants from the state and the *département*. This decision received strong support from the Inspecteur Primaire: on 14 June 1881, writing to the Inspecteur d'Académie at Blois, he said that the existing schoolhouse had many faults – it involved climbing a bad staircase of six or seven steps to get from one part of the building to another; its two yards were barely 50 cubic metres each and while one caught too much of the sun the other was too glacial and often wet. The Inspecteur Primaire argued that even after the classroom had been improved it would only be 46 square metres or 150 cubic metres, not large enough even for 40 pupils whereas there were 55–60 girls at Mesland of five to thirteen years of age. He strongly urged financial support to enable Mesland to have a new girls' school. After a year, on 30 April 1882, the Inspecteur Primaire reported to his superior, approving the site selected by Mesland council for its new girls' school. The site, of 7 ares 48 centiares, was said to be large enough for the 57 school-age girls in the commune (the space required by the Ministry was 10 square metres per pupil, so that for 57 children the site would have to be at least 5 ares 70 centiares). The inspector added that the proposed site, more than 200 metres from the cemetery, was southeast-facing, in an area of fields and farmhouses.

In May 1882 plans for the new building were considered by the Commission des Bâtiments scholaries of the *département*: in order to contain costs, it suggested that a door between the classroom and the living quarters of the schoolmistress was unnecessary, as was also the door at the front of the building (the back door, opening on to the yard, was deemed to be sufficient), views with which the Inspecteur d'Academie de Paris concurred in a letter to the prefect dated 8 June 1882. The commune council, on 23 July, decided it wanted to retain the plan for a door between the classroom and the private quarters of the schoolmistress, because it would be unpleasant for her to have to go round the outside of the building to get to the schoolroom, especially in winter after the adult evening classes and when it was raining or snowing. The council also wanted to retain the front door, which it was a '*un embellissement de la maison*' and which would be useful as a place from which to distribute prizes. The Inspecteur Primaire changed his mind on the latter issue, but was adamant about the former. During the autumn of 1882 the council considered the ways and means of financing the new school and in early 1883 the building works were put to tender. By May 1885 they were completed, four years after the mayor had first proposed the project (Fig. 12.7). On 28 May the council noted that the total cost had been almost 28,000F and that, after taking into account income of 7,000F from the sale of the old schoolhouse, of about 400F from interest, and of a grant of 19,000F from the state, there remained a deficit of almost 1,600F. Towards this the *département* awarded a grant of 400F (financed from income from fines rather than from any educational

Fig. 12.7 The girls' school of Mesland, completed in 1885.

resources) and the Ministry of Public Instruction authorized the commune to meet the rest of the deficit by raising a loan.

Thus, the late 1870s and early 1880s saw significant civil buildings added to the landscape of Mesland, with the boys' school and *mairie* completed for *la rentrée* of 1882 and the girls' school for that of 1885. Both buildings were substantial structures erected on the edge of the *bourg*. By 1882 primary schools in France had been made free, compulsory and secular. The two new school buildings in Mesland were unambiguous statements of the developing impact of the state upon the lives of the people of Mesland.[12]

Within the *canton* of Herbault, in which Mesland is situated, the proportion of illiterate conscripts fell from 20 per cent in 1860, to 11 per cent by 1870, to 10 per cent by 1880 and to 3 per cent by 1890 (with those conscript classes having left school some seven or more years earlier, in about 1853, 1863, 1873 and 1883 respectively).[13] By the early 1880s – when the new schools were being built at Mesland – almost all males of school-leaving age in the *canton* of Herbault were able to read and write. The new schools being constructed then were, it seems, affirmations of a process of schooling which had gradually been extended during the previous fifty years, since the *loi Guizot* of 1833 – and, indeed, before then in some places. The new school buildings consolidated the emergence of a literate, numerate and nationally

Fig. 12.8   The mixed school of the hamlet of Pré-Noir, completed in 1913.

aware population and asserted in the landscape the internal colonization of the locality by the French nation state. At Mesland, schooling seems at first to have developed slowly. Its first primary school was not functioning continuously until after 1846, or perhaps not until after 1848. In 1872 its adult population was still not very literate: 45 per cent of males and 63 per cent females over twenty years old – those who, for the most part, might have gone to school before, say, the mid-1860s – were said to be unable to read or write; those aged between six and twenty years – those who might have gone to school since the mid-1860s – were more literate but – with 20 per cent of boys and 27 per cent of girls said to be unable to read or write – Mesland was lagging behind the literacy levels of many other communes in the *canton* of Herbault in 1872.[14] There was, it seems, still basic work to be undertaken in the mixed school at Mesland in the 1870s and in its new and segregated schools from the 1880s.

Ecole mixte de Pré-Noir

Both of the new schools at Mesland had, of course, been built in the *bourg*. But in the early 1880s, some 70 per cent of the population of Mesland lived in its isolated farms and hamlets. Although by 1911 the outlying population had fallen to 64 per cent of the total, there were then still 510 people in the outlying settlements where there had been 506 in 1881. In late 1911 the commune council of Mesland decided to build a new school for boys and girls, at the hamlet of Pré-Noir, in the northern section of the commune, 3 kilometres from the *bourg*. The proposal received the approval of the *département* on 9 March 1912 and of the Ministry of Public Instruction on 3 May 1912. The architect in charge of the project expected the work to be completed by the end of 1913; contracts with a mason and a carpenter were agreed during the summer of 1913 but the project was interrupted by the outbreak of the war with Germany. It was not until the summer 1920 that the architect confirmed that all the building works for the *école mixte* contracted in 1913 had been completed (Fig. 12.8).

A new cemetery

Symbolic of the secularization of society and of the weakened role of the church within rural France was the removal of cemeteries beyond the built-up areas of villages. The political position of the church declined in Loir-et-Cher after 1830 and it was in May 1832 that the commune council of Mesland discussed the idea of removing the cemetery from what it then considered to 'an inconvenient location', although for centuries the cemetery had been immediately adjacent to the church itself and in the centre of the *bourg*. Displacement of the cemetery beyond the periphery of the *bourg* should be seen as an expression in the landscape of the growing marginalization of the church socially and politically. In November 1832 the prefect informed the mayor that he would approve removal of the cemetery to a new site, on condition that it was located in the northern section of the commune (and thus to the leeward side of the *bourg*'s prevailing winds) and at least 40 metres to 50 metres from any habitation. The council, however, had second thoughts: on 16 December the mayor reported to the prefect that the price being asked for the particular plot of land selected as a site for the new cemetery was too high. There the matter rested until the summer of 1836, when the council decided to purchase a different plot of land for the new cemetery but the prefect told the mayor such action would need to be approved by the government, after the completion of certain procedures. No further action was taken at that time.

The municipal council controlled the allocation of plots within the long-established cemetery and obtained an income from them. In February 1844, for example, it fixed the rates per square metre in the cemetery as follows: 45F for concessions in perpetuity; 15F for concessions for thirty years; and 7F50 for 'temporary' concessions, which were for between ten and thirty years. By August 1858 the council was concerned about the limited space remaining in the cemetery and wanted to limit the temporary concessions to ten years but the prefect pointed out that it would be illegal to do so.

In May 1863 the mayor reminded the council that it had long been its intention to remove the cemetery from *l'intérieur* of the *bourg*, where it was situated contrary to the law of 1804. The council agreed to seek authority to raise a special tax of 500F for the purchase of a site for a new cemetery and at its meeting in August agreed to the purchase of a plot of land, 15 ares 19 centiares, to the northeast of the *bourg* at a place called le Chateau Gaillard, at least 300 metres from the nearest houses. The plot would cost 375F and the balance of 125F would be used to pay for an enclosing wall. At 4 o'clock on the afternoon of 15 September 1863 the *canton's* justice of the peace held a public enquiry into the Proposal: he recorded the views of fifty-six people (of whom 31 per cent were unable to write their own names). Only three spoke in favour of moving the cemetery to the site proposed – because they considered it an appropriate site and because it was recommended by the commune council – while fifty-three spoke against it. The justice of the peace, in his summary of the enquiry, said that the commune's cemetery could not be left much longer at its present location in the centre of the *bourg*; he believed that the most appropriate location for a new cemetery would be a plot in the Clos de Hérisson or near by, to the northwest of the *bourg* – that was, he said the unanimous view of all the inhabitants, including the mayor and the *désservant*, because it had come to be recognized that the plot in the Bois Gaillard initially selected by the council was too far away from the *bourg* and not easily accessible, because of the steep slopes of the Cisse Valley. The preferred site was at a reasonable distance from the *bourg* and accessible on a good, quiet road. Moreover, most of the population lived on the western side of the Cisse Valley, whereas few lived on the eastern side where the Bois Gaillard was located.

Not much more was done for a couple of years. During the winter of 1865–6 some trial digging had been undertaken on a plot at Le Hérisson. In September 1866 the mayor reported that the plot, although retaining some moisture, could be effectively drained by ditches because it sloped; in November the council agreed to purchase the plot of 21 ares for 525F. It further agreed to have a wall built on three sides of the cemetery and a fence constructed on the fourth side, along which there was a deep ditch. With the cost of iron gates at the entrance and the setting up of a crucifix, the project was estimated to cost 2,000F, which the council agreed to meet by raising a loan to be repaid over five years. On 10 December 1866 the prefect approved the proposed transfer of the cemetery, noting that the proposed site was 135 metres from the nearest house. On 11 February 1867, having obtained an architect's estimates for the works involved, the mayor informed the council that to transfer the cemetery would cost 3,164F (including the cost of purchasing the land for it). The council agreed to seek authority to meet the costs by raising a loan of 2,283F to be repaid out of

Fig. 12.9   The new cemetery of Mesland, completed in 1868.

local taxation over three years and by seeking a grant of 869F (being one-third of the costs of the work in setting up the new cemetery) from the *département*. Not until a year later were the financial arrangements approved by the central authorities; the work was put in hand in May 1868 and completed by the autumn of that year, some five years after they had first been seriously considered by the council but some thirty years after the idea of transferring the cemetery at Mesland had first been discussed by its commune council (Fig. 12.9).

La place publique

A narrow space between the (old) cemetery and the entrance to the church served informally during the first three-quarters of the nineteenth century as public open space. It was here that the commune, in post-Revolutionary fervour, had planted its Liberty Tree which, on 11 January 1852 in a different, imperial, political climate, was saved by the council instead of being cut down, on the grounds that its location and its good condition meant that the tree was recognized as being an embellishment to the 'square'.

In November 1856 the mayor outlined a plan to enlarge *la place du bourg* which would involve severing a corner of the cemetery which for a dozen or so years had been of no use to the commune. The commune agreed that enlarging the *place publique* at the approach to the church was becoming more and more necessary; a corner of the cemetery, measuring 10 metres by 12 metres, had become unused twelve years previously when the remains of the de Montlivault family formerly buried there had been transferred to the commune of Montlivault and the council agreed to seek permission to use that part of the cemetery to enlarge the *place publique*. In February 1857 the prefect – the official representative of central authority – authorized the council – as the local authority – to add those 120 square metres to the *place publique*. The state's monitoring of the commune's initiatives was precise and detailed even over such a small plot of land. This incident illustrates the considerable dependence of the locality upon a central authority: some at least of the inhabitants of Mesland are likely to have seen the mark of central authority heavily imprinted on that small piece of land.

The principal move to create an imposing *place publique* was made on the initiative of the mayor in 1878. On 14 February he reported to the council that no burials had been made in the old cemetery since August 1868, when the commune's cemetery had been transferred to its new site; his proposal to convert the old site into a cemetery planted with trees was accepted by the council. At the public enquiry held into the proposal on 24 November 1878 no representations were made and its chairman (the Mayor of Seillac) recommended that the project should be authorized. This was noted by Mesland council on 1 December, when it decided to ask the prefect to allow the work to be put in hand as soon as possible and without putting the work out to tender, because it would not be expensive: the *place* could be prepared without cost and the construction of a *trottoir* (pavement) would only be necessary over a distance of about 50 metres. The council intended to meet the costs from its 1878 and 1879 budgets. On 24 December the prefect authorized the project as requested and approved plans, drawn up on 31 October by the *agent voyer* of the *canton*, which showed the alignment of the *trottoir* and the locations of twenty trees.

The work was carried out early in 1879, under the mayor's supervision, but not with the approval of all of the members of the council: on 1 April 1879 the prefect wrote to the *adjoint* at Mesland, reprimanding him for uprooting the trees and replanting them in a different arrangement and reminding him that it was the mayor's responsibility to look after such matters. On 9 April eight council members wrote to the prefect asking him to authorize a special meeting of the council to discuss the planting of the trees on the *place* by the mayor in a way approved by the majority of the council. The special meeting was held on 29 April: it noted that the mayor had had the trees planted without consulting most of the council and

Fig. 12.10   The *place publique* of Mesland, laid out in 1880.

decided to ask the prefect's permission to have them replanted in an arrangement approved by the council; but the mayor refused to undertake to supervize the replanting, so the council asked for it to be delegated to the *adjoint*, assisted by three other members of the council. This local disagreement simmered for some months. On 25 September 1879 the mayor suggested to the council that the *place* should be edged by *trottoirs* and – it seems for the first time – he showed the council members the plans drawn up by the cantonal *agent voyer* in October 1878. The council simply decided that no action should be taken unless it had the approval of a majority of the council. On 14 October the mayor informed the prefect that the two groups – those supporting him, and those opposing him – had agreed, in a spirit of conciliation, to follow his advice and had agreed unanimously to accept an opinion to be sought from the cantonal *agent voyer* about the problem of tree-planting and *trottoirs*. On 12 February 1880 the mayor placed before the council a plan prepared by the *agent des ponts et chaussées* for planting seventeen trees on the *place*; but the council, taking the view that this was not in accord with the advice of the prefect and the decision of the council taken on 13 October 1879, rejected the plan and adjourned debate of the issue. By May 1880 – following municipal elections – the council had a new mayor and agreed to go ahead with constructing the *trottoirs* and *demi-caniveaux* (gutters) around the *place*, at a cost of about 300F. The work was concluded in the autumn. What arrangement was finally agreed for the tree-planting remains unclear: the present layout of trees (Fig. 12.10) accords more with the plan drawn up in 1880 by the *agent des ponts et chaussées* than with the plan of 1878. This incident, of course, is more about internal tensions and conflicts between groups *within* Mesland than about a contest between local and central authorities. To some extent, it might indicate that the role of the council was coming to be more widely appreciated and its activities to be more democratized.

'Fire-station'

During the 1840s, 1850s, and 1860s many communes in Loir-et-Cher established *corps de sapeurs-pompiers* (fire-fighting brigades) but a *corps* was not established at Mesland until 1882.[15] A public appeal for funds to purchase a pump raised 1,541F, contributed to by fifty-one people (among them, the mayor, the *curé* and the schoolteacher, and also Théodore Chaput – but not François Collesse). On 17 April 1884 the commune council noted that 'les travaux relatifs au logements de la pompe' had been concluded, at a cost of 1,354F30, and decided to charge the work against funds which had earlier been set aside for repairs to the public fountain. A list of the commune's property insured in 1897 with the Société d'Assurances Mutuelles de Loir-et-Cher includes a slate-roofed stone building to the north of, and adjacent to, the *préau* of the boys' school. Thus the fire-station – a shed for the fire-fighting pump – was located on the western side of the school's playground, serving no doubt as a constant reminder to the boys of their duty to serve their community (the *corps* was made up of volunteers) and of the need to rely upon rational – rather than mystical – methods of protection against fire and similar hazards. By the 1880s fire was no longer accepted as an act of God; it was seen as man-made danger which could be fought by actions of the state, as represented locally by the commune council.

Post office

On 18 January 1887 the commune council considered and approved an estimate of 9,500F for building a *bureau de poste*. Given the heavy expenditure of the commune in recent years – on building two schools and on constructing and maintaining roads – the council decided to seek a grant from the *département* and to fund the remainder by raising a loan, to be repaid over thirty years. In practice, donations by twenty-four private individuals – giving sums ranging between 4F and 249F – raised 1,161F44 but no grant was forthcoming from the *département*; the thirty years' loan of 9,500F taken out in 1887 was, in 1893, converted into a ten years' loan of

Fig. 12.11   The *bureau de poste* of Mesland, completed in 1893.

5,000F, because the commune had been able to make several payments to reduce the size of the initial loan.

The *bureau de poste* was built at a central location, on the rue de l'Eglise at the southeastern corner of the *place publique* (Fig. 12.11). It was a prominent construction, with a waiting room for the public, the office itself, and a small kitchen and dining room on the ground floor; on the first floor was the main accommodation (two rooms and a *cabinet*) for the postmistress.

In 1904, in order to allow the installation of a telegraphic service in the commune, the council agreed to let the *bureau de poste* to the Directeur des Postes et Télégraphes de Loir-et-Cher for nine years at an annual rent of 280F, so that it could become, from 1 July 1905, *un bureau de poste, de télégraphe et téléphone*. In February 1906 the council considered the prefect's invitation to have Mesland included in the telephone network being planned for the *département* and agreed unanimously to pay 65F30 annually from 1906 for ten years in order to have a *bureau téléphonique* established in the commune.

Night shelter for poor travellers

Perceiving an increase in the number of poor people travelling through the commune and asking for overnight accommodation, the council in the summer of 1893 decided to construct a shelter. On 29 June the council approved an estimate of 861F for constructing such a shelter and decided to ask the *département* for a grant towards the cost; on 8 July the council noted that the prefect had arranged a grant of 450F and agreed to meet the balance from the commune's budget for 1893. The building, described in the estimates as 'un Abri de Nuit pour les étrangers de passage et sans ressources', was completed by November, at an actual cost of 634F16. The list of the commune's insured property in 1897 includes 'un petit local servant de poste et de logement des voyageurs indigents' and as being located 50 metres to the west of the boys' school. This slate-roofed stone building, a single room with a loft above, indicated the acceptance of a real, if limited, responsibility of the commune of Mesland for poor people travelling through its territory: again, a responsibility for centuries exercised by the church had been appropriated by the civil local authority.

Public washhouse

On 9 September 1910 the council approved plans and an estimate for the construction of a covered *lavoir* at the hamlet of Pré-Noir, some 3 kilometres to the north of the *bourg* of Mesland. Other than roads, this was the first public utility constructed by the council beyond the *bourg*. As we have already seen, plans for a mixed school et Pré-Noir were soon to follow. Work on the *lavoir*, completed by the autumn of 1911, cost 1,467F67. The contract for the work had stipulated that it would be undertaken 'suivant les règles de l'Art, mésures suivant leur dévéloppement réel ... les entrepreneurs renoncant aux usages locaux qui pourraient exister'. The slates were to come from the 'meilleur bancs des carrières d'Angers', while 'le Ciment sera du Portland artificiel de Boulogne'.

Conclusions

For much of the first half of the nineteenth century the public landscape of the commune of Mesland was one which it had largely inherited from the *paroisse* of the eighteenth century. Other than roads, streams and the fountain by the Cisse, the only 'public' components of the landscape were

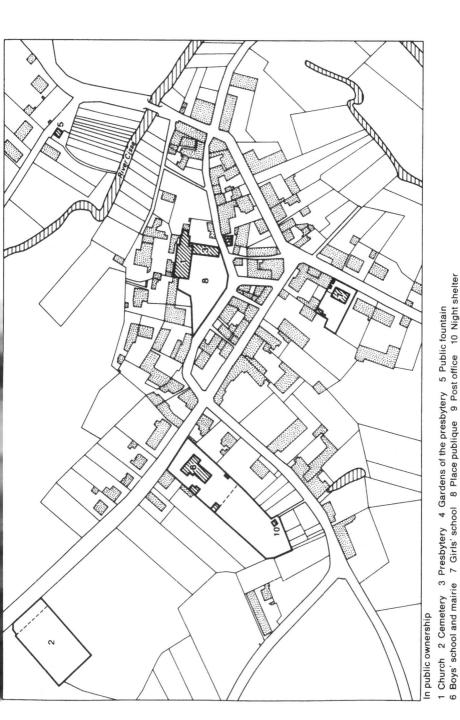

1 Church   2 Cemetery   3 Presbytery   4 Gardens of the presbytery   5 Public fountain
6 Boys' school and mairie   7 Girls' school   8 Place publique   9 Post office   10 Night shelter
—— Roads and bridges

Fig. 12.12   Ownership of the landscape of Mesland, c. 1914.

the church, the presbytery and the cemetery, and Mesland's only full-time 'public servant' was the *curé*: both fixed and ambulatory symbols of the potent ecclesiastical influence upon the *activités* and *mentalités* of the inhabitants of Mesland.

By the eve of the outbreak of the First World War the commune had acquired, through the agency of its municipal council acting both on its own initiative and in response to stimuli coming from central authority, a range of new landscape components (Fig. 12.12) and of public servants which was a forceful expression of the newly acquired power of the state, both central and local.[16] There were, of course, elements of continuity. The commune council assumed responsibility for building new roads and for maintaining and improving old ones. But the changing balance of power within the commune was significantly expressed in terms of the council's management of its inherited 'ecclesiastical' landscape and of the construction of a 'civil' landscape. The commune council, a municipal authority, came to assume responsibility for maintaining and repairing the church; the presbytery, after a short period in private hands, became the property of the commune; and the centuries-old cemetery in the centre of the *bourg* was displaced by the civil authority to the outskirts of the *bourg* and replaced by a *place publique*. Even more explicitly, the council proclaimed in the landscape its growing hegemony over the traditional, ecclesiastical source of socialization with Mesland. Two new schools (and plans for a third) and a post office were major statements of the growing power of the council; and a number of other additions to Mesland's landscape (the fire-station; the night shelter; the washhouse), although less striking, served to underscore those major statements. So, too, did the large cohort of public employees at Mesland by 1911.

The civil authority waged the battle for the minds of the people of Mesland in a host of other minor ways. With increased frequency, the national flag was flown from public buildings and used in street processions on public occasions. In the 1890s the *sapeurs-pompiers* were provided with uniforms by the commune council. By 1908, instead of ringing church bells to ward off thunder storms the people of Mesland were firing cannon into storm clouds to forestall the formation of hail stones. The *mentalite* of the people of Mesland had been transformed during the lifetimes of Thédore Chaput and François Collesse, and the construction and use of civic buildings and places had contributed to that transformation. The *événements* of Mesland were unique to that commune, but their underlying *conjonctures et structures* affected most communes of France. Consequently, the landscape symbolism recognizable within Mesland was being produced in many, indeed most, of France's rural communes. The ubiquity of the iconography of an emergent civic and national ideology was reinforced for the people of Mesland whenever they travelled (as they increasingly did) elsewhere in the French countryside. At the local cantonal centre of Herbault it was even more obvious than at Mesland; and it must have been most obvious if and when they went to Blois, the administrative centre of Loir-et-Cher, with its impressive array of nineteenth-century public buildings (notably the prefecture, the *palais de justice* and the *halles*), all within a stone's throw of its – much older, but now less powerful – cathedral.

ACKNOWLEDGEMENTS

I wish to thank M. Paul Roger, Mayor of Mesland, and M. Jacques Cartraud former Secretary of the Commune Council and schoolmaster at Mesland, for their friendship and advice during the period when this chapter was being prepared. I also wish to record again my considerable debt to the staff of the Archives Départementales de Loir-et-Cher for their assistance and advice, without which my research would simply not have been possible.

NOTES

1 This chapter is founded upon two substantial sets of unpublished documents deposited in the Archives Départementales du Loir-et-Cher (hereafter, AD) relating to the commune of Mesland: (i) the Minute Books of Mesland's Commune Council 1815–1913 (AD Régistres de délibérations du conseil municipal. Dépôts des Communes No. 510); (ii) the files of administrative correspondence of the Mayors of Mesland (AD 141:0⁶1 an XIII–1853; 0⁶2 1854–1883; 0⁶3 1881–1897; 0⁶4 1898–1921). Citations in this chapter of particular unpublished documents may (unless otherwise noted) be traced in one or other of these two manuscript collections: they are too numerous to be noted here individually.

2 E. Weber, *Peasants into Frenchmen* (London 1977) 310 and 110

3 T. Zeldin, *France 1848–1945* Vol. II. *Intellect, taste and anxiety* (Oxford 1977) 3–28; Weber, *Peasants*, 95–114. This view, it should be noted, is contested by P. Claval who argues that Zeldin and Weber are wrong in concluding that the 'idea of France' had not become part of popular consciousness before the mid nineteenth century. Claval contends that the idea had existed almost everywhere in France for a long time previously, although he does admit that its generalization was an achievement of the schooling system: see X. de Planhol and P. Claval, *Géographie historique de la France* (Paris 1988) 483 and P. Claval, The image of France and Paris in modern times: an historical-geographical problem, in A. R. H. Baker and M. Billinge (eds.), *Period and place: research methods in historical geography* (Cambridge 1982) 205–11. But the 'idea of France' ought not to be confused with French nationalism: Brian Jenkins has recently pointed out that for a century before the Revolution of 1789 'France had enjoyed a clear

territorial identity, had established a powerful centralized state, and its inhabit-
ants were slowly beginning to acquire a degree of collective consciousness ...
But it was not yet a "nation", a concept which had little currency before 1789':
B. Jenkins, *Nationalism in France: class and nation since 1789* (London 1990) 11

4 Weber, *Peasants*, 195–220; R. Price, *The modernization of rural France: communications networks and agricultural market structures in nineteenth century France*
(London 1983)

5 Weber, *Peasants*, 493–4

6 This paragraph draws upon an excellent account of French administrative
organisation in the nineteenth century: F. Ponteil, *Les Institutions de la France
de 1814 à 1870* (Paris 1966) esp. 30–3, 156–63, 282–5 and 378–81

7 Demographic and economic data for Mesland from 1831 to 1911 have been
drawn from the five-yearly manuscript census returns (AD 202M211 and
202M212)

8 AD Ancien cadastre et matrices des propriétés: Mesland, 1819–1825

9 In this regard, the experience of Mesland was perhaps precocious because it is
alleged that some communes, too poor to buy a flag, lacked that basic symbol of
the nation state until the mid nineteenth century: Weber, *Peasants*, 111

10 R. Price, *A social history of nineteenth-century France* (London 1987) 312;
J. Martin-Demézil, Analyse d'une grande enquête: l'enseignement primaire en
Loir-et-Cher, 1833 *Mémoires de la Société des Sciences et Lettres de Loir-et-Cher*
36 (1980) 59–76; P. Olivier, Les étapes de la scolarisation dans le Loir-et-Cher
de 1833 à 1910 *Mémoire, école normale supérieur* (Blois 1962)

11 This chapter focusses upon schools buildings as components of landscape and
does not concern itself with either the content of schooling or the quality of the
schoolteachers of Mesland but for some notes on these aspects, see: M. et Mme
Bourdin, Une École primaire vers 1860 *Centre Départemental pour le 3ᵉ Age en
Loir-et-Cher* (Blois 1980) 22–29

12 Cf. 'The schoolhouse at the centre of the village now came to challenge the
church as the place to which peasants came with their problems': B. Singer,
Village notables in nineteenth-century France: priests, mayors, schoolmasters
(Albany 1983) 110

13 Conscript illiteracy figures have been calculated from the military conscription
registers for the relevant years: AD Série R.1 Tableaux de Recrutement

14 Literacy standards are recorded for individuals in the 1872 population census
(AD 202M 114 and 211)

15 A. R. H. Baker, Fire-fighting fraternities? The *corps de sapeurs-pompiers* in
Loir-et-Cher during the nineteenth century *Journal of Historical Geography* 16
(1990) 121–39

16 Fig. 12.12 has been constructed by comparing the cadastral plan of 1936 with
the property registers of 1910

13

A contrast of Old World ideology:
Germans and Scotch-Irish in the Ozarks

RUSSEL L. GERLACH

The first European-descended settlers arrived in the Ozarks (Fig. 13.1) in the early nineteenth century, and by the mid-nineteenth century the region numbered among its inhabitants French, English, Scotch-Irish and Germans with smaller numbers representing a variety of other nationalities. By the mid nineteenth century two nationalities were dominant over much of the region – the Germans in areas bordering the Missouri and Mississippi Rivers and the Scotch-Irish in the less-accessible interior areas (Fig. 13.2)[1]. These two groups brought distinctive and contrasting ideologies to the Ozarks drawn from their experiences in the Old World, and their values and attitudes were reflected in the cultural landscapes that evolved in the region. This chapter examines the experiences of the Germans and Scotch-Irish in the Ozarks, and particularly their values and attitudes, and how these indicators of ideology were manifested in the cultural landscape.

Land selection

Scotch-Irish and German settlers selected different and contrasting environments in the Ozarks for a number of reasons. German settlement was highly organized for the most part, with many planned colonies established in the period after 1830. Those coming to Missouri directly from Germany were dominated by organized colonization societies and even those coming from areas in the eastern United States involved some degree of organization. The goals of these planned efforts were twofold. First, there was a desire to retain their Old World culture through compact group settlement. Secondly, virtually all sought locations where the possibilities for market agriculture and other forms of commerce were maximized. In particular, access and land quality were primary considerations. As most Germans arrived with some accumulated wealth, the costs of land acquisition were often of secondary concern. Employing land agents, and often working through land speculators many of whom were themselves

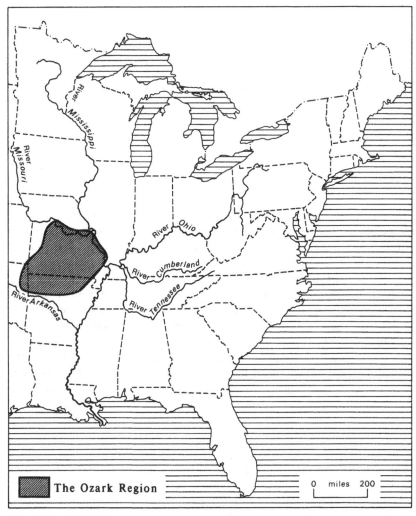

Fig. 13.1   The Ozark region.

German, the Germans were directed to the thinly settled Missouri and
Mississippi borders of the Ozarks which met their primary requirements for
settlement.[2] Some were attracted by the writings of earlier German pioneers in these areas which were in wide circulation in Germany by 1830.[3]
As a result, Germans generally acquired land of reasonable quality and
access, although instances were noted in which the desire of Germans to
establish colonies on unoccupied land led them to select sites in the Ozarks
that were either poorly located relative to rivers or of marginal quality.[4]
Through the process of chain migration, the areas expanded and filled in

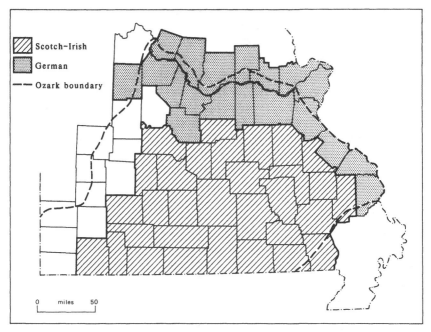

Fig. 13.2   Scotch-Irish and German settled counties in the Ozarks, 1860.

until there were few non-Germans remaining in these areas and land shortages limited further settlement.

The Scotch–Irish settlement of the Ozarks followed a pattern very different from that of the Germans. Those coming to Missouri had been in America since Colonial times. They were ethnically mixed, had moved several times since arriving in America, were generally of the lower classes and lacked any strong sense of group cohesion. Most came to the Ozarks from the hills of Kentucky and Tennessee where legal problems involving land titles and land shortages had prompted a relocation of many to areas farther west.[5] Lacking financial reserves, the Scotch–Irish were unable to compete for better land such as that acquired by the Ozark Germans. Beyond that consideration, their form of economy required neither quality land nor easy market access. They practised an extensive livestock-based grazing economy and, for that, the sparsely settled forested land of the interior Ozarks was well suited and, as it was government owned, there was little pressure to purchase or otherwise obtain a legal right to use the land. The land was there and no one else appeared interested in it. Lacking any real sense of group, Scotch–Irish settlement of the interior Ozarks was an individual effort, although in many cases it was kin-based with entire extended families locating in the same general area over a period of some years.

Land tenure

On the matter of land tenure Ozark Germans and Scotch–Irish differed in both attitude and practice. The Scotch–Irish were prone to occupy land without the formality of ownership, a tradition with strong Ulster antecedents.[6] Squatting, as the practice was known, was widespread throughout the Scotch–Irish settlements of the interior Ozarks. In their core in the Southern Courtois Hills in 1850 only seventy-five land entries had been filed for the more than 3,000 people then resident there,[7] and in Shannon County in the heart of this region, only 10 per cent of all individuals listing their occupation as farmer in 1850 owned any real estate.[8] In addition to attitudes among the Scotch–Irish that favoured the practice of squatting, several other factors encouraged its use. Population densities in the interior were low and settlement was widely dispersed. In addition, the majority of land in the interior Ozarks was owned by the government in the early years and absentee lumber interests in later years. In the early years squatting was not a problem, but in later years it did become an issue of some importance. Competition for land led to the problem and resulting tensions. The persistence of squatting into the twentieth century reflected the value system of the Scotch–Irish and their attitude towards outside authority.

Among the Ozark Germans, by contrast, squatting was a very uncommon practice. German settlements were organized and well ordered and this left little place for such a haphazard method of settlement as squatting. In Gasconade County in the northern Ozarks, more than 90 per cent of German farmers owned the land they worked in 1850.[9] Germans came from a tradition of peasant ownership of land, and most who came to the Ozarks possessed the means to acquire title to at least some land. Those that did not have the means were forced either to move temporarily to the city or hire their labour to their neighbours in order to acquire capital with which to buy land.

Open range

Open-range herding of livestock, like squatting, has long been associated with the Scotch–Irish.[10] They practised it in both Ulster and the eastern United States. The Spanish brought open-range herding to Missouri in the eighteenth century long before the arrival of the Scotch–Irish, and when Missouri adopted the English Common Law in 1816, open range was retained as the basic law regulating rangeland.[11] In the interior areas settled by the Scotch–Irish, livestock provided the quickest and least intensive form of income and with the use of open range, allowed land to be substituted for labour. The only land not available to ranging livestock

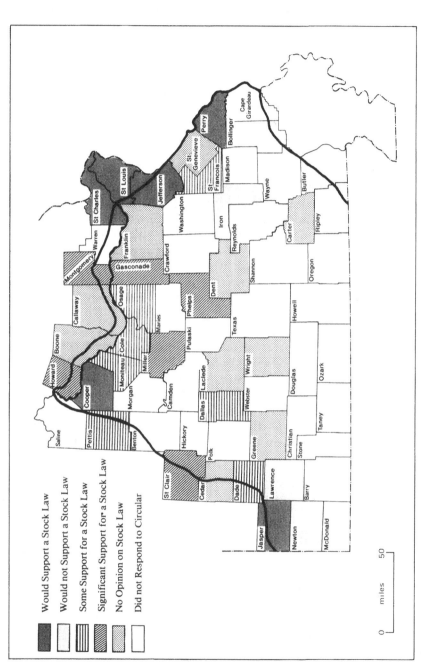

Would Support a Stock Law

Would not Support a Stock Law

Some Support for a Stock Law

Significant Support for a Stock Law

No Opinion on Stock Law

Did not Respond to Circular

0   miles   50

Fig. 13.3   Degree of support for a stock law in the Ozarks, 1872.

consisted of patches of cropland that were fenced in. In 1888 a brochure on Howell County in the southern Ozarks boasted that, 'For stock raising this county is preeminently one of the best in the state ... Not one acre in one hundred ... is yet fenced.'[12] Open range became a way of life for the Ozark Scotch–Irish and they resisted all efforts to control rangeland. A poll of all Missouri counties conducted by the Missouri State Board of Agriculture in 1872 indicated little support for stock laws (closed range) in the interior Ozarks (Fig. 13.3).[13]

For the Germans there were too many negatives associated with open range. It denied them access to the valuable manure from animals, an essential component in their agricultural system. After their initial years in the Ozarks, Germans relied less on livestock and more on grain, thus reducing their dependence on range, free or otherwise. Germans early recognized the difficulty of maintaining quality breeds in an open-range setting. Germans, whose settlements contained little public domain by 1860, were perhaps more sensitive to the destructive nature of open-range herding than were the Scotch–Irish, whose settlements were dominated by public domain. By the 1860s support for stock laws was apparent among the Germans (Fig. 13.3) and by 1874 stock laws were already in effect in some German settlements.[14] In 1883 an optional stock-law bill was passed by the state legislature.[15] Over the next several decades open range disappeared in the German areas but remained in the Scotch–Irish core of the interior Ozarks and by 1935 only the Scotch–Irish continued to support it (Fig. 13.4).[16] Finally, in 1967 the State of Missouri passed a law ending the practice of open range for the entire state, ostensibly to improve highway safety.[17] For the Scotch–Irish, the persistence of open range into the 1960s seemed to suggest a cultural affinity for the practice – something akin to a psychological need to possess the wilderness represented by the free range.

Attitudes toward farming

The Scotch–Irish and Germans brought different attitudes, interests and backgrounds regarding farming to the Ozarks. The Scotch–Irish were adaptable and took what the land offered under a given set of circumstances. They supplemented their farming with a variety of other activities, including mining, forestry and market hunting. The Germans, by contrast, never strayed from the pursuit of farming as their primary activity in the rural Ozarks. For many Germans, the only disruption in their tenure as farmers had been the time taken for their removal from Germany to the Ozarks. Thus, while the early Scotch–Irish were adaptable, exploiting several sectors of their habitat, the German, often poverty stricken, 'secured what little he had from the earth, the sweat of his brow and became correspondingly avaricious to its possession'.[18]

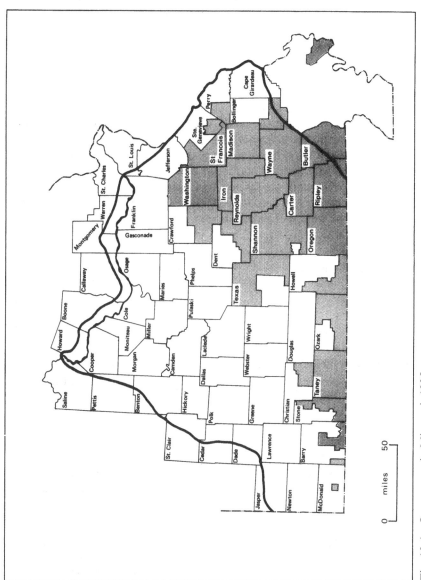

Fig. 13.4 Open range in Missouri, 1935.

It was often said of German farmers in America that they possessed agricultural skills far superior to those of neighbouring groups, and particularly settlers from the British Isles.[19] Describing an area in the northern Ozarks where the divide between the Germans and Scotch–Irish was fairly sharp, Curtis Marbut said, 'The cultivated land is better, the farms are more prosperous, and the farms are better equipped ... mainly because it is occupied to a very great extent by German farmers.'[20] A field worker for the Work Projects Administration in the 1930s stated, 'It is a safe bet that any county that is predominantly German is happy and prosperous.'[21] By contrast, an 1865 report on Crawford County in the core of the Scotch–Irish settled interior Ozarks noted 'Until our people are educated to a point where they can value a sheep higher than a dog, and agriculture ... better than opossum and coon hunting, I suppose our crop of nutritious grains will grow "to waste their fragrance on the desert air." '[22] Thus, it would appear that the Germans, as a group, had a greater commitment to farming than was the case for the Scotch–Irish in the Ozarks.

The Germans were motivated to some extent by a desire to succeed in purely material terms. Arthur Apprill said of the Germans in Gasconade County, 'Making money is their chief goal and aim in life, sometimes almost a passion, which may work to the detriment of higher interests and ideals.'[23] The Scotch–Irish seemed less concerned with material well-being. Throughout the interior Ozarks there was an even distribution of wealth; few were richer and few were poorer, and the Scotch–Irish were identified by, 'the apparent ease with which they satisfy their meager want and low level of expectations ... And it was said of them perpetually ... they were just not very ambitious.'[24]

Patterns of investment

Once established in the Ozarks, the Germans tended to move less than the Scotch–Irish. The locational stability of rural Germans and the greater mobility of the Scotch–Irish were characteristics associated with the two groups from an early date in the eastern United States. Commenting on this Maldwyn Jones said:

If the Germans appeared more industrious and frugal, that was partly because they generally remained where they first settled. Retaining the peasant's high regard for patrimonial property, they looked on their farms as legacies to be bequeathed to their children and gave a high priority to improvement. The Scotch–Irish, on the other hand, were slow to commit themselves to particular localities. Perhaps because migration had become second nature to them, they lacked any real attachment to the soil and tended to exploit their farms ruthlessly before moving on to others.[25]

In 1838 Eduard Zimmerman, in comparing the Ozark Germans and their neighbours, said of the Americans, as he called all non-Germans, that they 'never became attached to a given region. If they can sell their property to any given sort of advantage, they are certain to do so.'[26] Commenting on the German settlements in the Ozarks, Carl Sauer said, 'Stability remains the most distinguishing characteristic of the German stock. Where Germans have located in most cases they have remained ... Property is handed down from father to son, and in many cases the descendants of the original entrymen still retain the land.'[27]

Whether consciously or unconsciously this difference led the Germans to view land as a commodity to be developed, while the Scotch–Irish viewed land as a commodity to be exploited. The Germans invested time, effort and resources to upgrade the quality of their land while the Scotch–Irish adopted a system of land rotation as a means of maintaining productivity. In the German areas crop rotation, manuring and the use of cover crops were common practices. Arthur Cozzens said of the Ozark Germans, 'They have ... followed the typical German practice of applying all available manure to the land, liming the soil, and planting clover as part of their regular crop rotation.'[28] Carl Sauer observed that crop rotation was not commonly practised in the Scotch–Irish settled Ozark interior.[29] Commenting on the Scotch–Irish system of land rotation, a practice drawn from their Ulster heritage, a Forest Service study noted:

In order to survive they had to cut immature trees from the forest, overcrop good tillable land, clear land unsuited for cultivation, and burn the forests ... The overall income was too low to permit needed soil-building practices on the good tillable lands and the yields gradually decreased. Much of the cleared land was abandoned after a few years of use because of low productivity. Much of it reverted to a forest cover only to be cleared again ... Many of the people in this area have been going around in this vicious cycle ever since the original timber crop was cut.[30]

In German areas land abandonment was quite uncommon whereas in the Scotch–Irish areas it was the natural result of land rotation. Because of these differences, Germans tended to clear their cropland completely over a period of years whereas the Scotch–Irish only removed the larger trees. The Scotch–Irish relied heavily on the practice of burning both to clear land and to maintain productivity – a practice rarely used by the Germans. On the matter of conservation practices among the Scotch–Irish, Schroeder concluded that such practices were generally disregarded well into the twentieth century, and that they exploited the land as though there was no future to consider.[31]

Settlement characteristics

From the beginning both the German and Scotch–Irish settled areas of the Ozarks were characterized by dispersed rural settlement. Generally the

German settlements were more compact than were those of the Scotch–Irish and they were evidenced by a more regular or even distribution of land entries. There was a village system in the German areas that comingled geographically with the detached farms. Some of the Ozark German villages were, in fact, true farm villages, but even those not inhabited by farmers were an extension of the agricultural community, and the differences between dispersed rural settlements and villages were not always sharp. The German village in the Ozarks provided a sense of order and continuity in social, religious and ethnic matters as well as fulfilling its role as the economic centre of the farming community. The agglomerated settlements that evolved in the Scotch–Irish areas tended to serve primarily an economic function and were often established by outsiders who were later arrivals. In many cases they served non-agricultural interests, such as those of lumbering and mining, and thus had tenuous ties with the surrounding farming community.

The social organization of rural German communities in the Ozarks was tight. The community exerted an influence on all aspects of life and encouraged co-operative efforts in economic as well as social matters. In the Scotch–Irish areas of the Ozarks the farmer relied more on his own instincts, or 'personal ingenuity' as Cecil Gregory termed it.[32] His support group consisted of kin and neighbours who, like him, were also sheltered from outside influences, and, as Carl Sauer observed, 'Of cooperation the native [of the interior Ozarks] knows nothing beyond the relief of a neighbor in trouble.'[33]

The structural occupancy features of the Scotch–Irish and German areas of the Ozarks differed in a variety of ways. The majority of structures in German areas were constructed in the nineteenth century while very little remains of the nineteenth-century architectural heritage of the Scotch–Irish.[34] German structures were larger and more numerous than was the case for Scotch–Irish. All of this suggests that the Germans invested more heavily in farm improvements than did the Scotch–Irish; another indication of their intent to remain in place once located. The greater permanence of farm structures in the German areas appeared to be related to attitude and tradition.

The aesthetic qualities of farming were frequently raised concerning the Germans and Scotch–Irish in America. German farms were usually characterized as neat, tidy, clean, well ordered and, generally, reeking of prosperity.[35] Scotch–Irish farms, on the other hand, were often placed at the other end of the prosperity continuum. They were characterized as poor looking, disorganized and unkempt.[36] Robert Flanders addressed the more qualitative aspects of Scotch–Irish and German settlement features in the Ozarks. He wrote:

The German farms with their dwellings, barns and appurtenances tend to be 'snug'. Those of Scotch–Irish descent tend to be 'clarty'. 'Snug' barns and farmsteads ... simply means things 'lying close, warm, and comfortable; neat, trim, and convenient'. 'Clart' on the other hand ... means fouled and dirty, clogged [and] unarranged ... *Snugness* and *clartiness* are not mere manifestations of wealth vs. poverty, energy vs. sloth, sophistication vs. ignorance, good taste vs. no taste. They are, rather, manifestations of culture.[37]

The qualitative differences so often observed between farms of the Scotch–Irish and Germans may have been the result of contrasting cultural attitudes and values as much as any other factors. Kerby Miller suggested that the Ulster–American farmer's apparent destitution often masked material abundance and stemmed from his reluctance to display wealth.[38] Walter Schroeder observed, regarding the apparent prosperity of the Germans in Missouri, 'German-Missouri property has long been recognized for its cleanliness, orderliness, and neatness. Such care of buildings and surrounding grounds often gives a false impression of wealth and prosperity to some rural or semi-rural German-Missouri communities with highly marginal economies.'[39]

Conclusions

The Scotch–Irish and Germans in the rural Ozarks presented a striking contrast of cultures. For the Germans an ordered and predictable existence was of paramount importance, while for the Scotch–Irish a more chaotic life with all of its uncertainties was the norm. Permanence was a goal sought after by the Germans while the more opportunistic Scotch–Irish seemed content with life in constant transition where each new generation relived the eternal present. What resulted was a stable society among the Germans and one lacking stability among the Scotch–Irish. For both groups, their past was a major factor in explaining their very different adjustments to life in the rural Ozarks, and it was their cultural preadaptation to act in certain ways that truly made them distinctive, one from the other. Long after the overt signs and symbols of ethnicity had faded, the traditional values and attitudes of the Ozark Scotch–Irish and Germans continued to influence both groups, and the resulting contrasts in their cultural landscapes remain visible to the present.

NOTES

1 The source for Fig. 13.2 is Russel L. Gerlach, *Settlement patterns in Missouri* (Columbia Mo. 1985) map insert
2 The discussion of German settlement is taken from Russel L. Gerlach, *Immigrants in the Ozarks* (Columbia Mo. 1976), 33–43
3 Carl E. Schneider, *The German church on the American frontier* (St Louis 1939) 14
4 Carl O. Sauer, *The geography of the Ozark Highland of Missouri* (Chicago 1920) 167–8
5 John Solomon Otto, The migration of the Southern plain folk: an interdisciplinary synthesis *Journal of Southern History* 51 (1985) 196
6 See, for example, James G. Leyburn, *The Scotch–Irish: a social history* (Chapel Hill NC 1972) 192–94; William C. Lehmann, *Scottish and Scotch–Irish contributions to early American life and culture* (Port Washington, NY 1978) 54; Maldwyn Jones, Scotch–Irish in Stephen Thernstrom (ed.), *The Harvard encyclopedia of American ethnic groups* (Cambridge Mass. 1980) 900; Tyler Blethen and Curtis Wood, *The migration of the Scotch–Irish to southwestern North Carolina* (Cullowhee NC 1983) 22
7 Harbert L. Clendenen, Settlement morphology of the Southern Courtois Hills, Missouri, 1829–1860 (unpubl. Ph.D., Louisiana State University, 1973) 83
8 Based on a hand count from the manuscript schedules of population for Shannon County, Mo, 1850 federal census
9 *Ibid.* Gasconade County, Mo.
10 Grady McWhinney and Forrest McDonald, Celtic origins of southern herding practices *Journal of Southern History* 51 (1985) 168; Otto, The Migration of the Southern plain folk, 185; E. Estyn Evans, Cultural relics of the Ulster-Scots in the Old West of North America *Ulster Folklife* 11 (1965) 37
11 John H. Calvert, Fencing laws in Missouri – restraining animals *Missouri Law Review* 32 (1967) 524
12 *Howell County, Missouri: facts regarding climate, soil, agriculture, grazing, fruit and vine lands; forests, scenery, and game; reliable intelligence on other topics applicable to the necessities of the housekeeper, capitalist and tourist* (West Plains Mo. 1888) 10
13 Missouri State Board of Agriculture, *Eighth annual report for the year 1872* (Jefferson City Mo. 1873) 192–348. Replies were received from 87 of the 113 counties in Missouri
14 Records of Gasconade County, Mo., index to county court papers, 1841–1938; County court papers, box 5, bundle 3, paper 37
15 *Laws of Missouri*, Regular session (Jefferson City Mo. 1883) 26–38
16 Rudolph Bennitt, General ecology – free range in Missouri *Quarterly Report of the Missouri Cooperative Wildlife Research Unit and Associated Projects* (Columbia Mo. 1935) 13–15
17 *Revised statutes of the state of Missouri* 3 (Jefferson City Mo. 1969) 2184
18 Schneider, *The German church on the American frontier*, 26
19 Studies that portrayed German farmers as superior to others include Thomas J. Archdeacon, *Becoming American: an ethnic history* (New York 1983) 16; Walter

M. Killmorgan, A reconnaissance of some cultural–agricultural islands in the South *Economic Geography* 17 (1941) 409–30; M. Kollmorgan, Agricultural–cultural islands in the South – part 11 *Economic Geography* 19 (1943) 109–17; Richard Shryock, British versus German traditions in colonial agriculture *Mississippi Valley Historical Review* 26 (1939–40) 39–54. Studies that questioned the skills of some British farmers without reference to Germans include Charles M. Andrews, *The Colonial period in American History* (New Haven 1934) 209; Avery O. Craven, *Soil exhaustion as a factor in the agricultural history of Virginia and Maryland 1606–1860* (Urbana Ill. 1926) 37

20 Curtis F. Marbut, *Soil reconnaissance of the Ozark region of the Missouri and Arkansas* (Washington DC 1914) 84

21 US Work Projects Administration, Historical records survey, 1935–1942, Gasconade County, Mo. f.7126, Western historical manuscript collection (Columbia Mo.)

22 Missouri State Board of Agriculture, *First annual report* Appendix 1, *County agricultural reports* (Jefferson City, Mo. 1866) 59

23 Arthur W. Apprill, The culture of a German community in Missouri. (unpubl. MA thesis, University of Missouri–Columbia, 1935) 43

24 Robert Flanders, Alley, an Ozarks mill hamlet, 1890–1925, Report prepared for the National Park Service (Springfield Mo. 1985) 11; similar sentiments are expressed in Sauer, *Geography of the Ozark Highland*, 162; John S. Otto and Augustus M. Burns III, Traditional agricultural practices in the Arkansas highlands *Journal of American Folklore* 94 (1981) 186; Franklin G. Liming, Forest problem analysis and research program for the Missouri Ozarks, Central States Forest Experiment Station *Miscellaneous release no. 7* (St Paul Minn. 1951) 12.

25 Scotch–Irish, 900; see also Leyburn, *The Scotch–Irish*, 199–200; E. R. R. Green, Scotch–Irish immigration, an imperial problem *Western Pennsylvania Historical Magazine* 35 (1952) 209; Terry G. Jordan, *German seed in Texas soil* (Austin Tex. 1966) 37–8

26 Eduard Zimmerman, Travels into Missouri in October 1838 *Missouri Historical Review* 33 (1943) 243

27 *Geography of the Ozark Highland*, 173

28 Arthur B. Cozzens, Conservation in German settlements in the Missouri Ozarks *Geographical Review* 33 (1943) 294

29 *Geography of the Ozark Highland*, 195

30 Liming, Forest problem analysis and research program for the Missouri Ozarks, 12

31 Walter A. Schroeder, The eastern Ozarks: a geographic interpretation of the Rolla 1:250,000 topographic map, National Council for Geographic Education, *Special Publication Number 13* (Normal Ill. 1967) 39

32 C. E. Lively and C. L. Gregory, Rural social areas in Missouri, Missouri Agricultural Experiment Station, *Research Bulletin 305* (Columbia Mo. 1939) 33

33 Carl O. Sauer, The economic problem of the Ozark Highland *Scientific Monthly* 11 (1920) 223

34 The following discussion is based in part on 1,500 miles (2,400 kilometres) of

systematic road traverses, and included field observation of more than 4,500 farmsteads in the northern and eastern Ozarks

35 See, for example, Albert F. Faust, *The German element in the United States* 2 vols. (New York 1927) 1:445; Killmorgan, A reconnaissance of some cultural-agricultural islands in the South, 447

36 See, for example, David N. Doyle, *Ireland, Irishmen and revolutionary America, 1760–1820* (Dublin 1981) 84; John Fraser Hart, Land rotation in Appalachia *Geographical Review* 67 (1977) 148

37 Miller County survey, FY 80–81 Historic Preservation Program (Springfield Mo. n.d.) 46–7

38 K. Miller, *Emigrants and exiles: Ireland and the Irish exodus to North America* (New York 1985) 163

39 W. Schroeder, Rural settlement patterns in the German-Missouri cultural landscape, in Howard W. Marshall and James W. Goodrich, (eds.), *The German-American experience in Missouri* (Columbia Mo. 1986) 29

14

Ideology in the planned order upon the land: the example of Germany

DIETRICH DENECKE

The ordered, organized and hierarchical structures of past societies are reflected in the creation of planned settlements and landscapes, which persist in the contemporary landscape as relicts. These vestiges of past landscapes are artefacts that historical geographers have to analyse and explain in order to reconstruct the thoughts and ideologies of former people, governments and societies. However, by interpreting structures and changes in the landscape, we may decipher the activities, thoughts, ideas and ideologies which created the 'scenery'.

Ideology was, of course, only a single element in shaping the landscape. There is no doubt that natural resources, pragmatism, finance, social organization, skills and 'know-how' were also, to a greater or lesser extent, important. For all of these in the past, geography has adopted specific approaches, but the search for the ideology underlying the landscape is quite new. An ideological approach might reveal that its significance in influencing geographic phenomena is often greater than we realize.

In general, ideologies are created and developed to affect or guide people's thoughts, behaviour and activities. Even when ideology has a secondary or minimal impact on the physical landscape it may have a geographical relevance in many cases. Ideologically based activities which influenced the landscape were quite often demonstrative acts creating a visible public symbol with an appropriate frame to represent the ideology itself. However, even then it is rare that ideology alone is represented by the visible 'monument'; style, function, and technological development will all contribute to the final outcome in the landscape. Therefore, trying to detect an ideology behind a planned landscape is a difficult venture, subject to failure, as ideology is not always the obvious or dominant initiative in a particular planning process; although it often will provide the underlying structure.

Ideologies that may influence planning and landscape can be divided into four main categories; political, social and religious and ideologies

Ideologies in the landscape
or: ideology as manifest in the planned order of the land

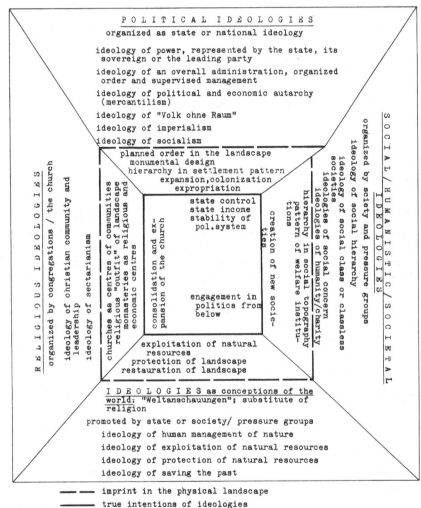

Fig. 14.1   Ideologies in the landscape, or ideologies as manifest in the planned
order of the land.

concerned with the conception of the world *Weltanschauungen*. In turn,
these four categories may be further subdivided (Fig. 14.1). The cultural
symbols and patterns in the landscape are widespread, manifold and,
especially when designed as monuments or monumental statements, quite
obvious. The ideological design behind the landscape features is a powerful
agent working to impress, manipulate or steer society.

Ideology is of great relevance in the *political context* – in the fields of land management, population policy and the fiscal strategies of government. The role of political ideologies is to legitimize, to explain or to justify the state political system to the territory, the nation and the people. Consequently the changing and shaping of the landscape is an important visible result of state ideology imposed on a people and its environment. Conversely the environment reflects the ideology and principles of the state, and also functions to guide and influence the actions, behaviour and thoughts of the people as a response. So ideology in the landscape acts as an active political instrument, steering a state-controlled nation. Thus having the favoured ideology represented in the management and planning of the landscape, in the economic setting and in the spatial organization of groups, communities and the people, the state had a vital interest in spatial organization, in planning and changing the environment of its citizens.

A number of different political ideologies are of specific importance (Fig. 14.1). Political ideologies may be identified in political programmes, in plans and projects, and also in the physical response to planning by the state. However, to discover and follow ideology in the landscape necessitates an understanding of the theoretical and political justification or legitimation of the political and economic aims of the state or government. To discover the character of the ideology guiding and mobilizing change it is necessary to employ methods used by detectives, as the instrument of change is generally disguised behind the plausible rhetoric of progress and individual benefits. The true implications of a political ideology are quite often unrecognized. The relationship between ideology and truth is complex, and in Germany this has been discussed by Hans Barth and Theodor Geiger.[1]

By the Middle Ages social-humanistic and welfare ideologies were already well developed and were related to the feudal or aristocratic hierarchy and to Christianity. The implementation of the ideology led to the construction of institutions such as hospitals, infirmaries, almshouses, hospices and roadside chapels, and also to some degree monasteries which were donated and founded by communities, the aristocracy and the church. Within Germany there was a fairly dense network of social-welfare institutions of this kind, particularly located along the main routes for trade and traffic.

During the nineteenth century social ideologies were increasingly assimilated by political ideologies based largely on socialist ideas, in order to win the political support of the lower and middle classes. This found physical expression in the landscape through social housing programmes.

Surprisingly, *religious ideologies* based on beliefs and doctrines have had little effect on the organization and physical structure of the cultural landscape in Germany, with the exception of the large numbers of churches, urban and rural monasteries and the numerous small religious institutions and monuments scattered across the country. Some authors, for example Goslar and Lubeck, have attempted to demonstrate that medieval symbols of Christianity and *Weltanschauungen* underlie the layout of some medieval towns, especially those towns having an oval shape, which has been interpreted as a mandorla.[2] Although we know that the church was involved in urban development in medieval times it is difficult to prove that Christian symbols were taken as models for town plans.[3]

Religious ideologies, based on beliefs and doctrines, did affect the physical structure of the cultural landscape insofar as the Christian churches normally were found in the centre of settlements and communities. This was the case even when churches were built in well-established pre-Christian settlements, as the village centre was chosen as the church site. The high church towers, symbols of the mighty God, were visible to all the congregation from far around. Monasteries were dominant landscape complexes, self-sufficient religious worlds which also played an important economic role during the Middle Ages.[4] However, we rarely find examples in Germany of settlements organized by religious groups or sects as mirrors of their own religious world; much better examples of such visible ideologies are found in North America, or in Asia.

Ideologies as concepts of the world are often substitutes for beliefs and, like political ideologies, are promoted by political powers, by leaders of society or by organized groups from below. Such ideologies often develop as movements, associated with new ideas and aim to reform the thoughts of man, and his activities. When guided solely by ideology projects often become rather utopian in character. This means that a project initially conceived and motivated by ideas and plans can easily run into technical and financial problems. So, ideology and utopia constitute another close relationship, characterizing ideology as a philosophy or belief which might not be easily turned into reality. In this respect, it would be worthwhile to undertake a comparative study of the great number of utopian projects in the world, in the past and present, during conception and planning, and in their final phases of failure and breakdown.

Finally, ideology also has close connections to idealism or to ideal forms and structures imposed on the cultural landscape. European town planning of the sixteenth century provides a very good example of this phenomenon (the 'baroque city' or *Barockstadt* of the Netherlands and Germany).[5]

The shaping and changing of the countryside and of settlement patterns are initiated by human thought. Thought, ideas or ideologies are behind action, transforming the visible cultural landscape in distinct periods. Human societies and ideologies have frequently a profound impact on the patterns of shape and change. One of the more effective impacts on the land is planning or, even more, replanning. The principle behind planning is order, functional arrangement and regular design, often derived from ideal

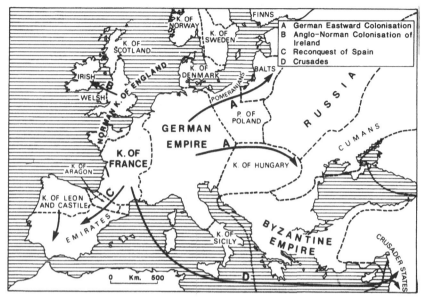

Fig. 14.2 Movements of colonization in twelfth-century Europe. *Source:*
A. Simms, Core and periphery in medieval Europe: the Irish experience in a wider
context, in W. J. Smith and K. Whelan (eds.), *Common ground: essays on the
historical geography of Ireland* (Cork 1988) 22–40.

models, while replanning is often directly steered by ideologies.[6] The most
important process and periods of intended change in the history of the
Central European cultural landscape will be discussed briefly, based on
selected examples.

The ideology of core-and-periphery and medieval expansion movements in Central Europe

Sovereigns and people in twelfth- and thirteenth-century Central Europe
considered themselves to be at the core of the Old World and as heirs of the
civilization of Latin Christendom. Land and economy were more developed and organized than in the periphery, especially compared to the Slav
territories in the east. There was a consciousness of superiority, an ideology
which stimulated expansion and colonization movements, as well as
economic push–pull factors.[7] The Anglo-Normans expanded towards
Ireland, France towards the emirates of the Iberian peninsula, the German
empire expanded towards the Slav territories in Hungary, Poland, Pomerania and the Baltic territories (Fig. 14.2).

The great number of sources and studies on the medieval expansion of
Europe have to be re-examined to find out more about the motives,

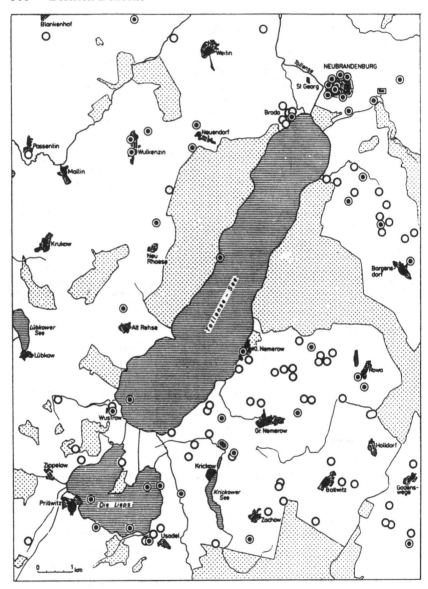

Fig. 14.3   Reorganization of the settlement pattern in Brandenburg, twelfth and thirteenth centuries. *Source:* Gringmuth-Dallmer, *Zeitschrift für Archäologie* 15 (1981) fig. 2, 248.

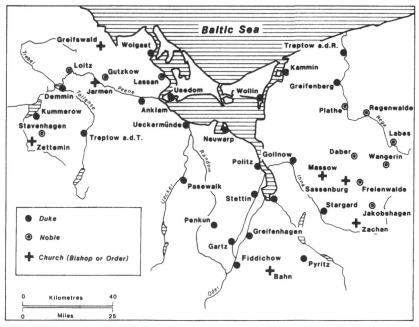

Fig. 14.4 Newly laid-out towns in medieval Western Pomerania and their lordships. *Source:* as Fig. 2, p. 32.

thoughts, ideas and ideologies which stimulated these movements and colonizations. Colonization often resulted in a reorganization of the settlement pattern, for example that which German settlers came upon in Brandenburg, Pomerania or Prussia.

New villages in a planned pattern (*Angerdörfer* 'green villages') and street villages (*Straßendörfer*) were laid out by the German colonists. Slav inhabitants were moved and a great number of the old Slav villages were deserted.[8] A new network of central places was laid out, which still forms the basis of the distribution pattern of towns and cities east of the Elbe (Fig. 14.3).

The driving spirit of the colonization process was the idea and the reality of abundant open space and the principle of expanding Christendom, its culture, power and economic development towards the periphery; it was also driven by the pressures of internal growth. This widespread movement also initiated the Crusades to Palestine and the establishment of castles and settlements there.[9]

Undoubtedly these colonization initiatives were spurred by ideologies developed by the leaders of these expansions, the Deutscher Orden or other orders and the aristocrats. They took over the lordship of the newly laid-out towns, thereby holding the key positions in the colonized territory (Fig. 14.4).

310   Dietrich Denecke

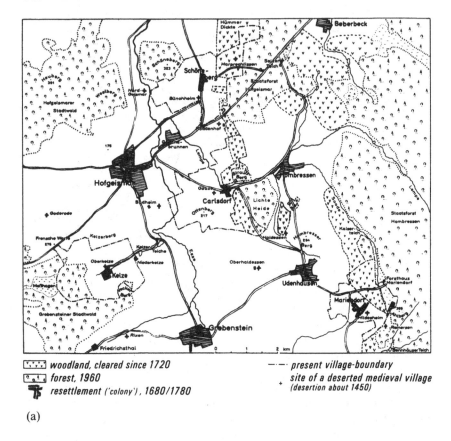

▦ *woodland, cleared since 1720*
▦ *forest, 1960*
🌲 *resettlement ('colony'), 1680/1780*

—·— *present village-boundary*
₊ *site of a deserted medieval village (desertion about 1450)*

(a)

The ideology of the core and periphery, linked to the perceived advantages of the periphery, helped to mobilize colonists from the core areas.

Mercantilist rearrangements of land use, economy and settlement systems: the ideology of political and economic autarchy

After the disaster and regression during the Thirty Years War in Central Europe (1618–48), the numerous territorial states recovered under the rule of sovereign lords, adopting mercantilist principles of a self-sufficient state with a closed economy. The method of implementing these economic principles and ideology was a systematic evaluation and the optimization of productivity of the land within the boundaries of the state's territory. This was organized by the sovereign himself in co-operation with trained specialists, often recruited from other countries, especially France and the Netherlands. Mercantilist ideology lay behind a great variety of planned landscape development:

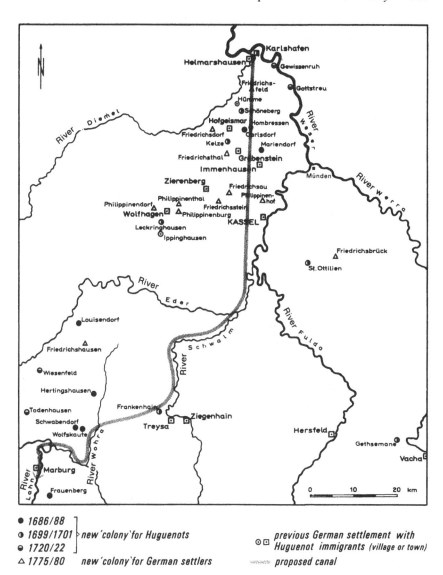

- *1686/88*
- *1699/1701* } *new 'colony' for Huguenots*
- *1720/22*
- △ *1775/80*   *new 'colony' for German settlers*

⊙▢ *previous German settlement with Huguenot immigrants (village or town)*

〰 *proposed canal*

(b)

Fig. 14.5 Eighteenth-century inner colonization in the territory of Hesse-Kassel. *Source:* L. Zogner, Hugenottendörfer in Nordhessen. Planung, Aufbau und Entwicklung von 17 französischen Emigrantenkolonien *Marburger geographischen Schriften* 28 (Marburg 1966).

1 The colonization of open space or unused land within the state territory (internal colonization).
2 The surveying of late-medieval deserted sites and fields with the aim of resettlement.
3 The distinct population policy of drawing trained people into the country for development, especially that of manufacturing.
4 The aim of initiating new manufacturing and of promoting commerce, supporting urbanization, infrastructural development (road and waterway construction).
5 Finally there was also a tendency to establish state-owned industries in order to organize production on a larger scale under state control.

The state of Hesse-Kassel during the late seventeenth and early eighteenth centuries is a good example. A great number of deserted sites were resettled by Huguenot refugees during the period between 1686 and 1722.[10] This scheme was organized by the sovereign but not much favoured by the German neighbours (Fig. 14.5).

Another, in this case, utopian project was to connect the rivers Weser and the Rhine with a 200-kilometre canal with more than 170 locks. In 1699 Duke Karl of Hesse-Kassel started to lay out a new town (Karlshafen) with a harbour in the centre with the intention of creating a thriving industrial place.[11] When Karl died thirty years later, the canal project was relinquished, after about 30 of the projected 200 kilometres had been completed. Only the town remained. The intention behind this project was to demonstrate the economic and technical power and the ability of the state; the achievement did not, however, match the initial pretensions and the heavy investment was fruitless.

Principles of rearrangement of villages and towns after fire: the ideology of administrative order and management

A great number of villages and towns in Germany developed fairly irregular settlement patterns. Fires very often destroyed substantial parts of settlements, so that decisions had to be made about reconstruction. As houses were in most cases reduced to ground level this created the possibility of laying out a completely new pattern. However, up to the seventeenth century it was normal to keep to the old property boundaries.

During the seventeenth and eighteenth centuries, the development of administration in the territorial states caused a major change. There are many examples of new street and property patterns being imposed on the previous plan. The reconstruction not only imposed a new order in the formal pattern, but also in the functional and social pattern of the settlement. Straight lines, right angles, standardized measures, symmetrical order, separation of functions, a differential arrangement of a social hier-

Planned order upon the land in Germany

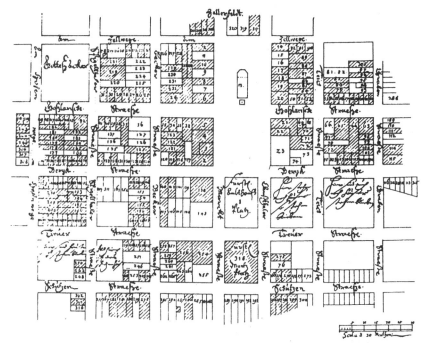

Fig. 14.6   The mining town Zellerfeld, replanned 1674 in a regular pattern after fire in 1672. *Source:* H. G. Griep, Das Bürgerhaus der Oberharzer Bergstädte *Das deutsche Bürgerhaus* 19 (Tübingen 1975).

archy are obvious elements in rebuilt towns and villages of that period. What lies behind these sophisticated arrangements? It is not merely a mode of design or a period of style. Interpreting the additional plans and papers of the redevelopment projects, we realize that administration ideals are primarily behind this ideology of order. Reconstruction was organized in detail by officials, following strict rules and regulations. Power was given to this hierarchy of official representatives, trained as financiers in administration to manage this public work of reconstruction.

Reorganization after fire is one of the outstanding examples of the ideology of administrative order and management during the Early Modern period. It plays an important role in the periods of the birth of public regulations and restrictions. The well-documented procedure, from the initial report of the disaster to the first provisional aid, and then the step-by-step plans of redevelopment give an illustrative insight into the spatial impact of the ideas, concepts and organization of administrative principles applied by state officials. Studying this phenomenon in different European countries would be a valuable comparative exercise.

There is not yet any comparative study of redevelopments and reconstructions of towns after fires in Germany, however there is a wealth of
interesting, well-documented examples giving an insight into the idea and
even the ideology of a new order and regulation behind the physical
structure of replanning. An early example of a replanned town is the
mining town of Zellerfeld in the Harz Mountains, which burned down
nearly completely in 1672. An entirely new town plan was laid out by the
skilled surveyors of the mining administration (*Markscheider*) in the form
of a regular grid, disregarding the uneven, sloping topography of the site
among some mining operations and integrating some buildings and sections which survived (Fig. 14.6).[12] The church, administrative building and
a market place formed a main axis around which the accommodation of the
mining population was arranged in a clear hierarchical order from the
centre to the periphery of the town plan. The intention of the idea and
ideology was to stress the hierarchical order of this highly specialized
community, and to express the order of this society within the regular
layout of the town plan. Another good example is the small medieval town
of Moringen (Lower Saxony) which was almost completely razed by fire in
1737 and which was reorganized, replanned and expanded in a regular
rectangular pattern which again reflected the social hierarchy of the community (Fig. 14.7). In the substantial records of the replanning process we
can often perceive the underlying idea of administrative order, regulation
and rearrangement. Even villages, with an irregular nucleated pattern
dating from medieval times, were replanned with a regular, often symmetrical layout, and with a standardized house type, arranged and differentiated
according to the social hierarchy of eighteenth-century rural society, if fire
destroyed the settlement (Fig. 14.8).

**Ideology behind the organization of building societies and working-class
housing: the ideology of social concern**

Until the nineteenth century, building and house ownership in towns and
cities was a privilege of the townspeople with *Bürger* ('citizens") rights.
There were usually some garden land and fields in addition to the house.
With the expansion of the cities in Germany during the second half of the
nineteenth century, builders, construction firms and private investors
started to build not just single homes and villas for the upper class but also
tenement houses or 'tenement villas' with flats to rent. Almost all the
building activities in German towns were private enterprises, especially in
the period 1870–90.

This began to change during the 1870s when building societies were
formed with an underlying ideology of social concern. The idea was to raise
funds by organizing communities of interests, mainly of middle and lower

————— *Medieval townwall*
⁄⁄⁄⁄⁄⁄ *Quarter which survived the fire 1737*

Replanned streetpattern and properties
■ *New gatehouses*

Fig. 14.7   The replanned town of Moringen (Lower Saxony) after fire, 1737.

316   Dietrich Denecke

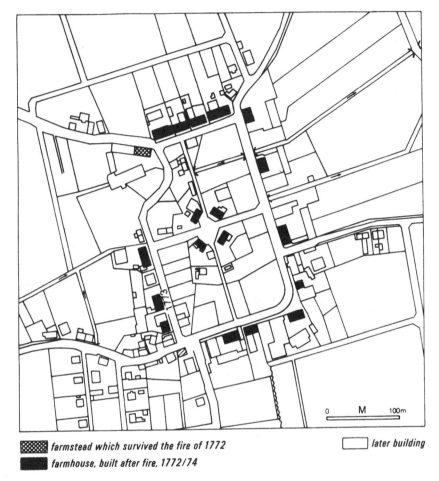

farmstead which survived the fire of 1772   ☐ later building
farmhouse, built after fire, 1772/74

Fig. 14.8   Replanned village after fire: Immensen (Landkr. Northeim, southern Lower Saxony).

class social and occupational groups. Social reforms were reflected in these societies; large numbers of publications, committed to eradicating appalling living and housing conditions, called on social concern to solve the housing problems for the rapidly growing urban populations. In the big cities, especially in Berlin, progressive architects were at work, promoting human aspects in middle-class housing and town planning. An expression of the ideology of social concern was the distinct quarters or neighbourhoods called *Siedlung*, *Stadtrandsiedlung*, *Vorstadt*, or *Gartenstadt* (settlement, city-edge-settlement, suburb, garden city). The spatial arrangement of the houses and their unity of style reflects the ideology of a homogeneous

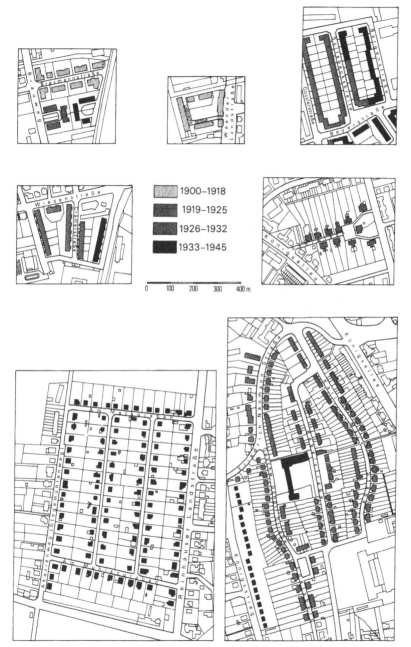

Fig. 14.9   Types of neighbourhood planned by building societies in the town of Göttingen (Lower Saxony), 1900–1950.

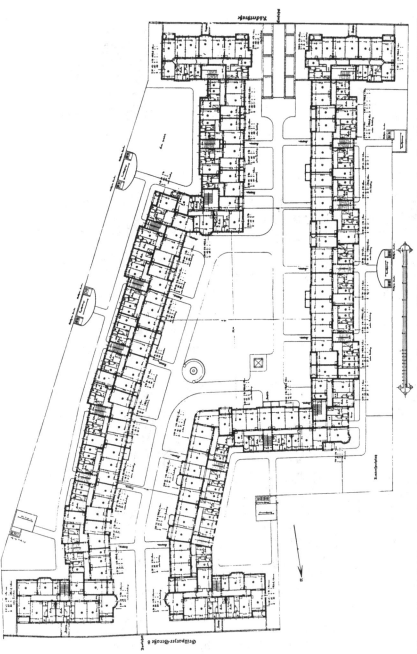

Fig. 14.10   Neighbourhood project (*Wohnstrasse*) by the Beamten Wohnungsverein, Berlin-Steglitz, Fritschweg, 1907–1908; architect Paul Mebes. *Source*: J. Posener, *Berlin auf dem Wege zu einer neuen Architektur – Das Zeitalter Wilhelm II* (Munich 1979) 353.

social structure in these neighbourhoods (Figs. 14.9 and 14.10). These neighbourhoods undoubtedly had a powerful impact in forming a united community and a feeling of solidarity.

The garden land and small stable attached to these dwellings was another element of the social ideology and environment of working-class quarters. All this might be seen as the German version of the garden-city movement, expressed by socially committed architects such as Wagner, Schmitthenner and others.[13]

This social movement and building activity came to its peak after the First World War, during the 1920s and early 1930s, when a great number of small societies, often operating just one scheme, were founded. Most of these societies only existed for a short time, coming to an end during the second half of the 1930s.[14] This type of architecturally and socially homogeneous quarter still exists today, showing a remarkable stability in social structure. The quarters usually form closely knit communities, even within the framework of the growing urbanization and mobility of present times, when the ideology of the founding generation has long since evaporated.

The ideology of the German Nazi regime (fascism) and its schemes of spatial reorganization

Ambitious and rigorous concepts of spatial reorganization were implemented under the ideology of the German Nazi regime. Activity was focussed on the big cities. Architecture and urban design were vital instruments of the state ideology, symbols of power, state control and authoritarian behaviour, meant to create a collective national and socialist mentality and solidarity of the people (*Volk*). The physical result consisted of central monumental buildings for public use, an egalitarian pattern of residential quarters and hierarchical arrangements. Systematic programmes were ordered from above to reorganize the centres of the bigger German cities to demonstrate the new national-socialist (*Nationalsozialistische Partei*, effectively the Nazi Party) ideology in the urban environment.[15]

These aims were legalized in 1937 in the law covering the renewal of German towns.[16] A special group of *Führerstädte* was privileged, obtaining funds for the remodelling schemes direct from the government.[17] The programme of reorganization involved changes on a gigantic scale (Fig. 14.11a and b). However, most of the plans, drawn up by Hitler himself and by the architects of the Reich (Harter, Gutschow, Hentrich, Heuser, Graubner Hinsch, Dyrssen) never materialized, because the war reduced building activities. The enormous input required to demolish the existing building fabric and to raise the gigantic new structures was also underestimated, covered by ideological ambition.

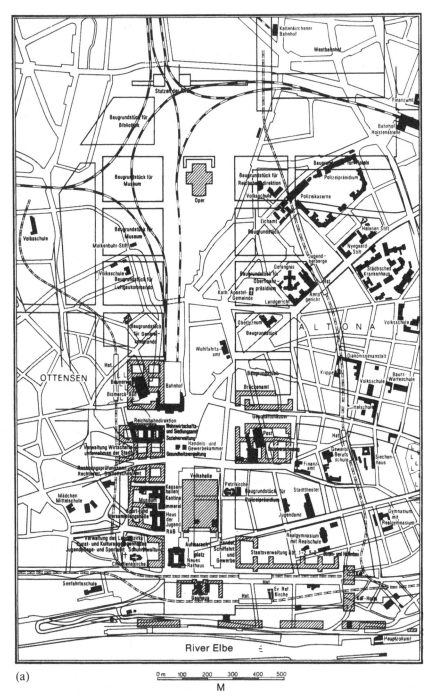

(a)

Fig. 14.11   The proposed replanning of Hamburg: (a) plan by K. Gutschow, 1940;
(b) plan by H. Hentrich and H. Heuser, 1944. *Source:* J. Lafrenz, Planung der

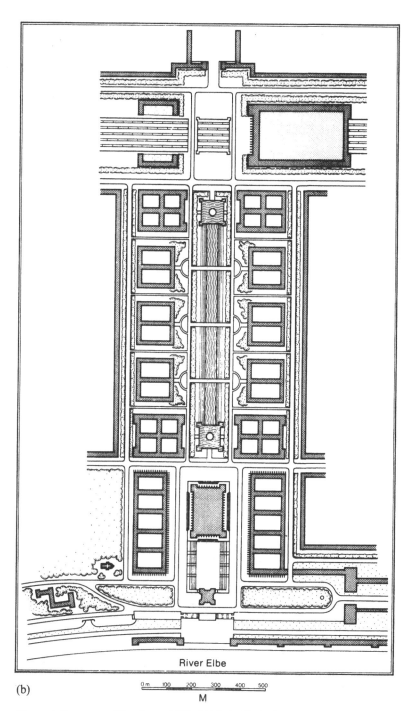

River Elbe

(b)

0 m   100   200   300   400   500

M

Neugestaltung von Hamburg 1933–1945, in H. Heineberg (ed.), Innerstädtische Differenzierung und Prozesse im 19. und 20. Jahrhundert *Städteforschung A* 25 (Cologne 1987) 397f.

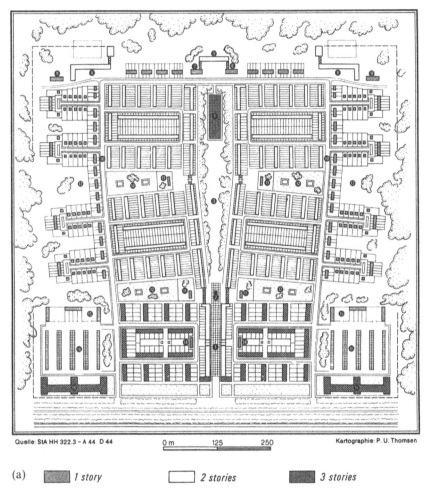

(a)   ▨ *1 story*   ☐ *2 stories*   ▨ *3 stories*

Fig. 14.12   Models of (a) an *Ortsgruppe* and (b) a *Hauptortsgruppe*, 1944, by W. Hinsch and F. Dyrssen. *Source:* as Fig. 11, pp. 428–9.

As well as these programmes to reorganize city centres, there were plans to establish a new hierarchy of central places, especially for those territories in the east which came under German control during the war. For these plans Walter Christaller was engaged as an expert.[18]

A key idea was to arrange urbanized regions after a feudal hierarchical order. The city was seen as an organism, consisting of a group of cells, corresponding with the structure of the political organization. Decentralization on the one hand, hierarchy in rank and size on the other, were the main ideas behind the pattern of newly laid-out urbanized areas. The

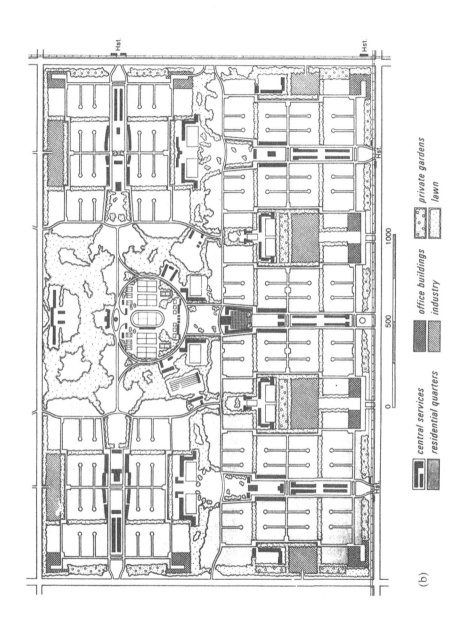

(b)

central services

residential quarters

office buildings

industry

private gardens

lawn

0   500   1000

Hst.
Hst.
Hst.
Hst.

smallest urban unit, called *Ortsgruppe*, for about 5,000–6,000 inhabitants, consisted of five or six cells, grouped together in a symmetrical order around a broad arterial route or highway orientated towards the main building of the party and an open square for political gatherings (Fig. 14.12a and b). The next unit was the *Hauptortsgruppe*, which consisted of five or six of the basic units and had about 30,000 inhabitants.

Other projects affecting the landscape and settlement were small and large-scale colonization schemes for new farming colonies in different parts of Germany.[19]

The socialist ideology of the German Democratic Republic

In the eastern part of Germany after 1945 the new socialist government began to establish the socialist or Marxist ideology and political system. The main tool was expropriation and nationalization of property and production. With it the leading socialist party took over the responsibility for and management of the economy and planning. The result was the creation of production pools, a state-controlled production process, a state-controlled market and a closed trade and monetary system. Settlements and fields were remodelled. Most barns and stables fell into disuse, new facilities for specialized large-scale agriculture were erected, usually on the periphery of the former village. Individual plots were amalgamated to become large co-operatively worked fields. The process of expropriation was a long one. The new ideology was not well received, because farmers were forced to give up their land.

City planning was also based on the new socialist ideology. In East Berlin, the capital of the East German state, the old city centre with the king's residence was cleared to lay out a large open square for political gatherings, the Karl-Marx-Platz, and the main axis leading to the east, then called Stalinallee, was completely renewed in a monumental style.

A socialist planning strategy was developed in 1950.[20] An important aim was to combine industrial development and production with the erection of new residential blocks, ten to twenty storeys high, constructed from prefabricated concrete panels. Building and construction work became industrialized. Four completely new cities were erected as demonstrative symbols of the new ideology, all combined with heavy industry. On the other hand, the old cities were expanded by widespread uniform, prefabricated residential blocks, in the form of block units or neighbourhoods (Fig. 14.13).

During the 1970s and 1980s more emphasis was placed on remodelling the old centres of smaller towns, in general by complete replacement, retaining only some elements of the old street pattern and the boundary of the town wall (Fig. 14.14). Almost complete expropriation of houses and

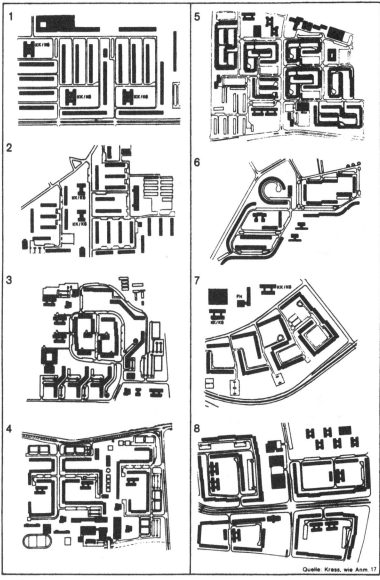

1 Rostock – Lütten Klein
2 Berlin – Hans - Loch - Viertel
3 Erfurt – Riethstraße
4 Leipzig – Grünau

5 Schwerin – Großer Dreesch
6 Gera – Lusan
7 Berlin – Am Tierpark
8 Magdeburg – Neustädter See

Fig. 14.13   New residential quarters under the concept of the socialist city. *Source:*
P. Schöller, Stadtumbau und Stadterhaltung in der DDR, in H. Heineberg (ed.),
Innerstädtische Differenzierung und Prozesse im 19. und 20. Jahrhundert *Städte-forschung A* 25 (Cologne 1987) 451.

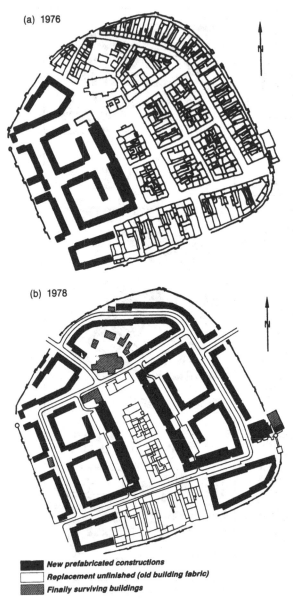

(a) 1976

(b) 1978

New prefabricated constructions
Replacement unfinished (old building fabric)
Finally surviving buildings

Fig. 14.14   Remodelling medieval town centres as a concept of socialist town planning: Bernau-Brandenburg. *Source:* as Fig. 14.13, p. 456.

property in towns led to total decay in the old town centres, so that the process of replacement is still going on. Most of the old building fabric of many old towns in East Germany will be gone within the next twenty years, though the idea and possibility of preservation is now in progress.

Conclusion

The aim of this chapter has been to identify and examine distinct periods in German history during which ideology was a basic factor in promoting landscape change and reorganization. These periods mainly coincided with times of absolutist, feudal, dictatorial or socialist forms of government. Different ideologies producing a spatial impact may be identified, and their influence upon the landscape usually took the form of remodelling and replacement. Reorganization is a fundamental element of ideology, and its application may manifest itself in the physical restructuring of landscapes. Landscape change was often a result of fundamental and directed planning; however, the initial aims were rarely achieved during the reign of the power group which created the ideology. Thus to search for and to identify legacies of ideology within landscapes should be an important aspect of historical geography.

NOTES

1 Hans Barth, *Wahrheit und Ideologie* (Zürich 1945); Theodor Geiger, *Ideologie und Wahrheit. Eine soziologische Kritik des Denkens* (Stuttgart 1953)
2 Mandorla or vesica: pointed almondlike oval, used as the aureole around the head or figure of a saint or martyr in medieval sculpture and painting
3 Otto Brendel, Origin and meaning of the mandorla *Gazette des Beaux-Arts* ser. 6, vol. 25 (1944) 5–24; Hannes Thorhauer, *Goslar – Kriterien der Grund- und Aufrißgestaltung einer mittelalterlichen Stadt*, Diss. Archit. Braunschweig (1986) 66–86
4 Dominant in this respect was the order of the Cistercians: Louis J. Lekai, *The Cistercians: ideals and reality* (Kent 1977); Ambrosius Schneider *et al.* (eds.), *Die Cistercienser – Geschichte, Geist, Kunst* (Cologne 1983); especially in this context: Winfried Schenk, Mainfräkische Kulturlandschaft unter klösterlicher Herrschaft – Die Zisterzienserabtei Ebrach als raumwirksame Institution vom 16. Jahrhundert bis 1803 *Würzburger Geographische Arbeiten* 71 (Würzburg 1988)
5 Ed Taverne, *In't land van belofte: in de nieue stadt – Ideaal en werkelijkheid van de stadsuitleg in de Republiek 1580–1680* (Maarssen 1978)
6 Order and regulation is also related to political philosophies; see for example: Hans Barth, *Die Idee der Ordnung – Beiträge zu einer politischen Philosophie* (Erlenbuch 1958); Kultur und Norm, *Schriften zur wissenschaftlichen Weltorientierung* 2 (Berlin 1957)

7 D. Hay, *Europe: the emergence of an idea* (Edinburgh 1957); J. R. S. Phillips, *The medieval expansion of Europe* (Oxford 1988); Walter Schlesinger, *Die deutsche Ostsiedlung des Mittelalters als Problem der Europäischen Geschichte* (Sigmaringen 1975); Anngret Simms, Core and periphery in medieval Europe: the Irish experience in a wider context, in William J. Smyth and Kevin Whelan (eds.), *Common ground – essays on the historical geography of Ireland* (Cork 1988) 22–40

8 Günter Mangelsdorf, Zur Verbreitung mittelalterlicher Ortswüstungen im Bezirk Potsdam *Zeitschrift für Archäologie* 17 (1983) 59–84; Günter Mangelsdorf, Mittelalterliche Vüstungen zwischen Havel und Flämingnoirdrand *Veröffentlichungen des Museums für Ur- und Frühgeschichte Potsdam* 17/18 (1983) 231–60

9 J. Kirtland Wright, *The geographical lore of the time of the crusades* (New York 1965); J. K. Hyde, Real and imaginary journeys in the later Middle Ages *Bulletin of the John Rylands Library* 65 (1982/3); J. Prawer, *The Latin kingdom of Jerusalem: European colonialism in the Middle Ages* (London 1972); M. Benvenisti, *The crusaders in the Holy Land* (New York 1972)

10 Lothar Zögner, Hugenottendörfer in Nordhessen. Planung, Aufbau und Entwicklung von 17 französischen Emigrantenkolonien *Marburger Geographishcen Schriften* 28 (Marburg 1966)

11 Rüdiger Recknagel, Karlshafen. Fragment einer städtebaulichen Portalanlage um 1700 *Hessische Forschungen zur geschichtlichen Landes- und Volkskunde 2* (Kassel 1958)

12 Hans-Günter Griep, Das Bürgerhaus der Oberharzer Bergstädte *Das deutsche Bürgerhaus* 19 (Tübingen 1975)

13 Julius Posener, *Berlin auf dem Wege zu einer neuen Architektur – Das Zeitalter Wilhelm II* (Munich 1979)

14 See as an example for the town of Göttingen, Dietrich Denecke, *Göttingen – Materialien zur historischen Stadtgeographie und zur Stadtplanung* (Göttingen 1979) 74f.

15 Recently there is a growing interest in the planning concepts and the ideas of replanning during the Nazi period; see for example J. Petsch, *Baukunst und Stadtplanung im Dritten Reich* (Munich 1976); J. Dülffer, J. Thies and J. Henke, *Hitlers Stadtebaupolitik im Dritten Reich, eine Dokumentation* (Cologne 1978); Jochen Thies, Hitler's European building programme *Journal of Contemporary History* 13 (1978) 413–43; Jürgen Lafrenz, Planung der Neugestaltung von Hamburg 1933–1945, in Heinz Heineberg (ed.), Innerstädtische Differenzierung und Prozesse im 19. und 20. Jahrhundert *Städteforschung A 25* (Cologne 1987) 385–437; important collection of papers: Stadterneuerung in der Weimarer Republik und im National-sozialismus – Beiträge sur stadtbaugeschichtlichen Forschung *Arbeitsberichte des Fachbereichs Stadtplanung und Landschaftsplanung der Gesamthochschule Kassel* 73 (Kassel 1987)

16 Jost Dülffer, NS – Herrschaftssystem und Stadtgestaltung: Das Gesetz zur Neugestaltung deutscher Städte vom 4. Oktober 1937 in: Stadterneuerung in der Weimarer Republik (see note 15) 192–221

17 Jochen Thies, Nationalsozialistische Städteplanung: Die 'Führerstädte' *Die alte Stadt* 5 (1978) 23–38

18 Walter Christaller, Grundsätzliches zur Neubildung des Deutschen Reiches und seiner Verwaltungsbezirke *Geographische Wochenschrift* (1933) 913–19; Walter Christaller, Grundgedanken zur Siedlungs- und Verwaltungsbau im Osten *Neues Bauerntum* (1940) 305–12

19 J. G. Smit, *Neubildung deutschen Bauerntums – Innere Kolonisation im Dritten Reich. Fallstudien in Schleswig-Holstein* (Kassel 1983)

20 Frank Werner, *Stadt, Städtebau und Architektur in der DDR* (Erlangen 1981); Peter Schöller, Stadtumbau und Stadterhaltung in der DDR. In Heinz Heineberg (ed.), Innerstädtische Differenzierung und Prozesse im 19. und 20. Jahrhundert *Städteforschung A* 25 (Cologne 1987) 439–71

15

Parading: a lively tradition in early Victorian Toronto

PETER G. GOHEEN

The visit of His Excellency Sir Charles Bagot to Toronto, has been the signal for general rejoicing, and the cordial welcome with which His Excellency has been greeted on this occasion, is in the highest degree honourable to the inhabitants ... Besides the large number that formed the procession, the crowd of spectators was immense, and the windows of the houses were occupied by the beauty and fashion of the city, – all eager to witness the triumphal approach of their distinguished visitor.[1]

Our city was on Tuesday night the scene of one of those disgraceful and disgusting displays of Hell-engendered malice which ... gives such an unenviable distinction to ... Orangemen. About 11 o'clock ... in the vicinity of King Street ... a ragged and drunken mob, who paraded the street ... proceeded, unhindered, to the Hon. W. W. Baldwin's residence, where they attempted to ... burn ... effigies ... amid cries of 'traitors' and all manner of Orange ribaldry and execration.[2]

Parades were a staple of public life in the cities of colonial Canada. They marked any occasions, official and unofficial, and like mood music provided a congenial ambience for expressing a wide range of emotions. They were a popular, practised form of publicly enacted ceremony; few inhabitants of the towns would have disagreed with the contention that they were 'far more than drapery'.[3] They claimed attention because as performances they were distinguished from ordinary movement in the city by their imposed order and by their presentation of music, scenery, costumed participants and banners. The symbols displayed were clearly presented and widely understood. The parade was a comprehensible ritual, a 'rule-governed activity of a symbolic character which draws the attention of its participants to objects of thought and feeling which they hold to be of special significance'.[4]

Toronto entered the Victorian era as an almost newly incorporated city which had already experienced decades of parades in its streets. Processions were integral to the functioning of its founding institutions: the military and the government. The 1840s were a period of maturation for the colonial city as it grew into its role as a regional commercial capital which had yet to experience the social and economic transformation that would accompany the arrival of large-scale factory industry. The times were not quiescent, however. The unrest that flared into rebellion in 1837 left its trace in troops stationed in the city, and in political heats that cooled slowly. The arrival of impoverished Irish, mostly Catholic, immigrants changed the religious and, to some degree, the ethnic composition of the population. The immediate significance of this change would be difficult to exaggerate in a city and region where the conduct of colonial society still relied intimately on inherited distinctions transferred from Europe. Toronto housed more Irish than English. The new environment worked its changes, but these were slow adaptations. At the end of the 1840s the Governor-General signed the Rebellion Losses Bill, stirring deep passions that emphasized the incipient antagonism between anglophone and francophone, and the open differences separating Tory and reformer.

The parade as ritual

The parades which occupied Toronto's streets indicate to the modern observer the social valuation of time and space, especially of the public space of the streets where they occurred. For parades were foremost among the organized events to occupy the streets, and thereby to manipulate the symbolic value of time and space to meet their own aims. Their organizers well understood that 'social space ... is never neutral, never homogeneous. Some sites have more power and significance than others.'[5] Neither was timing a matter of indifference; a commemorative calendar comprised both inherited and recently affixed dates. Likewise, the week and the hour of parading held meaning and were deliberately set. Interpreted in this context parades can offer insight into the social valuation of public space and social time in the city; they can serve to identify the groups involved, the purposes served and the response of the community to them. And they let us know how the reaction was reflected in the conduct of parades themselves.

   The streets of early Victorian Toronto were its most accessible public open spaces. The repertoire of parades varied from a civic procession to honour the presence of a visiting Governor-General to a riotous parade marking Guy Fawkes' Day. These events, described in the opening paragraphs of the chapter, took place in April 1842 and November 1843 respectively. They claimed attention because their participants took command, even briefly, of the streets. The city's inhabitants acted on the assumption that they possessed the right of unhindered egress to this space. Their attitude to the streets can be described in words recently applied to eighteenth-century English towns: 'the street ... was ... a coherent and lively arena for social peregrination, perception, challenge and engagement ... [A]ll comers had access to the public terrain.'[6]

The parade offered the most carefully orchestrated of a variety of presentations regularly to be enacted on the streets. Other, more casual, forms of collective behaviour included perambulations through the principal streets to mark notable occasions such as the birth of an heir to the throne. Such celebrations were likely to feature special displays and fireworks as amusements. Hangings, routinely conducted in open air, drew crowds into adjoining thoroughfares. Political meetings were often held where they spilled into the streets. Voting took place out of doors, in full view of the street. Voters could be required to assert their right to the franchise and had openly to declare their choice in the public glare of observers. These activities implied a crowd composed, as was often noted by contemporaries, of country as well as city folk, of all manner of people. Like parades, they could be witnessed by all. Nevertheless, in Toronto the parade was the most frequent and popular street performance.

Urban parades were already an old, adaptable and well-understood form by the early nineteenth century. They came to North America as cultural baggage brought by the immigrants. They marked occasions of many varieties in Toronto: funerals, political victories, religious holy days, military celebrations, visits by distinguished figures, sporting events and commemorations by sodalities. Processions planned for official occasions were the centrepieces of the most elaborate and stately public events of the time. When staged without the sponsorship of governments they could still be impressive and even spectacular events. They depended for their vitality on no single segment of society. Parades legitimated a wide variety of claims, being regarded as effective devices for expressing both consensus and conflict within the communities. Street processions require organization and imply a measure of orderliness, but public disorder was a none too infrequent by-product of their enactment. Resulting violations of the peace were normally confined to minor infractions, described, with reference to Protestant–Catholic clashes, as a 'pattern of restrained ritual riot'.[7] Traditional provocations practised in Toronto in the 1840s included carrying figures in effigy and displaying taunting slogans on banners.

Toronto's streets remained open to the whole breadth of society and its multifarious purposes in the first years of the Victorian era. The population could still practise such rights in part because the city possessed only weak instruments for controlling access to the streets. The city lacked both a professional police force and any concept of administering it independently of control by elected politicians. The latter, it seems, preferred to exercise their power over the streets through informal rather than formal means.

The impact of this practice was demonstrated as early as the 'types riot' of June 1826. On this occasion a local editor and publisher suffered the vandalizing of his printing shop by a mob among whom were several persons intimately involved with the town's administrative and professional elite. A clearer demonstration that equal protection under law was unavailable in the town could scarcely have been imagined. It was, in the words of a scholar of the early legal system, 'only the first episode in a pattern of political violence that was to prevail . . . until the late 1840s'.[8] Parades were the focus for much of this violence, which frequently involved Orangemen. In the 1840s criminal justice and public order were still administered with a keen discrimination for whose interests were involved; they were treated as 'intensely personal' affairs.[9] The streets were the arena for many different processions and demonstrations, most of which were unpoliced and peaceful. Police intervention, when it occurred, reflected the bias of the political establishment and was felt unequally by the various groups affected.

The management of collective behaviour in city streets assumed increasing importance as strains on urban social order increased, in part as a result of a more diverse urban political. Recent research amounts to a consensus view that the 'ordering of urban public space' became associated with its privatization, partially 'to reduce . . . the complexity of living in a world of strangers'.[10] The distinctive meaning of public and private space blurred for those elites which retreated from their former engagement in open collective ceremonies in which the whole population could participate.[11] The street was to lose its symbolic power to host ceremonies capable of representing the community as a whole, we are told. The withdrawal of some segments of the urban population from active involvement with organized life on the streets robbed public ceremonies of formerly important actors. Parades – 'as images of social relations' – registered well such 'ways of interpreting and knowing a changing city'.[12] The 'decline of parading as a full public rite, capable of enrolling a wide spectrum of social groups' was the consequence of a series of decisions made by increasing numbers of groups over time.[13] What, if any, evidence of this trend surfaced in Toronto, especially as a consequence of a crisis of management of parades in the streets that overtook normal practice in 1843?

Parading as celebration and contention

The visit of Sir Charles Bagot on 21 April 1842 offered official Toronto an unsurpassed opportunity to enact an elaborate civic ritual. The Governor-General personified the imperial connection, a powerful symbol in a society which cherished reflections of monarchical and aristocratic life in Britain.[14] A parade suitably formed the centrepiece of the welcoming ceremonies. It required careful organization, involving the co-operation of many groups, and took time to prepare. The process of its planning followed the standard pattern for the time: the occasion rendered a routine procedure special.

Two Police Constables – mounted
(with their staves)

Band

Marshal of the Day – mounted
Fire Companies
(in order of precedence)

St Andrew's Society
(with banners)

St Patrick's Society
(with banners)

St George's Society
(with banners)

The Mechanics' Institute

Two Police Constables – mounted
(with their staves)

The Corporation
(in carriages)

The Bishop – The Mayor
The GOVERNOR

The Chief Justice
Judges and College Council

The Sheriff – The Warden

Members of the Learned Profession
(in carriages or mounted)

Other inhabitants generally

The Hega Bailiff – mounted
(with his baton)

Fig. 15.1   Order of the Toronto procession to greet the Governor-General, 21 April 1842, published in the *British colonist*, 20 April 1842.

The official announcement of Bagot's visit precipitated a well-practised chain of events. A 'requisition, respectably signed, was presented to His Worship the Mayor, requesting a public meeting be called, to make arrangements for giving His Excellency a public reception'.[15] The mayor obliged, and hosted the organizing meeting at City Hall on Monday, 11 April at 3.00 pm. The 'respectable' of the city were represented in strength and passed motions to arrange a reception and present an address to His Excellency. They had prepared the address, read at the meeting by the sheriff, moved by the president of the Board of Trade, seconded by the president of the Bank of Upper Canada, adopted by the meeting, and ordered to be signed by the mayor on behalf of the inhabitants. Then the

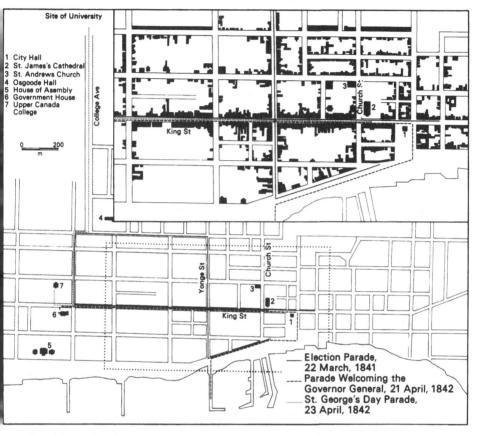

Fig. 15.2   Map showing routes taken through Toronto city centre by the election parade of 22 March 1841, and the arrival and parade associated with the Governor-General in April 1842. Based on the following map housed in the collection of the Metropolitan Toronto Reference Library: 'Topographical plan of the City and liberties of Toronto in the Province of Canada', surveyed, drawn and published by James Cane, Topographical Engineer, in 1842 and dedicated to Sir Charles Bagot. Information on parade routes is from the following sources: election parade information from the *British colonist*, 24 March 1841; parade welcoming Bagot to Toronto from the *British colonist*, 27 April 1842; and parade on 23 April 1842 from *The mirror*, 29 April 1842 and the *British colonist*, 27 April 1842.

meeting appointed a committee of its peers to make the necessary preparations.

A week later, on 18 April, the Programme of the Procession was announced. The participation of national societies, of local fire companies and the constabulary, of the judiciary, the learned societies, the educational and religious establishments and of the corporation had been co-ordinated and the order of precedence assigned (Fig. 15.1). Thursday morning, the day of the visit, the principal streets through which the parade would move received their decorations. These included two triumphal arches bearing flags, one constructed where the guest was to land. Care in planning extended to the timing. The signal for the procession to form at the City Hall, to march to the wharf to greet Sir Charles, would be the ringing of the town bells. This would commence when the steamer came into sight.[16]

His Excellency approached at the hour appointed, awaited by an honour guard and an assembled crowd of spectators. The welcoming dignitaries and the ordered ranks of the parade completed the scene as he disembarked. He landed 'under a royal salute, the band of the 43rd playing the national anthem ... the loud and continued cheering of the assembled multitude ... indicated the heartfelt gratification experienced by all present'.[17] The procession reformed, now joined by the visitor, and paraded through the city centre (Fig. 15.2).[18] The desire to participate reached all strata of society: one journalist observing 'a respectable number of coloured inhabitants who joined the procession, with banners; and also a considerable body of Indians'.[19] Others, including women who were seldom seen in parades, lined the streets and observed from windows of buildings along the route. 'Never in our recollection, did Toronto witness so large an assemblage of people as on this occasion; there could not have been less than from four or five thousand persons present.'[20] The Governor-General's destination was Government House, where the mayor read the address of the inhabitants and the guest delivered his reply. Following this ceremony the procession reformed, retracing its steps to City Hall where it disbanded. On Friday His Excellency received addresses from the various institutions and societies of the city and vicinity, and attended a levee, tickets for which, at 30 shillings, gave the elite an opportunity to mingle with the Queen's representative.

Saturday, 23 April, St George's Day, provided the principal event of the visit to Toronto. As Chancellor of King's College, Bagot was to lay the foundation stone for its new premises. An even larger and longer parade than that of Thursday conveyed him to the site of the intended new building (Fig. 15.2). These arrangements followed a morning of tribute to St George. This was accomplished in the customary fashion, with a parade by the national societies – devoted to St George, St Patrick and St Andrew – through the centre of town to a 10.30 am church service in the Anglican cathedral. Following the service, society members marched to Upper Canada College where they joined the company of the others who were to process to the new grounds of King's College. At 1.00 pm Bagot arrived and the parade formed after some necessary preliminary remarks addressed in Latin to the crowd. Like all such parades, it was accompanied by the sound of music, it offered a display of banners, and a panoply of costume.

'The Governor's rich Lord-Lieutenant's dress, the Bishop's sacerdotal robes, the Judicial ermine of the Chief Justice, the splendid Convocation robes of Dr McCaul, the gorgeous uniforms of the [Governor-General's] suite', impressed at least one contemporary who composed this 'highly wrought description'. The anonymous author considered it 'all formed one moving picture of civic pomp, one glorious spectacle which can never be remembered but with satisfaction'.[21] The parade drew a crowd estimated at from 10,000 to 15,000, impressive for a city of less than 15,000 people.[22]

There could have been no better opportunity than this for the elite to stage an event to reflect its preferred view of society. Its order, display of symbols and performance created impressive ceremony, which 'gives an image of how things should be ... of ideal harmony and order'.[23] Order was a matter of meticulous planning. Bagot assumed pride of place in both parades, the status of other participants varying with their remoteness from his place in the procession. Both parades reflect the ritual inversion so often attributed to them. Lower-ranking participants preceded the higher ranking, until the Governor-General's appearance. Top prestige went to the leaders of the principal institutions: the courts, the government, the church and the city. Members of the national societies, the Mechanics' Institute and professional men were granted recognition of a less exalted degree. Citizens who could claim no titles or affiliations found their place in the least prestigious positions. The difference in precedence in the two parades was a matter of small degree; the general ranking of privilege was not to be disputed. Every participant and observer understood the importance of the order of the procession. Costume, banners and other symbols on display clearly identified the groups and the offices being represented. There was no more dramatic or visible way than a parade to convey the message that a well-ordered society respected privilege and honoured hierarchical principles. The assigned place for every person was to be respected.

Parades likewise served well as tools for contention, understood as a peculiar form more of order than disorder. It is, Charles Tilly instructs us, 'order created by the rooting of collective action in the routines and organization of everyday social life, and by its involvement in a continuous process of signaling, negotiation and struggle with other parties whose interests the collective action touches'.[24] Among those groups most continuously testing the boundaries of acceptable collective behaviour through public demonstrations was the Orange Order. It paraded on several anniversaries, including that marking Guy Fawkes' Day, 1843. At 11.00 pm on Tuesday, 7 November 1843, its members formed a procession and proceeded noisily through King Street carrying

a sort of canvass box with the following illuminated sentences:

'Baldwin, Sullivan and Hincks.'
'Gallows Hill, and no Surrender.'
'Corporation of Toronto.'
'No Repeal for Ireland.' and
'To H–ll with the Pope.'[25]

They bore aloft effigies of the three men named, each of whom was Irish-born, wealthy and a prominent leader of Reform politics in the colony. Their destination was the Hon. W. W. Baldwin's residence where they attempted to burn the effigies. In targeting reformers and in carrying effigies they recall similar processions in England dating to the late eighteenth century and the political tensions arising from the French Revolution.[26]

This incident illustrates the unfettered access to the streets permitted some rowdy throngs by the authorities. The Catholic editor of the *Mirror* complained bitterly that:

[t]he Police authorities were of course attending to their *duty*, but to whom? Not to the Citizens, who *pay* them to protect their property from harm, and themselves from insult! – but to the Corporation who hire them to advance, as well, *their* party objects. The duty of the Policemen on this occasion, was to keep out of the way, or more correctly speaking, to assist those dastardly ruffians in disturbing the peace ... We understand the principal actors in this despicable piece of business, were subordinate officers of the Corporation.[27]

The Protestant editor of the *British colonist* referred briefly to the event as 'a ridiculous scene ... a very silly business'.[28]

Repertoire of parades

By the 1840s the parade's repertoire in Toronto was expanding, as this most popular form of attracting public attention proved to be eminently adaptable. Processions had been a part of the city's life since it emerged from the wilderness with its garrison and government functions. The opening and closing of Parliament always required the formality of a procession, and troops paraded in the streets with some regularity. Other civic occasions such as the opening of public works or the funeral of a high state servant called for parades from the earliest days of town life.[29] Similar events continued in the 1840s to bring forth processions, notably for visiting Governors-General – as in 1838, 1839, 1840, 1842, 1843 and so forth. Religion motivated many an occasion for holding parades. The participants included many who would also have found their place in civic and governmental processions, but there would have been no mistaking their roles. The institutions represented in each were different, as the banners displayed, costumes worn and music played made plain.

As Toronto grew and matured socially an increasing number of parades relied less and less on religious or official ceremony. The long-established processions by national societies on their saints' feast days bore indelible marks of religious celebration. Other institutions, however, joined the calendar of parades with rituals more strictly non-sectarian in their design. The parades associated with the athletic games known as the 'Gathering' were among the earliest. First held in September 1839, they commenced in the only appropriate fashion imaginable:

The slumbers of the Inhabitants were disturbed at an early hour this morning by . . . [tunes] played thro' the City, by four Pipers of the gallant 95th, gaily attired in the garb of Clan Gregor, reminding us that the Gathering had begun. – At nine o'clock crowds of spectators, and competitors, from all quarters moved toward Caer Howell.[30]

Celebration of the Queen's birthday, which Toronto began soon after her accession in 1837, involved no religious element. Neither was it an occasion for official pomp and circumstance; the preferred form was a procession of citizens without distinction of rank or affiliation. The first 'tee-total' (*sic*) procession, bringing out 500 marchers in October 1841, is an early example of secular parading.[31] The ritual of election parades included chairing the successful candidate through the town, an ancient practice that diffused to the Canadian colonies.[32] It was practised regularly when contests for the provincial legislature were announced. Public meetings called to nominate candidates for election occasionally precipitated parades as well.

Parading was far from an incidental feature of the corporate life of many associations. It was, for instance, a corollary of membership by all who joined the Orange Order.[33] Its performance was the focus of annual celebrations by the national societies and by other fraternal groups. Not surprisingly, these organizations sponsored a significant proportion of the parades seen in Toronto's streets each year. In 1841, the number was six out of a total of twenty-one recorded in the newspapers consulted (Fig. 15.3). In that year, an equal number of parades was associated with the activities of the military who, following the rebellion of 1837, continued to be visible on the city's streets. The tramp of regimental boots to the accompaniment of military music was soon to become a rarer event in the city, although civilian organizations continued to prize the skills of military bandsmen when they were permitted to offer musical accompaniment for processions. In 1841, an election year in the colony, successful candidates assumed their traditional rights of staging victory parades. Non-sectarian and civic occasions contributed four parades, three of these being associated with sporting events. Only one funeral is reported, its significance arising from the unhappy conditions of the death in a riot that disrupted an election parade.

Sponsors or purpose

Fraternal societies	6
Military	6
Election	4
Civic or social	4
Funeral	1

Timing

Monday	7
Tuesday	1
Wednesday	3
Thursday	4
Friday	4
Saturday	1
Sunday	0
Day unknown	1
Total	21

Fig. 15.3   Parades held in Toronto, 1841. Information on the purpose and timing of parades comes from the *British colonist* and *The mirror*.

Normally the press reported funerals and processions only when public figures died. It is difficult to know whether the ordinary citizen was laid to rest with the same ceremonial passage through the streets of a cortège as attended the prominent to the grave, although occasional references to such events recur in the press during the 1840s.

The ritual calendar of early Victorian Toronto established the timing of a number of important parades (Fig. 15.4). Signifying the colonial society in which the celebrations occurred, all of the events thus commemorated referred to origins outside of North America. Some, like the feast days of the patron saints, had been long established. The Queen's birthday, on the other hand, had only begun to be celebrated after her accession in 1837; likewise, the celebration by the 'coloured inhabitants' of the anniversary of their emancipation from slavery within the British empire was but recently added to the calendar.

It perhaps bears reflection that none of these dates marks major religious holy days; yet religious services were incorporated as an element of ritual in most of these celebrations. That this role was almost entirely in the hands of the Protestant denominations is a fact that it would be difficult to exaggerate. The only challenge raised to this uniformity of practice concerned 17 March and the annual St Patrick's Day procession. The usual procedure was followed on Saturday, 17 March 1838 when 'the sons of St Patrick, held their annual celebration. They walked in procession, accompanied by the sister societies, St George and St Andrew, to the Episcopal Church, where a suitable sermon was preached.'[34] The procession on

17 March	St Patrick's Day
23 April	St George's Day
24 May	Queen's birthday
12 July	Battle of the Boyne – Orange Order
1 August	Coloured people celebrate abolition
30 November	St Andrew's Day
28 December	St John's Day – Freemasons

Fig. 15.4   Early Victorian Toronto's ritual calendar of parades, as published in the *British colonist, The globe* and *The mirror*.

Sunday, 17 March 1839 to the Catholic church, accompanied as usual by members of the other societies, was unusual.[35] When next we encounter reference to a St Patrick's Day parade destined for the Catholic church, in 1842, there were two competing groups in the streets: Protestant and Catholic, and two societies parading: the St Patrick's Society and the St Patrick's Benevolent Society.

The latter organization had been established in Toronto in 1841 to create 'a bond of union between all sects and denominations of Irishmen', something the existing St Patrick's Society, a Catholic editor complained, did not achieve. The new society would not aim, readers were assured, 'to be [in] the glare of spectators by its procession', but would focus on assisting the needy.[36] There was, though, no escaping the urge to establish identity and measure significance in the traditional fashion: by asserting the right to parade on an appropriate occasion. In his report of the 1842 parade, the same editor who penned the words just quoted proudly noted that the Benevolent Society 'turned out a multitude' comprising no less than 800 or 900 in its parade, 'a larger number than ever before turned out in Toronto on similar occasions. – The utmost order and decorum were observed ... [proving] that the masses of Irishmen, when unmolested, can be as observant of the law and of good order as any other body of men.'[37] The process of winning community and self-respect, by staging a large and orderly parade, was unmistakable. A rival, Protestant, paper noted the innovation of having two parades, regretting 'this public demonstration of the split amongst Irishmen, but, viewing the two processions, we cannot conceal the fact, that the Society dissenting from that first formed, was by far the most numerous'.[38]

The timing of parades in 1841 shows no preference for the weekend which became so prominent later in the century. Indeed, there was an avoidance of Sunday and Saturday, the busiest market day of the week, was also unpopular. The evidence for 1841 would suggest a continuation of 'St Monday' as a day when workers declined to labour in order to claim time for such social purposes as taking part in processions.[39] The tendency was less pronounced in other years. Evidence of timing, often incomplete, suggests that parades happened at a wide variety of hours through the late morning and afternoon and into the evening. No basis was found to conclude that they were deliberately scheduled for non-working hours. A more reasonable inference would be that time discipline was not so rigidly enforced in the workplace to prevent those who wished once or twice a year to join a procession at any hour of the day. On special occasions such as visits by Governors-General, editors printed requests by the city for businesses to close in order to permit a respectful turnout of the population.

Claiming symbolic space

Parades claimed privileges in certain streets where prestige accrued by monopolizing their use for the passing minute or hour. The public space they chose to occupy lent recognition to the sponsoring organization and helped legitimize its existence. The intention was to associate with the power and to appropriate the status therein represented.[40] The physical elements, as one historian has put it, 'had a meaning so clear ... that ... the targets glitter in the eye of history as signs of the conception of the nature of society'.[41]

In early Victorian Toronto the spatial definition of prestige was clearly articulated for all to see and understand. Within easy walk along a few blocks of the main streets in the centre of the city were to be found all of the elite institutions of the city, and the residences of many of their leaders. Here focussed the economic, religious, political and social life of the city; here was the ceremonial heart of the city also, where power and status were to be won, measured and displayed. King Street was the fulcrum of this authority and power, and of the processions which sought to appropriate it. In such circumstances no sharp boundaries yet separated the prestige territory devoted to business and residence. There is much evidence to suggest that the city's inhabitants understood clearly the constellation of values that cleaved to this district, and used their knowledge when they sought by parading to command this ceremonial landscape.

The ability to combine order with large numbers of units in a parade guaranteed accessibility and participation. In parades designed to demonstrate consensus and civic solidarity, size meant success; the largest were among the most memorable, attracting big and appreciative crowds of spectators. Of this, contemporary journalists' accounts leave no doubt. Maintenance of happy relations among parade participants rested on the organizers' abilities to control the arrangement of units comprising the whole, to choose the order of the march. Order was significant, its meaning broadly understood. So, likewise, was the separation of the various contingents in the parade. To put it slightly differently, '[t]he genius of the parade was that it allowed many contending constituencies of the city to line up and move through the streets without ever encountering one another face to face, much less stopping to play specified roles in one co-ordinated pageant.'[42]

Other parades commemorated partisanship, emphasizing criteria for participation both of inclusion and exclusion. Such events served to divide rather than to unite the city; they highlighted a process of contention. Precisely because of the organization they entailed and the routine nature of their timing and performance, parades were ideal vehicles for expressing contention: they were deliberate and calculated acts. Not inherently disorderly, they could nevertheless serve as catalysts for just this reaction. However ritualized and within whatever limits such riots normally were contained, serious disorder was always a possibility. In Toronto in the 1840s Protestant–Catholic antipathy, especially as it had been expressed in Ireland, lay at the root of many contentious parades. The recurring Orange demonstration of 12 July was its most visible signal, but not its only expression.

The 1840s in Toronto

The first election to the new United Parliament of the Province of Canada (Ontario and Quebec) commenced with the candidates and their supporters walking in procession, with banners and music, to the hustings. This happened at 9.00 am, Monday, 15 March 1841 – the first of six consecutive days of polling. On Wednesday, 17 March – St Patrick's Day – the electors of the Catholic congregation, having attended mass to honour their saint's day, processed as a body to the polls. Results were announced after each day's voting, and the winners declared on Saturday. Victory brought the two successful candidates a parade; supported by their electors and with a band playing and banners on display they made their way for over a mile through the streets of central Toronto (Fig. 15.2). They not unexpectedly drew the attention of their opposition, the unsuccessful Tories among whom Orangemen, 'vehement partisans' fearful of a new association of the winners with French Catholic Canada, were prominent.[43]

Contention had accompanied the whole election process, as unexceptional an occurrence for the times as processions. Harassment of candidates at their meetings, attempts to plant party banners against the hustings, and even the flagrant partiality of city constables in favour of one of the parties constituted for Toronto a predictable routine, a contentious order.[44] The events of Monday, 22 March surpassed these bounds and forced a reassessment of a habit of tolerating the unimpeded access to the streets of any group with the ability to organize a parade.

The day began badly. Even before the parade had been formed, a piper was attacked by a mob in the street. He was beaten with clubs and his pipes seized. The attackers tore off the tartan scarves worn by his two companions and trampled them on the ground. The mayor, having been informed of the incident by an eyewitness who volunteered to identify the assailants, refused to dispatch a constable, intimidating the complainant in the process. The parade formed at 1.00 pm, the leaders in carriages with others mounted and on foot. Under the control of the marshals it proceeded 'in the most orderly manner' until it arrived at the corner of King and Church Streets, the heart of the ceremonial city. Here it was set upon by 'ruffians' armed with stones and cudgels. Among the attackers was a brother and one of the defeated candidates who was also the son of an ex-judge. The city magistrates (council members) were prepared, it seems, having taken 'the precaution to close their shop windows' and secure themselves indoors. They made no effort to stem the violence, leaving one editor to conclude that they 'countenanced' it.[45] The procession paused to defend itself and then proceeded eastward on King Street until it was met by another mob at a tavern flying Orange banners. From an upper window several shots were fired from a musket, killing one of the parade members and severely wounding three others. At this, order disappeared as the marchers attacked the premises. In the absence of any intervention by the police, no magistrates being found to make the requisition, troops of the 34th Regiment eventually appeared, read the riot act and ordered the crowd to disperse. An alderman having arrived, he entered the tavern and, with the assistance of the military, arrested a number of the perpetrators.[46] The 'Orange paper', the Tory *Patriot*, reported the 'terrible provocation' given by those parading to the tavern's inmates.[47]

A three-man committee appointed by the Governor-General to enquire into election violence in Toronto held hearings in the city in June, laying its report before the House of Assembly in August. In Toronto its contents 'astonished' the members of the Common Council who, having declared it to be one-sided, sought to appoint a committee of their own. Only one council member had been willing to testify before the commission, and as an editor put it: '[t]he impression abroad is, that the sting of the report is in its truth'.[48] This assessment is consistent with the opinion expressed a year later by Charles Dickens on visiting the city, that the incident was an outrage committed under the banner of the Orange Order.[49]

Whereas divers persons in considerable numbers distinguished by ribbons, favors and other emblems expressive of party feelings, are in the practice of meeting and marching in procession ... upon certain Festivals, Anniversaries and other occasions in celebration of certain Political Events; and whereas such celebrations under whatever pretence held, are found to give great offence ... and to occasion heats and perpetuate animosities, injurious to social order and dangerous to the Public Peace ... it is therefore expedient entirely to prohibit the same.[50]

The 'Act to restrain Party Processions in certain cases' sought to end parades that created social friction by declaring them unlawful assemblies. Its scope was limited only by the clause that it would not apply to religious processions that took place in the course of public worship 'or in the celebration of any religious rite enjoined or ordered by any ... Church ... which shall be accompanied by the Clergy.'[51] It became law on receiving Royal Assent on 9 December 1843.[52] A parallel bill to discourage secret societies was presented to the Legislative Assembly and passed but was reserved, hence failing to become law. A third bill, introduced in the same session, passed and given Royal Assent on 9 December 1843, sought to guarantee the orderly conduct of public meetings.[53]

The Party Processions Act aimed not to ban all commemorative parades. As debate in the Legislative Assembly revealed, the intention was to avoid interfering with parades of the national societies, unless their banners or other trappings provoked religious feuds or contributed to breaches of the peace by use of offensive weapons. The Orange Order, further questioning made clear, was the central target of the bill.[54] What would be the effect of such a sweeping challenge to well-established practice?

Processions organized by the national societies continued to appear on the streets of Toronto, as the Act intended. The immediate impact of the legislation was felt by less-well-entrenched groups. The Freemasons failed to stage their usual procession of 28 December 1843. In the years that followed the press carried fewer reports of the city's socio-religious groups celebrating in the streets. Numerous groups responded by curtailing their use of the streets for parades. The civic or state parade, organized to mark official occasions, survived and even incorporated contingents from the societies whose freedom to march in their own stead was denied by law.

Nonetheless, in its aim of eliminating Orange parades the legislation was notably unsuccessful. The malevolent expression of religious differences between Protestants and Catholics that Orange parades had become was too highly prized to be easily abandoned. For religion was '[a]n emotional tinder so carefully kept dry [that it] required only a small spark to burst into flame.'[55] The Orange parade provided the ignition. And so Orange parades were advertised under different labels and were even exported to distant places in order to circumvent their legal prohibition. A parade on 12 July had only one purpose; it could scarcely be meant to mark any other anniversary.

In 1844 signs appeared throughout the city advertising a 'Conservative Excursion' to Niagara Falls on 12 July. About 400 persons, male and female, embarked by steamer on the appointed morning. They were met near the Falls by 1,500 Catholic labourers, mostly Irish, who were engaged in constructing the Welland Canal. These men, by way of precaution . . . armed themselves with such weapons of defence as were at hand, viz., sticks, pitchforks, scythes, muskets, pistols, &c. They thought that if Orangemen had the assurance to come in the very teeth of the law, to insult them, their country, their religion, they would be prepared to resent the indignity, and to teach them a different tune.

The party of excursionists from Toronto was met by a Catholic priest who advised them to lay aside their Orange badges and ribbons. This they did, and forewent playing party tunes or sending O'Connell's effigy over the Falls.[56] Meanwhile, on the streets of Toronto, 200 persons, 'as ragged, dirty ruffianly a set as any one need wish to look upon', marched in procession through its main streets displaying Orange flags, ribbons and badges. Marching illegally, and upon being ordered to disperse, they refused, to an ensuing contained riot.[57] The episode was characteristic of the continuous testing of the boundaries of permissible demonstration in which Orangemen were engaged throughout the decade. The reaction depended upon one's position. C. Donlevy, the Irish Catholic editor of the *Mirror*, expressed his contempt in the language of his paper. Aldermen and magistrates in the city were, on the other hand, always willing to defend the marchers in court and, on at least one occasion to resign their posts in order to participate in an illegal Orange march.

The tranquillity of the community was still disturbed on occasion by contentious parading. The Orangemen of the city flouted the prohibition in subsequent years, under various titles. Their parade of 1849 was as large as any yet seen within the city, and featured a display of flags and banners. Enforcing the law would have been 'no easy matter' on account of the parade's size and because the authorities, according to one interpretation, had permitted breaches of the same law during the past year. Indeed, two magistrates resigned their posts as aldermen a few days prior to 12 July to enable them to march in the parade free of responsibility to enforce the law they were breaking.[58]

Even the official parade could become a contentious event when the community was deeply divided, as it was in 1849 over the decision by the Governor-General, Elgin, to sign the Rebellion Losses Bill. This act, parallel to an earlier motion passed with reference to Canada West (Ontario), compensated persons in Canada East (Quebec) who had suffered damage in the rebellion of 1837, save only those who had been convicted by the courts or banished. It was received by Tories in Toronto with 'horror' and 'an angry readiness to riot'.[59] They indulged themselves on the nights of 22 March and 2 May. When Elgin's intention to visit the city was made known in late summer its residents were deeply divided in their attitude towards him. Editors' columns reflected nervous anticipation. The ordinary procedure of a public meeting to select a reception committee and to present an address that the city council could approve for presentation had met with delay and hostility. Elgin, having postponed his visit, arrived on Tuesday, 9 October. Members of the city corporation assembled at the wharf to greet him, and he proceeded as part of a small procession the few blocks to his accommodation, a hotel on King Street. Those citizens who concurred in the sentiments expressed in the city's address to the Governor-General were invited to assemble to greet his arrival. The Tory paper, the *British colonist*, reported that the 'multitude was of a mixed character . . . The reception was not such as the representative of Her Majesty usually received in this city.' An absence of 'enthusiasm' was noted, as was an egg thrown at the carriage carrying the visitor. 'The movement of the crowd along the street was very irregular, exhibiting nothing like an orderly procession.' The address was read in front of the hotel, where Elgin read his reply amid great noise.[60] This was threadbare ceremony.

The order that 'every respectable citizen' wished to be maintained broke down Tuesday night about 10.00 pm when a torch-light procession, composed of 'a small party of the worst characters about town', formed. The fire-bell was rung to call out the demonstrators who formed their ranks in Church Street and paraded into King Street, moving in the direction of Elgin's lodgings, carrying his effigy which they intended to burn. The police, being on the alert, stopped the procession as it approached Yonge Street. Some who escaped arrest reassembled opposite the hotel, shouting for some time before they dispersed.[61] On Friday an orderly parade of members of the St Andrew's Society made its way along the same route to present His Excellency the congratulations of the society. The Scots editor of the *Globe* could report with pride on the three pipers in full highland costume followed by the officers and 200 society members who marched two abreast, flags and banners proudly displayed.[62] The timing of his departure not being public knowledge, Elgin left with few to see him off, around 8.00 am Monday morning.

On Wednesday, 15 October 1851 a procession assembled at City Hall that was not remarkably different from many preceding it, except that it was even more inclusive than most. In addition to the presence of contingents from the national societies, other fraternal organizations represented included the Orange Order, the Odd Fellows, the York Agricultural Society and the Mechanics' Institute. Governments, schools and the learned professions provided contingents. Toronto, then the seat of government, was privileged to have the attendance at the ceremony of the Governor-General and Lady Elgin. The well-practised form of corporate celebration was to honour the commencement of the railway age in Toronto. The occasion was the turning of the first sod.[63] The event was significant in Toronto's life because it was without any reference to inherited holidays, actions taken elsewhere or transferred cultural practices.

What was being celebrated was entirely Canadian: the economic vitality and future of Toronto, which was to be enhanced by the construction of its first railway, to connect it with Georgian Bay. The importance of the event, novel in its purpose, was clarified for its audience by its reference to understandable symbolic performance. The ritual of parades in colonial cities involved the adaptation in new settings of activities which refer to accepted forms of behaviour assumed to have deep roots in cultural practice. This is what Eric Hobsbawm calls the invention of tradition.[64]

The procession came to America as a tradition; its enactment recalled a long experience of such public demonstrations. It must not be thought of as an invariant form, for it was no such thing. The circumstances of staging parades in the raw frontier environment of colonial Toronto would, if nothing else, have imposed some changes. Colonial society was busy adapting understood genres of collective behaviour to fit its own purposes. Its challenges, too, were distinct. The failure of legislation in 1843 to prevent the insistent demonstration of Orangeism in the streets informs us that society – in this case civic government and a powerful fraternal organization – valued parading too highly to permit any draconian legislation to interfere significantly with a right well established in practice. It was a vital and malleable practice, adaptable to the new circumstances of celebrating the city's economic future in 1851.

Conclusion

The ritual of parading was well practised in Toronto not only by civic and provincial officials wishing to dignify their function or office, but by persons from every class and station in society. They paraded in the company of each other, in enormous processions that were intended to enhance the city's prestige as well as to honour occasions or distinguished guests. They paraded separately, as members of fraternal or national societies or as supporters of temperance or the abolition of slavery. They paraded to create consensus or to incite contention. Whatever their purpose or constituency, they targeted the same urban space. This space, that to which maximum prestige adhered, conveyed the greatest benefit to those engaged in its 'symbolic capture'.[65] Parades, as a moving pageant displaying symbols and offering music, created a 'time and place of withdrawal from normal modes of social action', the quality of liminality.[66] They offered an experience difficult to ignore. They were times of intensified awareness, defined for participant and spectator alike by the ritual of the procession and by the unmistakable values conveyed by the place where they were enacted.

NOTES

1 *British colonist*, 27 April 1842
2 *The mirror*, 10 November 1843
3 Quentin Skinner, The world as a stage *New York Review of Books*, 28, 6 (16 April 1981) 36
4 Steven Lukes, *Essays in social theory* (London 1977) 54; W. Lloyd Warner, *The living and the dead: a study of the symbolic life of Americans* (New Haven 1959) Part II
5 Hilda Kuper, The language of sites in the politics of space *American Anthropologist* 74 (1972) 421
6 Penelope J. Corfield, Walking the city streets: the urban odyssey in eighteenth-century England *Journal of Urban History* 16 (1989–90) 133–4
7 Gregory S. Kealey, *Toronto workers respond to industrial capitalism, 1867–1892* (Toronto 1980) 115
8 Paul Romney, From the types riot to rebellion: elite ideology, anti-legal sentiment, political violence, and the rule of law in Upper Canada *Ontario History* 79 (1987) 139; see also Michael Cross, Stony Monday, 1849: the rebellion losses riots in Bytown *Ontario History* 63 (1971)
9 John Weaver, Crime, public order and repression: the Gore district in upheaval, 1832–1851 *Ontario History* 78 (1986) 204
10 Lyn H. Lofland, *A world of strangers: order and action in urban public space* (New York 1973) 118
11 François Bedarida and Anthony Sutcliffe, The street in the structure and life of the city: reflections on nineteenth-century London and Paris *Journal of Urban History* 6 (1979–80)
12 Susan G. Davis, *Parades and power: street theatre in nineteenth-century Philadelphia* (Philadelphia 1986) 164; see also Sallie A. Marston, Public rituals and community power: St Patrick's Day parades in Lowell, Massachusetts, 1841–1874 *Political Geography Quarterly* 8 (1989); Timothy J. Meagher, 'Why should we care for a little trouble or a walk through the mud': St Patrick's and Columbus Day parades in Worcester, Massachusetts, 1845–1915 *New England Quarterly* 58 (1985)
13 Mary Ryan, The American parade: representations of the nineteenth-century social order, in Lynn Hunt (ed.), *The new cultural history* (Berkeley 1989) 147; see also Kathleen Neils Conzen, Ethnicity as festive culture, in Werner Sollors (ed.), *The invention of ethnicity* (New York 1989) 52–3
14 Arthur R. M. Lower, *Colony to nation: a history of Canada* (Toronto 1961) 256
15 *British colonist*, 13 April 1842
16 *British colonist*, 20 and 27 April 1842
17 *British colonist*, 27 April 1842
18 Fig. 15.2 is based on the following map housed in the collection of the Metropolitan Toronto Reference Library: 'Topographical plan of the city and liberties of Toronto in the Province of Canada', surveyed, drawn and published by James Cane, Topographical Engineer, in 1842 and dedicated to Sir Charles Bagot. Information on parade routes: The election parade, *British colonist* 24 March 1841. The parade to welcome Bagot, *British colonist* 27 April 1842, and

the parade on 23 April 1842, *The Mirror* 29 April 1842; *British colonist* 27 April 1842

19 *Ibid.*

20 *The mirror*, 22 April 1842

21 The anonymous contemporary report is quoted in Henry Scadding, *Toronto of old*, F. H. Armstrong (ed.) (Toronto 1966) 234

22 *British colonist*, 27 April 1842; *Census of the Canadas, 1851–2*, vol. I, xiii

23 John Skorupski, *Symbol and theory: a philosophical study of theories of religion in social anthropology* (Cambridge 1976) 91

24 Charles Tilly, *The contentious French* (Cambridge Mass. 1986) 4

25 *The mirror*, 10 November 1843

26 E. P. Thompson, *The making of the English working class* (New York 1963) 74–5, 112–13

27 *The mirror*, 10 November 1843

28 *British colonist*, 9 November 1843

29 Edith G. Firth (ed.), *The town of York, 1815–1834: a further collection of documents of early Toronto* (Toronto 1966); Scadding, *Toronto of old*

30 *British colonist*, 18 September 1839

31 *The mirror*, 22 October 1841

32 Peter Burke, *Popular culture in early modern Europe* (London 1978) 196–7

33 Gregory S. Kealey, The Orange Order in Toronto: religious riot and the working class, in Gregory S. Kealey and Peter Warrian (eds.), *Essays in Canadian working class history* (Toronto 1976) 16; Cecil J. Houston and William J. Smyth, *The sash Canada wore: a historical geography of the Orange Order in Canada* (Toronto 1980)

34 *British colonist*, 22 March 1838

35 *British colonist*, 20 March 1839

36 *The mirror*, 5 February 1841

37 *The mirror*, 18 March 1842

38 *British colonist*, 23 March 1842

39 Davis, *Parades and power*, 40; Mark Harrison, *Crowds and history: mass phenomena in English towns, 1790–1835* (Cambridge 1988) 119–25; Harrison, The ordering of the urban environment: time, work and the occurrence of crowds, 1790–1835 *Past and Present* 110 (February, 1986) 157

40 Mona Ozouf, Le cortège et la ville. Let itineraires parisiens des fêtes révolutionnaires *Annales, ESC* 26 (1971)

41 William M. Reddy, The textile trade and the language of the crowd at Rouen, 1752–1871 *Past and Present* 74 (February, 1977) 84; Monz Ozouf, *Festivals and the French Revolution*, translated by Alan Sheridan (Cambridge, Mass. 1988) Chapter 6

42 Ryan, American parade, 137

43 J. M. S. Careless, *The union of the Canadas: the growth of Canadian institutions, 1841–1845* (Toronto 1967) 45

44 *British colonist*, 17 March 1841; *The mirror*, 20 March 1841

45 *British colonist*, 24 March 1841

46 *British colonist*, 24 and 31 March 1841; *The mirror*, 26 March 1841

47 Quoted in *The mirror*, 2 April 1841

48 *British colonist*, 11, 18 and 25 August 1841
49 Charles Dickens, *American notes for general circulation*, John S. Whitley and Arnold Goldman (eds.) (Harmondsworth 1972) 248
50 *Provincial Statutes of Canada*, vol. III. 3rd Session, 1st Parliament, 1843, 17
51 *Ibid.*
52 The Act was repealed in 1851
53 Province of Canada, *Journals of the Legislative Assembly*, vol. III, Session, 1843
54 *British colonist*, 27 October 1843
55 John Webster Grant, *A profusion of spires: religion in nineteenth-century Ontario* (Toronto 1988) 205; Gregory S. Kealey, Orangemen and the Corporation: the politics of class during the union of the Canadas, in Victor l. Russell (ed.) *Forging a consensus: historical essays on Toronto* (Toronto 1984)
56 *The mirror*, 19 July 1844
57 *Ibid.*; Ruth Bleasdale, Class conflict in the canals of Upper Canada in the 1840s *Labour/Le Travailleur* 7 (Spring 1981)
58 *British colonist*, 13 July 1849
59 J. M. S. Careless, *Brown of the globe Vol. I. Voice of Upper Canada, 1818–1859* (Toronto 1989) 89
60 *British colonist*, 12 October 1849; see also *The Globe*, 2, 9 and 11 October 1849
61 *The Globe*, 11 October 1849; see also *British Colonist*, 12 October 1849
62 *The globe*, 13 October 1849
63 *The mirror*, 17 October 1851
64 Eric Hobsbawm, Introduction: inventing traditions, in Eric Hobsbawm and Terence Ranger (eds.), *The invention of tradition* (Cambridge 1983); P. Jackson, Street life: the politics of carnival *Society and Space* 6 (1988)
65 John Berger, The nature of mass demonstrations *New Society* 11, 295 (23 May 1968) 755; Temma Kaplan, Civic rituals and patterns of resistance in Barcelona, 1890–1930, in Pat Thane, Geoffrey Crossick and Roderick Floud (eds.), *The power of the past: essays for Eric Hobsbawm* (Cambridge 1984); Michael Korovkin and Guy Lanoue, On the substantiality of form: interpreting symbolic expression in the paradigm of social organization *Comparative Studies in Society and History* 30 (1988) 642
66 Victor W. Turner, *The ritual process: studies in structure and anti-structure* (Chicago 1969) 156; Rob Shields, The 'system of pleasure': liminality and the carnivalesque at Brighton *Theory, Culture and Society* 7 (1990)

Index

Titles marked with an asterisk are available in paperback